D0025482

Applied Discrete Structures

Part 1 - Fundamentals

Applied Discrete Structures

Part 1 - Fundamentals

Al Doerr
University of Massachusetts Lowell

Ken Levasseur
University of Massachusetts Lowell

May 13, 2019

Edition: 3rd Edition - version 6

Website: faculty.uml.edu/klevasseur/ADS2

©2017   Al Doerr, Ken Levasseur

Applied Discrete Structures by Alan Doerr and Kenneth Levasseur is licensed under a Creative Commons Attribution-NonCommercial-ShareAlike 3.0 United States License. You are free to Share: copy and redistribute the material in any medium or format; Adapt: remix, transform, and build upon the material. You may not use the material for commercial purposes. The licensor cannot revoke these freedoms as long as you follow the license terms.

ISBN 978-1-365-93358-5

9 781365 933585

To our families

Donna, Christopher, Melissa, and Patrick Doerr

Karen, Joseph, Kathryn, and Matthew Levasseur

Acknowledgements

We would like to acknowledge the following instructors for their helpful comments and suggestions.

- Tibor Beke, UMass Lowell
- Alex DeCourcy, UMass Lowell
- Vince DiChiacchio
- Matthew Haner, Mansfield University (PA)
- Dan Klain, UMass Lowell
- Sitansu Mittra, UMass Lowell
- Ravi Montenegro, UMass Lowell
- Tony Penta, UMass Lowell
- Jim Propp, UMass Lowell

I'd like to particularly single out Jim Propp for his close scrutiny, along with that of his students, who are listed below.

I would like to thank Rob Beezer, David Farmer, Karl-Dieter Crisman and other participants on the pretext-xml-support group for their guidance and work on MathBook XML, which has now been renamed PreTeXt. Thanks to the Pedagogy Subcommittee of the UMass Lowell Transformational Education Committee for their financial assistance in helping getting this project started.

Many students have provided feedback and pointed out typos in several editions of this book. They are listed below. Students with no affiliation listed are from UMass Lowell.

- Ryan Allen
- Rebecca Alves
- Junaid Baig
- Anju Balaji
- Carlos Barrientos
- Chris Berns
- Raymond Berger, Eckerd College
- Brianne Bindas
- Nicholas Bishop
- Nathan Blood
- Cameron Bolduc
- Sam Bouchard
- Eric Breslau
- Rachel Bryan
- Rebecca Campbelli
- Eric Carey
- Emily Cashman
- Cora Casteel
- Rachel Chaiser, U. of Puget Sound
- Sam Chambers
- Hannah Chiodo
- Sofya Chow
- David Connolly
- Sean Cummings

- Alex DeCourcy
- Ryan Delosh
- Hillari Denny
- John El-Helou
- Adam Espinola
- Josh Everett
- Anthony Gaeta
- Lisa Gieng
- Holly Goodreau
- Lilia Heimold
- Kevin Holmes
- Alexa Hyde
- Michael Ingemi
- Eunji Jang
- Kyle Joaquim
- Devin Johnson
- Jeremy Joubert
- William Jozefczyk
- Antony Kellermann
- Yorgo A. Kennos
- Thomas Kiley
- Cody Kingman
- Leant Seu Kim
- Jessica Kramer
- John Kuczynski
- Justin LaGree
- Daven Lagu
- Kendra Lansing
- Gregory Lawrence
- Pearl Laxague
- Kevin Le
- Matt LeBlanc
- Maxwell Leduc
- Ariel Leva
- Robert Liana
- Tammy Liu
- Laura Lucaciu
- Andrew Magee
- Matthew Malone
- Logan Mann
- Sam Marquis

- Amy Mazzucotelli
- Adam Melle
- Jason McAdam
- Nick McArdle
- Christine McCarthy
- Shelbylynn McCoy
- Conor McNierney
- Albara Mehene
- Joshua Michaud
- Max Mints
- Charles Mirabile
- Timothy Miskell
- Genevieve Moore
- Mike Morley
- Zach Mulcahy
- Tessa Munoz
- Zachary Murphy
- Logan Nadeau
- Carol Nguyen
- Hung Nguyen
- Shelly Noll
- Steven Oslan
- Harsh Patel
- Beck Peterson
- Donna Petitti
- Paola Pevzner
- Samantha Poirier
- Ian Roberts
- John Raisbeck
- Adelia Reid
- Derek Ross
- Tyler Ross
- Jacob Rothmel
- Zach Rush
- Steve Sadler, Bellevue College (WA)
- Doug Salvati
- Chita Sano
- Noah Schultz
- Ben Shipman
- Florens Shosho
- Jonathan Silva
- Joshua Simard
- Mason Sirois

- Sana Shaikh
- Joel Slebodnick
- Greg Smelkov
- Andrew Somerville
- Alicia Stransky
- James Tan
- Bunchhoung Tiv
- Joanel Vasquez
- Rolando Vera

- Anh Vo
- Ryan Wallace
- Uriah Wardlaw
- Steve Werren
- Laura Wikoff
- Henry Zhu
- Several students at Luzurne County Community College (PA)

Preface

This is part 1 of *Applied Discrete Structures*. It contains the first ten chapters of the full version, suitable for a single semeseter course..

Applied Discrete Structures is designed for use in a university course in discrete mathematics spanning up two semesters. Its original design was for computer science majors to be introduced to the mathematical topics that are useful in computer science. It can also serve the same purpose for mathematics majors, providing a first exposure to many essential topics.

We embarked on this open-source project in 2010, twenty-one years after the publication of the 2nd edition of *Applied Discrete Structures for Computer Science* in 1989. We had signed a contract for the second edition with Science Research Associates in 1988 but by the time the book was ready to print, SRA had been sold to MacMillan. Soon after, the rights had been passed on to Pearson Education, Inc. In 2010, the long-term future of printed textbooks is uncertain. In the meantime, textbook prices (both printed and e-books) had increased and a growing open source textbook movement had started. One of our objectives in revisiting this text is to make it available to our students in an affordable format. In its original form, the text was peer-reviewed and was adopted for use at several universities throughout the country. For this reason, we see *Applied Discrete Structures* as not only an inexpensive alternative, but a high quality alternative.

An initial choice of Mathematica for "source code" was based on the speed with which we could do the conversion. However, the format was not ideal, with no viable web version available. The project had been well-received in spite of these issues. Validation through the listing of this project on the American Institute of Mathematics was very helpful. The current version of *Applied Discrete Structures* has been developed using *PreTeXt*, a lightweight XML application for authors of scientific articles, textbooks and monographs initiated by Rob Beezer, U. of Puget Sound. When the PreTeXt project was launched, it was the natural next step. The features of PreTeXt make it far more readable, with easy production of web, pdf and print formats.

The current computing landscape is very different from the 1980's and this accounts for the most significant changes in the text. One of the most common programming languages of the 1980's was Pascal. We used it to illustrate many of the concepts in the text. Although it isn't totally dead, Pascal is far from the mainstream of computing in the 21st century. The open source software movement was just starting in the late 1980's and in 2005, the first version of Sage (later renamed SageMath), an open-source computer algebra system, was first released. In Applied Discrete Structures we have replaced "Pascal Notes" with "SageMath Notes."

Ken Levasseur
Lowell MA

Contents

Chapter 1

Set Theory

Goals for Chapter 1. We begin this chapter with a brief description of discrete mathematics. We then cover some of the basic set language and notation that will be used throughout the text. Venn diagrams will be introduced in order to give the reader a clear picture of set operations. In addition, we will describe the binary representation of positive integers and introduce summation notation and its generalizations.

1.1 Set Notation and Relations

1.1.1 The notion of a set

The term set is intuitively understood by most people to mean a collection of objects that are called elements (of the set). This concept is the starting point on which we will build more complex ideas, much as in geometry where the concepts of point and line are left undefined. Because a set is such a simple notion, you may be surprised to learn that it is one of the most difficult concepts for mathematicians to define to their own liking. For example, the description above is not a proper definition because it requires the definition of a collection. (How would you define "collection"?) Even deeper problems arise when you consider the possibility that a set could contain itself. Although these problems are of real concern to some mathematicians, they will not be of any concern to us. Our first concern will be how to describe a set; that is, how do we most conveniently describe a set and the elements that are in it? If we are going to discuss a set for any length of time, we usually give it a name in the form of a capital letter (or occasionally some other symbol). In discussing set A, if x is an element of A, then we will write $x \in A$. On the other hand, if x is not an element of A, we write $x \notin A$. The most convenient way of describing the elements of a set will vary depending on the specific set.

Enumeration. When the elements of a set are enumerated (or listed) it is traditional to enclose them in braces. For example, the set of binary digits is $\{0, 1\}$ and the set of decimal digits is $\{0, 1, 2, 3, 4, 5, 6, 7, 8, 9\}$. The choice of a name for these sets would be arbitrary; but it would be "logical" to call them B and D, respectively. The choice of a set name is much like the choice of an identifier name in programming. Some large sets can be enumerated without actually listing all the elements. For example, the letters of the alphabet and the integers from 1 to 100 could be described as $A = \{a, b, c, \ldots, x, y, z\}$, and $G = \{1, 2, \ldots, 99, 100\}$. The three consecutive "dots" are called an ellipsis. We use them when it is clear what elements are included but not listed. An ellipsis is used in two other situations. To enumerate the positive integers, we would write $\{1, 2, 3, \ldots\}$, indicating that the list goes on infinitely. If we want to list a more general set such as the integers between 1 and n, where n is some undetermined positive integer, we might write $\{1, \ldots, n\}$.

Standard Symbols. Sets that are frequently encountered are usually given symbols that are reserved for them alone. For example, since we will be referring to the positive integers throughout this book, we will use the symbol $\mathbb{P}$ instead of writing $\{1, 2, 3, \ldots\}$. A few of the other sets of numbers that we will use frequently are:

- $(\mathbb{N})$: the natural numbers, $\{0, 1, 2, 3, \ldots\}$

- $(\mathbb{Z})$: the integers, $\{\ldots, -3, -2, -1, 0, 1, 2, 3, \ldots\}$

- $(\mathbb{Q})$: the rational numbers

- $(\mathbb{R})$: the real numbers

- $(\mathbb{C})$: the complex numbers

Set-Builder Notation.   Another way of describing sets is to use set-builder notation. For example, we could define the rational numbers as

$$\mathbb{Q} = \{a/b \mid a, b \in \mathbb{Z}, b \neq 0\}.$$

Note that in the set-builder description for the rational numbers:

- a/b indicates that a typical element of the set is a "fraction."

- The vertical line, $|$, is read "such that" or "where," and is used interchangeably with a colon.

- $a, b \in \mathbb{Z}$ is an abbreviated way of saying a and b are integers.

- Commas in mathematics are read as "and."

The important fact to keep in mind in set notation, or in any mathematical notation, is that it is meant to be a help, not a hindrance. We hope that notation will assist us in a more complete understanding of the collection of objects under consideration and will enable us to describe it in a concise manner. However, brevity of notation is not the aim of sets. If you prefer to write $a \in \mathbb{Z}$ and $b \in \mathbb{Z}$ instead of $a, b \in \mathbb{Z}$, you should do so. Also, there are frequently many different, and equally good, ways of describing sets. For example, $\{x \in \mathbb{R} \mid x^2 - 5x + 6 = 0\}$ and $\{x \mid x \in \mathbb{R}, x^2 - 5x + 6 = 0\}$ both describe the solution set $\{2, 3\}$.

A proper definition of the real numbers is beyond the scope of this text. It is sufficient to think of the real numbers as the set of points on a number line. The complex numbers can be defined using set-builder notation as $\mathbb{C} = \{a + bi : a, b \in \mathbb{R}\}$, where $i^2 = -1$.

In the following definition we will leave the word "finite" undefined.

Definition 1.1.1 Finite Set. A set is a finite set if it has a finite number of elements. Any set that is not finite is an infinite set.   ◇

Definition 1.1.2 Cardinality. Let A be a finite set. The number of different elements in A is called its cardinality. The cardinality of a finite set A is denoted $|A|$.   ◇

As we will see later, there are different infinite cardinalities. We can't make this distinction until Chapter 7, so we will restrict cardinality to finite sets for now.

1.1.2 Subsets

Definition 1.1.3 Subset. Let A and B be sets. We say that A is a subset of B if and only if every element of A is an element of B. We write $A \subseteq B$ to denote the fact that A is a subset of B. ◊

Example 1.1.4 Some Subsets.

(a) If $A = \{3, 5, 8\}$ and $B = \{5, 8, 3, 2, 6\}$, then $A \subseteq B$.

(b) $\mathbb{N} \subseteq \mathbb{Z} \subseteq \mathbb{Q} \subseteq \mathbb{R} \subseteq \mathbb{C}$

(c) If $S = \{3, 5, 8\}$ and $T = \{5, 3, 8\}$, then $S \subseteq T$ and $T \subseteq S$.

□

Definition 1.1.5 Set Equality. Let A and B be sets. We say that A is equal to B (notation $A = B$) if and only if every element of A is an element of B and conversely every element of B is an element of A; that is, $A \subseteq B$ and $B \subseteq A$. ◊

Example 1.1.6 Examples illustrating set equality.

(a) In Example 1.1.4, $S = T$. Note that the ordering of the elements is unimportant.

(b) The number of times that an element appears in an enumeration doesn't affect a set. For example, if $A = \{1, 5, 3, 5\}$ and $B = \{1, 5, 3\}$, then $A = B$. Warning to readers of other texts: Some books introduce the concept of a multiset, in which the number of occurrences of an element matters.

□

A few comments are in order about the expression "if and only if" as used in our definitions. This expression means "is equivalent to saying," or more exactly, that the word (or concept) being defined can at any time be replaced by the defining expression. Conversely, the expression that defines the word (or concept) can be replaced by the word.

Occasionally there is need to discuss the set that contains no elements, namely the empty set, which is denoted by $\emptyset$. This set is also called the null set.

It is clear, we hope, from the definition of a subset, that given any set A we have $A \subseteq A$ and $\emptyset \subseteq A$. If A is nonempty, then A is called an **improper subset** of A. All other subsets of A, including the empty set, are called **proper subsets** of A. The empty set is an improper subset of itself.

Note 1.1.7 Not everyone is in agreement on whether the empty set is a proper subset of any set. In fact earlier editions of this book sided with those who considered the empty set an improper subset. However, we bow to the emerging consensus at this time.

1.1.3 Exercises for Section 1.1

1. List four elements of each of the following sets:

    (a) $\{k \in \mathbb{P} \mid k - 1 \text{ is a multiple of } 7\}$

    (b) $\{x \mid x \text{ is a fruit and its skin is normally eaten}\}$

    (c) $\{x \in \mathbb{Q} \mid \frac{1}{x} \in \mathbb{Z}\}$

(d) $\{2n \mid n \in \mathbb{Z}, n < 0\}$

(e) $\{s \mid s = 1 + 2 + \cdots + n \text{ for some } n \in \mathbb{P}\}$

2. List all elements of the following sets:

(a) $\{\frac{1}{n} \mid n \in \{3, 4, 5, 6\}\}$

(b) $\{\alpha \in \text{ the alphabet } \mid \alpha \text{ precedes } \mathbf{F}\}$

(c) $\{x \in \mathbb{Z} \mid x = x + 1\}$

(d) $\{n^2 \mid n = -2, -1, 0, 1, 2\}$

(e) $\{n \in \mathbb{P} \mid n \text{ is a factor of } 24 \}$

3. Describe the following sets using set-builder notation.

(a) $\{5, 7, 9, \ldots, 77, 79\}$

(b) the rational numbers that are strictly between -1 and 1

(c) the even integers

(d) $\{-18, -9, 0, 9, 18, 27, \ldots\}$

4. Use set-builder notation to describe the following sets:

(a) $\{1, 2, 3, 4, 5, 6, 7\}$

(b) $\{1, 10, 100, 1000, 10000\}$

(c) $\{1, 1/2, 1/3, 1/4, 1/5, \ldots\}$

(d) $\{0\}$

5. Let $A = \{0, 2, 3\}$, $B = \{2, 3\}$, and $C = \{1, 5, 9\}$. Determine which of the following statements are true. Give reasons for your answers.

(a) $3 \in A$

(b) $\{3\} \in A$

(c) $\{3\} \subseteq A$

(d) $B \subseteq A$

(e) $A \subseteq B$

(f) $\emptyset \subseteq C$

(g) $\emptyset \in A$

(h) $A \subseteq A$

6. One reason that we left the definition of a set vague is Russell's Paradox. Many mathematics and logic books contain an account of this paradox. Two references are [40] and [35]. Find one such reference and read it.

1.2 Basic Set Operations

1.2.1 Definitions

Definition 1.2.1 Intersection. Let A and B be sets. The intersection of A and B (denoted by $A \cap B$) is the set of all elements that are in both A and B. That is, $A \cap B = \{x : x \in A \text{ and } x \in B\}$. $\diamond$

Example 1.2.2 Some Intersections.

- Let $A = \{1, 3, 8\}$ and $B = \{-9, 22, 3\}$. Then $A \cap B = \{3\}$.

- Solving a system of simultaneous equations such as $x + y = 7$ and $x - y = 3$ can be viewed as an intersection. Let $A = \{(x, y) : x + y = 7, x, y \in \mathbb{R}\}$ and $B = \{(x, y) : x - y = 3, x, y \in \mathbb{R}\}$. These two sets are lines in the plane and their intersection, $A \cap B = \{(5, 2)\}$, is the solution to the system.

- $\mathbb{Z} \cap \mathbb{Q} = \mathbb{Z}$.

- If $A = \{3, 5, 9\}$ and $B = \{-5, 8\}$, then $A \cap B = \emptyset$.

$\square$

Definition 1.2.3 Disjoint Sets. Two sets are disjoint if they have no elements in common. That is, A and B are disjoint if $A \cap B = \emptyset$. $\diamond$

Definition 1.2.4 Union. Let A and B be sets. The union of A and B (denoted by $A \cup B$) is the set of all elements that are in A or in B or in both A and B. That is, $A \cup B = \{x : x \in A \text{ or } x \in B\}$. $\diamond$

It is important to note in the set-builder notation for $A \cup B$, the word "or" is used in the inclusive sense; it includes the case where x is in both A and B.

Example 1.2.5 Some Unions.

- If $A = \{2, 5, 8\}$ and $B = \{7, 5, 22\}$, then $A \cup B = \{2, 5, 8, 7, 22\}$.

- $\mathbb{Z} \cup \mathbb{Q} = \mathbb{Q}$.

- $A \cup \emptyset = A$ for any set A.

$\square$

Frequently, when doing mathematics, we need to establish a universe or set of elements under discussion. For example, the set $A = \{x : 81x^4 - 16 = 0\}$ contains different elements depending on what kinds of numbers we allow ourselves to use in solving the equation $81x^4 - 16 = 0$. This set of numbers would be our universe. For example, if the universe is the integers, then A is empty. If our universe is the rational numbers, then A is $\{2/3, -2/3\}$ and if the universe is the complex numbers, then A is $\{2/3, -2/3, 2i/3, -2i/3\}$.

Definition 1.2.6 Universe. The universe, or universal set, is the set of all elements under discussion for possible membership in a set. We normally reserve the letter U for a universe in general discussions. $\diamond$

1.2.2 Set Operations and their Venn Diagrams

When working with sets, as in other branches of mathematics, it is often quite useful to be able to draw a picture or diagram of the situation under consideration. A diagram of a set is called a Venn diagram. The universal set U is represented by the interior of a rectangle and the sets by disks inside the rectangle.

Example 1.2.7 Venn Diagram Examples. $A \cap B$ is illustrated in Figure 1.2.8 by shading the appropriate region.

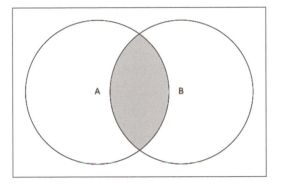

Figure 1.2.8: Venn Diagram for the Intersection of Two Sets

The union $A \cup B$ is illustrated in Figure 1.2.9.

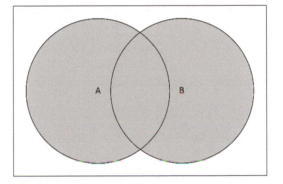

Figure 1.2.9: Venn Diagram for the Union $A \cup B$

In a Venn diagram, the region representing $A \cap B$ does not appear empty; however, in some instances it will represent the empty set. The same is true for any other region in a Venn diagram.   □

Definition 1.2.10 Complement of a set. Let A and B be sets. The complement of A relative to B (notation $B - A$) is the set of elements that are in B and not in A. That is, $B - A = \{x : x \in B \text{ and } x \notin A\}$. If U is the universal set, then $U - A$ is denoted by A^c and is called simply the complement of A. $A^c = \{x \in U : x \notin A\}$.   ◇

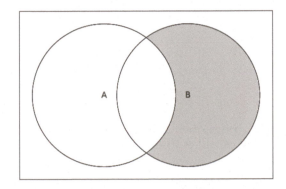

Figure 1.2.11: Venn Diagram for $B - A$

Example 1.2.12 Some Complements.

(a) Let $U = \{1, 2, 3, ..., 10\}$ and $A = \{2, 4, 6, 8, 10\}$. Then $U - A = \{1, 3, 5, 7, 9\}$ and $A - U = \emptyset$.

(b) If $U = \mathbb{R}$, then the complement of the set of rational numbers is the set of irrational numbers.

(c) $U^c = \emptyset$ and $\emptyset^c = U$.

(d) The Venn diagram of $B - A$ is represented in Figure 1.2.11.

(e) The Venn diagram of A^c is represented in Figure 1.2.13.

(f) If $B \subseteq A$, then the Venn diagram of $A - B$ is as shown in Figure 1.2.14.

(g) In the universe of integers, the set of even integers, $\{\ldots, -4, -2, 0, 2, 4, \ldots\}$, has the set of odd integers as its complement.

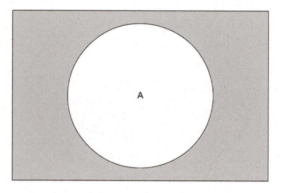

Figure 1.2.13: Venn Diagram for A^c

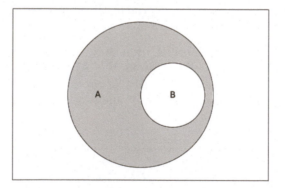

Figure 1.2.14: Venn Diagram for $A - B$ when B is a subset of A

□

Definition 1.2.15 Symmetric Difference. Let A and B be sets. The symmetric difference of A and B (denoted by $A \oplus B$) is the set of all elements that are in A and B but not in both. That is, $A \oplus B = (A \cup B) - (A \cap B)$. ◊

Example 1.2.16 Some Symmetric Differences.

(a) Let $A = \{1, 3, 8\}$ and $B = \{2, 4, 8\}$. Then $A \oplus B = \{1, 2, 3, 4\}$.

(b) $A \oplus \emptyset = A$ and $A \oplus A = \emptyset$ for any set A.

(c) $\mathbb{R} \oplus \mathbb{Q}$ is the set of irrational numbers.

(d) The Venn diagram of $A \oplus B$ is represented in Figure 1.2.17.

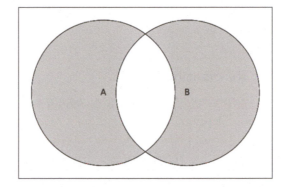

Figure 1.2.17: Venn Diagram for the symmetric difference $A \oplus B$

$\square$

1.2.3 SageMath Note: Sets

To work with sets in Sage, a set is an expression of the form Set(*list*). By wrapping a list with Set( ), the order of elements appearing in the list and their duplication are ignored. For example, L1 and L2 are two different lists, but notice how as sets they are considered equal:

```
L1=[3,6,9,0,3]
L2=[9,6,3,0,9]
[L1==L2, Set(L1)==Set(L2) ]
```

```
[False,True]
```

The standard set operations are all methods and/or functions that can act on Sage sets. *You need to evaluate the following cell to use the subsequent cell.*

```
A=Set(srange(5,50,5))
B=Set(srange(6,50,6))
[A,B]
```

```
[{35, 5, 40, 10, 45, 15, 20, 25, 30}, {36, 6, 42, 12, 48, 18,
    24, 30}]
```

We can test membership, asking whether 10 is in each of the sets:

```
[10 in A, 10 in B]
```

```
[True, False]
```

The ampersand is used for the intersection of sets. Change it to the vertical bar, |, for union.

```
A & B
```

```
{30}
```

Symmetric difference and set complement are defined as "methods" in Sage. Here is how to compute the symmetric difference of A with B, followed by their differences.

```
[A.symmetric_difference(B),A.difference(B),B.difference(A)]
```

```
[{35, 36, 5, 6, 40, 42, 12, 45, 15, 48, 18, 20, 24, 25, 10},
 {35, 5, 40, 10, 45, 15, 20, 25},
 {48, 18, 36, 6, 24, 42, 12}]
```

1.2.4 Exercises for Section 1.2

1. Let $A = \{0, 2, 3\}$, $B = \{2, 3\}$, $C = \{1, 5, 9\}$, and let the universal set be $U = \{0, 1, 2, ..., 9\}$. Determine:

    (a) $A \cap B$ (e) $A - B$ (i) $A \cap C$

    (b) $A \cup B$ (f) $B - A$ (j) $A \oplus B$

    (c) $B \cup A$ (g) A^c

    (d) $A \cup C$ (h) C^c

2. Let A, B, and C be as in Exercise 1, let $D = \{3, 2\}$, and let $E = \{2, 3, 2\}$. Determine which of the following are true. Give reasons for your decisions.

    (a) $A = B$ (e) $A \cap B = B \cap A$

    (b) $B = C$ (f) $A \cup B = B \cup A$

    (c) $B = D$ (g) $A - B = B - A$

    (d) $E = D$ (h) $A \oplus B = B \oplus A$

3. Let $U = \{1, 2, 3, ..., 9\}$. Give examples of sets A, B, and C for which:

    (a) $A \cap (B \cap C) = (A \cap B) \cap C$ (d) $A \cup A^c = U$

    (b) $A \cap (B \cup C) = (A \cap B) \cup (A \cap C)$ (e) $A \subseteq A \cup B$

    (c) $(A \cup B)^c = A^c \cap B^c$ (f) $A \cap B \subseteq A$

4. Let $U = \{1, 2, 3, ..., 9\}$. Give examples to illustrate the following facts:

    (a) If $A \subseteq B$ and $B \subseteq C$, then $A \subseteq C$.

    (b) There are sets A and B such that $A - B \neq B - A$

    (c) If $U = A \cup B$ and $A \cap B = \emptyset$, it always follows that $A = U - B$.

5. What can you say about A if $U = \{1, 2, 3, 4, 5\}$, $B = \{2, 3\}$, and (separately)

    (a) $A \cup B = \{1, 2, 3, 4\}$

    (b) $A \cap B = \{2\}$

    (c) $A \oplus B = \{3, 4, 5\}$

6. Suppose that U is an infinite universal set, and A and B are infinite subsets of U. Answer the following questions with a brief explanation.

    (a) Must A^c be finite?

    (b) Must $A \cup B$ be infinite?

    (c) Must $A \cap B$ be infinite?

7. Given that U = all students at a university, D = day students, M = mathematics majors, and G = graduate students. Draw Venn diagrams illustrating this situation and shade in the following sets:

(a) evening students

(b) undergraduate mathematics majors

(c) non-math graduate students

(d) non-math undergraduate students

8. Let the sets D, M, G, and U be as in exercise 7. Let $|U| = 16,000$, $|D| = 9,000$, $|M| = 300$, and $|G| = 1,000$. Also assume that the number of day students who are mathematics majors is 250, 50 of whom are graduate students, that there are 95 graduate mathematics majors, and that the total number of day graduate students is 700. Determine the number of students who are:

(a) evening students

(b) nonmathematics majors

(c) undergraduates (day or evening)

(d) day graduate nonmathematics majors

(e) evening graduate students

(f) evening graduate mathematics majors

(g) evening undergraduate non-mathematics majors

1.3 Cartesian Products and Power Sets

1.3.1 Cartesian Products

Definition 1.3.1 Cartesian Product. Let A and B be sets. The Cartesian product of A and B, denoted by $A \times B$, is defined as follows: $A \times B = \{(a, b) \mid a \in A \text{ and } b \in B\}$, that is, $A \times B$ is the set of all possible ordered pairs whose first component comes from A and whose second component comes from B. $\diamond$

Example 1.3.2 Some Cartesian Products. Notation in mathematics is often developed for good reason. In this case, a few examples will make clear why the symbol $\times$ is used for Cartesian products.

- Let $A = \{1, 2, 3\}$ and $B = \{4, 5\}$. Then $A \times B = \{(1, 4), (1, 5), (2, 4), (2, 5), (3, 4), (3, 5)\}$. Note that $|A \times B| = 6 = |A| \times |B|$.

- $A \times A = \{(1, 1), (1, 2), (1, 3), (2, 1), (2, 2), (2, 3), (3, 1), (3, 2), (3, 3)\}$. Note that $|A \times A| = 9 = |A|^2$.

$\square$

These two examples illustrate the general rule that if A and B are finite sets, then $|A \times B| = |A| \times |B|$.

We can define the Cartesian product of three (or more) sets similarly. For example, $A \times B \times C = \{(a, b, c) : a \in A, b \in B, c \in C\}$.

It is common to use exponents if the sets in a Cartesian product are the same:

$$A^2 = A \times A$$

$$A^3 = A \times A \times A$$

and in general,

$$A^n = \underbrace{A \times A \times \ldots \times A}_{n \text{ factors}}.$$

1.3.2 Power Sets

Definition 1.3.3 Power Set. If A is any set, the power set of A is the set of all subsets of A, denoted $\mathcal{P}(A)$.   ◊

The two extreme cases, the empty set and all of A, are both included in $\mathcal{P}(A)$.

Example 1.3.4 Some Power Sets.

- $\mathcal{P}(\emptyset) = \{\emptyset\}$

- $\mathcal{P}(\{1\}) = \{\emptyset, \{1\}\}$

- $\mathcal{P}(\{1,2\}) = \{\emptyset, \{1\}, \{2\}, \{1,2\}\}.$

We will leave it to you to guess at a general formula for the number of elements in the power set of a finite set. In Chapter 2, we will discuss counting rules that will help us derive this formula.   □

1.3.3 SageMath Note: Cartesian Products and Power Sets

Here is a simple example of a cartesian product of two sets:

```
A=Set([0,1,2])
B=Set(['a','b'])
P=cartesian_product([A,B]);P
```

The cartesian product of ({0, 1, 2}, {'a', 'b'})

Here is the cardinality of the cartesian product.

```
P.cardinality()
```

6

The power set of a set is an iterable, as you can see from the output of this next cell

```
U=Set([0,1,2,3])
subsets(U)
```

<generator **object** powerset at 0x7fec5ffd33c0>

You can iterate over a powerset. Here is a trivial example.

```
for a in subsets(U):
    print(str(a)+ "␣has␣" +str(len(a))+"␣elements.")
```

```
[] has 0 elements.
[0] has 1 elements.
[1] has 1 elements.
[0, 1] has 2 elements.
[2] has 1 elements.
[0, 2] has 2 elements.
[1, 2] has 2 elements.
[0, 1, 2] has 3 elements.
```

```
[3] has 1 elements.
[0, 3] has 2 elements.
[1, 3] has 2 elements.
[0, 1, 3] has 3 elements.
[2, 3] has 2 elements.
[0, 2, 3] has 3 elements.
[1, 2, 3] has 3 elements.
[0, 1, 2, 3] has 4 elements.
```

1.3.4 EXERCISES FOR SECTION 1.3

1. Let $A = \{0, 2, 3\}$, $B = \{2, 3\}$, $C = \{1, 4\}$, and let the universal set be $U = \{0, 1, 2, 3, 4\}$. List the elements of

   (a) $A \times B$ (e) $A \times A^c$

   (b) $B \times A$ (f) B^2

   (c) $A \times B \times C$ (g) B^3

   (d) $U \times \emptyset$ (h) $B \times \mathcal{P}(B)$

2. Suppose that you are about to flip a coin and then roll a die. Let $A = \{HEADS, TAILS\}$ and $B = \{1, 2, 3, 4, 5, 6\}$.

   (a) What is $|A \times B|$?

   (b) How could you interpret the set $A \times B$?

3. List all two-element sets in $\mathcal{P}(\{a, b, c, d\})$

4. List all three-element sets in $\mathcal{P}(\{a, b, c, d\})$.

5. How many singleton (one-element) sets are there in $\mathcal{P}(A)$ if $|A| = n$?

6. A person has four coins in his pocket: a penny, a nickel, a dime, and a quarter. How many different sums of money can he take out if he removes 3 coins at a time?

7. Let $A = \{+, -\}$ and $B = \{00, 01, 10, 11\}$.

   (a) List the elements of $A \times B$

   (b) How many elements do A^4 and $(A \times B)^3$ have?

8. Let $A = \{\bullet, \square, \otimes\}$ and $B = \{\square, \ominus, \bullet\}$.

   (a) List the elements of $A \times B$ and $B \times A$. The parentheses and comma in an ordered pair are not necessary in cases such as this where the elements of each set are individual symbols.

   (b) Identify the intersection of $A \times B$ and $B \times A$ for the case above, and then guess at a general rule for the intersection of $A \times B$ and $B \times A$, where A and B are any two sets.

9. Let A and B be nonempty sets. When are $A \times B$ and $B \times A$ equal?

1.4 Binary Representation of Positive Integers

1.4.1 Grouping by Twos

Recall that the set of positive integers, $\mathbb{P}$, is $\{1, 2, 3, ...\}$. Positive integers are naturally used to count things. There are many ways to count and many ways to record, or represent, the results of counting. For example, if we wanted to count five hundred twenty-three apples, we might group the apples by tens. There would be fifty-two groups of ten with three single apples left over. The fifty-two groups of ten could be put into five groups of ten tens (hundreds), with two tens left over. The five hundreds, two tens, and three units is recorded as 523. This system of counting is called the base ten positional system, or decimal system. It is quite natural for us to do grouping by tens, hundreds, thousands, ... since it is the method that all of us use in everyday life.

The term positional refers to the fact that each digit in the decimal representation of a number has a significance based on its position. Of course this means that rearranging digits will change the number being described. You may have learned of numeration systems in which the position of symbols does not have any significance (e.g., the ancient Egyptian system). Most of these systems are merely curiosities to us now.

The binary number system differs from the decimal number system in that units are grouped by twos, fours, eights, etc. That is, the group sizes are powers of two instead of powers of ten. For example, twenty-three can be grouped into eleven groups of two with one left over. The eleven twos can be grouped into five groups of four with one group of two left over. Continuing along the same lines, we find that twenty-three can be described as one sixteen, zero eights, one four, one two, and one one, which is abbreviated 10111_{two}, or simply 10111 if the context is clear.

1.4.2 A Conversion Algorithm

The process that we used to determine the binary representation of 23 can be described in general terms to determine the binary representation of any positive integer n. A general description of a process such as this one is called an algorithm. Since this is the first algorithm in the book, we will first write it out using less formal language than usual, and then introduce some "algorithmic notation." If you are unfamiliar with algorithms, we refer you to Section A.1

(1) Start with an empty list of bits.

(2) Step Two: Assign the variable k the value n.

(3) Step Three: While k's value is positive, continue performing the following three steps until k becomes zero and then stop.

　(a) divide k by 2, obtaining a quotient q (often denoted k div 2) and a remainder r (denoted (k mod 2)).

　(b) attach r to the left-hand side of the list of bits.

　(c) assign the variable k the value q.

Example 1.4.1 An example of conversion to binary. To determine the binary representation of 41 we take the following steps:

- $41 = 2 \times 20 + 1$　$List = 1$

- $20 = 2 \times 10 + 0 \quad List = 01$

- $10 = 2 \times 5 + 0 \quad List = 001$

- $5 = 2 \times 2 + 1 \quad List = 1001$

- $2 = 2 \times 1 + 0 \quad List = 01001$

- $1 = 2 \times 0 + 1 \quad List = 101001$

Therefore, $41 = 101001_{two}$ □

The notation that we will use to describe this algorithm and all others is called pseudocode, an informal variation of the instructions that are commonly used in many computer languages. Read the following description carefully, comparing it with the informal description above. Appendix B, which contains a general discussion of the components of the algorithms in this book, should clear up any lingering questions. Anything after // are comments.

Algorithm 1.4.2 Binary Conversion Algorithm. *An algorithm for determining the binary representation of a positive integer.*

Input: a positive integer n.

Output: the binary representation of n in the form of a list of bits, with units bit last, twos bit next to last, etc.

(1) $k := n$ //initialize k

(2) $L := \{ \}$ //initialize L to an empty list

(3) While $k > 0$ do

    (a) $q := k \ div \ 2$ //divide k by 2

    (b) $r := k \ mod \ 2$

    (c) $L := prepend \ r \ to \ L$ //add r to the front of L

    (d) $k := q$ //reassign k

Here is a Sage version of the algorithm with two alterations. It outputs the binary representation as a string, and it handles all integers, not just positive ones.

```
def binary_rep(n):
    if n==0:
        return '0'
    else:
        k=abs(n)
        s=''
        while k>0:
            s=str(k%2)+s
            k=k//2
        if n < 0:
            s='-'+s
        return s

binary_rep(41)
```

'101001'

Now that you've read this section, you should get this joke.

Figure 1.4.3: With permission from Randall Munroe

1.4.3 Exercises for Section 1.4

1. Find the binary representation of each of the following positive integers by working through the algorithm by hand. You can check your answer using the sage cell above.

    (a) 31   (c) 10

    (b) 32   (d) 100

2. Find the binary representation of each of the following positive integers by working through the algorithm by hand. You can check your answer using the sage cell above.

    (a) 64   (c) 28

    (b) 67   (d) 256

3. What positive integers have the following binary representations?

    (a) 10010   (c) 101010

    (b) 10011   (d) 10011110000

4. What positive integers have the following binary representations?
   (a) 100001   (c) 1000000000

   (b) 1001001   (d) 1001110000

5. The number of bits in the binary representations of integers increases by one as the numbers double. Using this fact, determine how many bits the binary representations of the following decimal numbers have without actually doing the full conversion.
   (a) 2017   (b) 4000   (c) 4500   (d) 2^{50}

6. Let m be a positive integer with n-bit binary representation: $a_{n-1}a_{n-2}\cdots a_1a_0$ with $a_{n-1} = 1$ What are the smallest and largest values that m could have?

7. If a positive integer is a multiple of 100, we can identify this fact from its decimal representation, since it will end with two zeros. What can you say about a positive integer if its binary representation ends with two zeros? What if it ends in k zeros?

8. Can a multiple of ten be easily identified from its binary representation?

1.5 Summation Notation and Generalizations

1.5.1 Sums

Most operations such as addition of numbers are introduced as binary operations. That is, we are taught that two numbers may be added together to give us a single number. Before long, we run into situations where more than two numbers are to be added. For example, if four numbers, a_1, a_2, a_3, and a_4 are to be added, their sum may be written down in several ways, such as $((a_1 + a_2) + a_3) + a_4$ or $(a_1 + a_2) + (a_3 + a_4)$. In the first expression, the first two numbers are added, the result is added to the third number, and that result is added to the fourth number. In the second expression the first two numbers and the last two numbers are added and the results of these additions are added. Of course, we know that the final results will be the same. This is due to the fact that addition of numbers is an associative operation. For such operations, there is no need to describe how more than two objects will be operated on. A sum of numbers such as $a_1 + a_2 + a_3 + a_4$ is called a series and is often written $\sum_{k=1}^{4} a_k$ in what is called *summation notation*.

We first recall some basic facts about series that you probably have seen before. A more formal treatment of sequences and series is covered in Chapter 8. The purpose here is to give the reader a working knowledge of summation notation and to carry this notation through to intersection and union of sets and other mathematical operations.

A *finite series* is an expression such as $a_1 + a_2 + a_3 + \cdots + a_n = \sum_{k=1}^{n} a_k$ In the expression $\sum_{k=1}^{n} a_k$:

- The variable k is referred to as the *index*, or the index of summation.

- The expression a_k is the *general term* of the series. It defines the numbers that are being added together in the series.

- The value of k below the summation symbol is the *initial index* and the value above the summation symbol is the *terminal index*.

- It is understood that the series is a sum of the general terms where the index start with the initial index and increases by one up to and including the terminal index.

Example 1.5.1 Some finite series.

(a) $\sum_{i=1}^{4} a_i = a_1 + a_2 + a_3 + a_4$

(b) $\sum_{k=0}^{5} b_k = b_0 + b_1 + b_2 + b_3 + b_4 + b_5$

(c) $\sum_{i=-2}^{2} c_i = c_{-2} + c_{-1} + c_0 + c_1 + c_2$

□

Example 1.5.2 More finite series. If the general terms in a series are more specific, the sum can often be simplified. For example,

(a) $\sum_{i=1}^{4} i^2 = 1^2 + 2^2 + 3^2 + 4^2 = 30$

(b)

$$\sum_{i=1}^{5} (2i - 1) = (2 \cdot 1 - 1) + (2 \cdot 2 - 1) + (2 \cdot 3 - 1) + (2 \cdot 4 - 1) + (2 \cdot 5 - 1)$$
$$= 1 + 3 + 5 + 7 + 9$$
$$= 25$$

□

1.5.2 Generalizations

Summation notation can be generalized to many mathematical operations, for example, $A_1 \cap A_2 \cap A_3 \cap A_4 = \bigcap_{i=1}^{4} A_i$

Definition 1.5.3 Generalized Set Operations. Let $A_1, A_2, \ldots, A_n$ be sets. Then:

(a) $A_1 \cap A_2 \cap \cdots \cap A_n = \bigcap_{i=1}^{n} A_i$

(b) $A_1 \cup A_2 \cup \cdots \cup A_n = \bigcup_{i=1}^{n} A_i$

(c) $A_1 \times A_2 \times \cdots \times A_n = \bigtimes_{i=1}^{n} A_i$

(d) $A_1 \oplus A_2 \oplus \cdots \oplus A_n = \bigoplus_{i=1}^{n} A_i$

◊

Example 1.5.4 Some generalized operations. If $A_1 = \{0, 2, 3\}$, $A_2 = \{1, 2, 3, 6\}$, and $A_3 = \{-1, 0, 3, 9\}$, then

$$\bigcap_{i=1}^{3} A_i = A_1 \cap A_2 \cap A_3 = \{3\}$$

and

$$\bigcup_{i=1}^{3} A_i = A_1 \cup A_2 \cup A_3 = \{-1, 0, 1, 2, 3, 6, 9\}.$$

With this notation it is quite easy to write lengthy expressions in a fairly compact form. For example, the statement

$$A \cap (B_1 \cup B_2 \cup \cdots \cup B_n) = (A \cap B_1) \cup (A \cap B_2) \cup \cdots \cup (A \cap B_n)$$

becomes

$$A \cap \left(\bigcup_{i=1}^{n} B_i \right) = \bigcup_{i=1}^{n} (A \cap B_i).$$

$\square$

1.5.3 Exercises for Section 1.5

1. Calculate the following series:

(a) $\sum_{i=1}^{3}(2 + 3i)$

(c) $\sum_{j=0}^{n} 2^j$ for $n = 1, 2, 3, 4$

(b) $\sum_{i=-2}^{1} i^2$

(d) $\sum_{k=1}^{n}(2k - 1)$ for $n = 1, 2, 3, 4$

2. Calculate the following series:

(a) $\sum_{k=1}^{3} k^n$ for $n = 1, 2, 3, 4$

(b) $\sum_{i=1}^{5} 20$

(c) $\sum_{j=0}^{3} (n^j + 1)$ for $n = 1, 2, 3, 4$

(d) $\sum_{k=-n}^{n} k$ for $n = 1, 2, 3, 4$

3.

(a) Express the formula $\sum_{i=1}^{n} \frac{1}{i(i+1)} = \frac{n}{n+1}$ without using summation notation.

(b) Verify this formula for $n = 3$.

(c) Repeat parts (a) and (b) for $\sum_{i=1}^{n} i^3 = \frac{n^2(n+1)^2}{4}$

4. Verify the following properties for $n = 3$.

(a) $\sum_{i=1}^{n} (a_i + b_i) = \sum_{i=1}^{n} a_i + \sum_{i=1}^{n} b_i$

(b) $c \left(\sum_{i=1}^{n} a_i \right) = \sum_{i=1}^{n} c a_i$

5. Rewrite the following without summation sign for $n = 3$. It is not necessary that you understand or expand the notation $\begin{pmatrix} n \\ k \end{pmatrix}$ at this point.

$$(x + y)^n = \sum_{k=0}^{n} \begin{pmatrix} n \\ k \end{pmatrix} x^{n-k} y^k.$$

6.

(a) Draw the Venn diagram for $\bigcap_{i=1}^{3} A_i$.

(b) Express in "expanded format": $A \cup \left(\bigcap_{i=1}^{n} B_i \right) = \bigcap_{i=1}^{n} (A \cup B_n)$.

7. For any positive integer k, let $A_k = \{x \in \mathbb{Q} : k - 1 < x \le k\}$ and $B_k = \{x \in \mathbb{Q} : -k < x < k\}$. What are the following sets?

(a) $\bigcup_{i=1}^{5} A_i$ (c) $\bigcap_{i=1}^{5} A_i$

(b) $\bigcup_{i=1}^{5} B_i$ (d) $\bigcap_{i=1}^{5} B_i$

8. For any positive integer k, let $A_k = \{x \in \mathbb{Q} : 0 < x < 1/k\}$ and $B_k = \{x \in \mathbb{Q} : 0 < x < k\}$. What are the following sets?

(a) $\bigcup_{i=1}^{\infty} A_i$ (c) $\bigcap_{i=1}^{\infty} A_i$

(b) $\bigcup_{i=1}^{\infty} B_i$ (d) $\bigcap_{i=1}^{\infty} B_i$

9. The symbol Π is used for the product of numbers in the same way that Σ is used for sums. For example, $\prod_{i=1}^{5} x_i = x_1 x_2 x_3 x_4 x_5$. Evaluate the following:

(a) $\prod_{i=1}^{3} i^2$ (b) $\prod_{i=1}^{3} (2i + 1)$

10. Evaluate

(a) $\prod_{k=0}^{3} 2^k$ (b) $\prod_{k=1}^{100} \frac{k}{k+1}$

Chapter 2

Combinatorics

Enumerative Combinatorics

Enumerative combinatorics
Date back to the first prehistorics
Who counted; relations
Like sets' permutations
To them were part cult, part folklorics.

Michael Toalster, The Omnificent English Dictionary In Limerick Form

Throughout this book we will be counting things. In this chapter we will outline some of the tools that will help us count.

Counting occurs not only in highly sophisticated applications of mathematics to engineering and computer science but also in many basic applications. Like many other powerful and useful tools in mathematics, the concepts are simple; we only have to recognize when and how they can be applied.

2.1 Basic Counting Techniques - The Rule of Products

2.1.1 What is Combinatorics?

One of the first concepts our parents taught us was the "art of counting." We were taught to raise three fingers to indicate that we were three years old. The question of "how many" is a natural and frequently asked question. Combinatorics is the "art of counting." It is the study of techniques that will help us to count the number of objects in a set quickly. Highly sophisticated results can be obtained with this simple concept. The following examples will illustrate that many questions concerned with counting involve the same process.

Example 2.1.1 How many lunches can you have? A snack bar serves five different sandwiches and three different beverages. How many different lunches can a person order? One way of determining the number of possible lunches is by listing or enumerating all the possibilities. One systematic way of doing this is by means of a tree, as in the following figure.

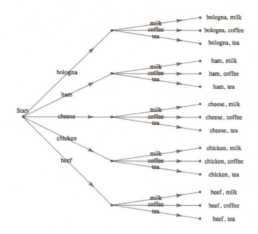

Figure 2.1.2: Tree diagram to enumerate the number of possible lunches.

Every path that begins at the position labeled START and goes to the right can be interpreted as a choice of one of the five sandwiches followed by a choice of one of the three beverages. Note that considerable work is required to arrive at the number fifteen this way; but we also get more than just a number. The result is a complete list of all possible lunches. If we need to answer a question that starts with "How many . . . ," enumeration would be done only as a last resort. In a later chapter we will examine more enumeration techniques.

An alternative method of solution for this example is to make the simple observation that there are five different choices for sandwiches and three different choices for beverages, so there are $5 \cdot 3 = 15$ different lunches that can be ordered. $\square$

Example 2.1.3 Counting elements in a cartesian product. Let $A = \{a, b, c, d, e\}$ and $B = \{1, 2, 3\}$. From Chapter 1 we know how to list the elements in $A \times B = \{(a, 1), (a, 2), (a, 3), ..., (e, 3)\}$. Since the first entry of each pair can be any one of the five elements a, b, c, d, and e, and since the second can be any one of the three numbers 1, 2, and 3, it is quite clear there are $5 \cdot 3 = 15$ different elements in $A \times B$. $\square$

Example 2.1.4 A True-False Questionnaire. A person is to complete a true-false questionnaire consisting of ten questions. How many different ways are there to answer the questionnaire? Since each question can be answered in either of two ways (true or false), and there are ten questions, there are

$$2 \cdot 2 \cdot 2 \cdot 2 \cdot 2 \cdot 2 \cdot 2 \cdot 2 \cdot 2 \cdot 2 = 2^{10} = 1024$$

different ways of answering the questionnaire. The reader is encouraged to visualize the tree diagram of this example, but not to draw it! $\square$

We formalize the procedures developed in the previous examples with the following rule and its extension.

2.1.2 The Rule Of Products

If two operations must be performed, and if the first operation can always be performed p_1 different ways and the second operation can always be performed p_2 different ways, then there are $p_1 p_2$ different ways that the two operations can be performed.

Note: It is important that p_2 does not depend on the option that is chosen in the first operation. Another way of saying this is that p_2 is independent of the first operation. If p_2 is dependent on the first operation, then the rule of products does not apply.

Example 2.1.5 Reduced Lunch Possibilities. Assume in Example 2.1.1, coffee is not served with a beef or chicken sandwiches. Then by inspection of Figure 2.1.2 we see that there are only thirteen different choices for lunch. The rule of products does not apply, since the choice of beverage depends on one's choice of a sandwich. □

Extended Rule Of Products. The rule of products can be extended to include sequences of more than two operations. If n operations must be performed, and the number of options for each operation is p_1, $p_2, \ldots, p_n$ respectively, with each p_i independent of previous choices, then the n operations can be performed $p_1 \cdot p_2 \cdots \cdots p_n$ different ways.

Example 2.1.6 A Multiple Choice Questionnaire. A questionnaire contains four questions that have two possible answers and three questions with five possible answers. Since the answer to each question is independent of the answers to the other questions, the extended rule of products applies and there are $2 \cdot 2 \cdot 2 \cdot 2 \cdot 5 \cdot 5 \cdot 5 = 2^4 \cdot 5^3 = 2000$ different ways to answer the questionnaire. □

In Chapter 1 we introduced the power set of a set A, $\mathcal{P}(A)$, which is the set of all subsets of A. Can we predict how many elements are in $\mathcal{P}(A)$ for a given finite set A? The answer is yes, and in fact if $|A| = n$, then $|\mathcal{P}(A)| = 2^n$. The ease with which we can prove this fact demonstrates the power and usefulness of the rule of products. Do not underestimate the usefulness of simple ideas.

Theorem 2.1.7 Power Set Cardinality Theorem. *If A is a finite set, then* $|\mathcal{P}(A)| = 2^{|A|}$.

Proof. Proof: Consider how we might determine any $B \in \mathcal{P}(A)$, where $|A| = n$. For each element $x \in A$ there are two choices, either $x \in B$ or $x \notin B$. Since there are n elements of A we have, by the rule of products,

$$\underbrace{2 \cdot 2 \cdots \cdots 2}_{n \text{ factors}} = 2^n$$

different subsets of A. Therefore, $\mathcal{P}(A) = 2^n$. ■

2.1.3 Exercises

1. In horse racing, to bet the "daily double" is to select the winners of the first two races of the day. You win only if both selections are correct. In terms of the number of horses that are entered in the first two races, how many different daily double bets could be made?

2. Professor Shortcut records his grades using only his students' first and last initials. What is the smallest class size that will definitely force Prof. S. to use a different system?

3. A certain shirt comes in four sizes and six colors. One also has the choice of a dragon, an alligator, or no emblem on the pocket. How many different shirts could you order?

4. A builder of modular homes would like to impress his potential customers with the variety of styles of his houses. For each house there are blueprints for three different living rooms, four different bedroom configurations, and two different garage styles. In addition, the outside can be finished in cedar shingles or brick. How many different houses can be designed from these plans?

5. The Pi Mu Epsilon mathematics honorary society of Outstanding University wishes to have a picture taken of its six officers. There will be two rows of three people. How many different way can the six officers be arranged?

6. An automobile dealer has several options available for each of three different packages of a particular model car: a choice of two styles of seats in three different colors, a choice of four different radios, and five different exteriors. How many choices of automobile does a customer have?

7. A clothing manufacturer has put out a mix-and-match collection consisting of two blouses, two pairs of pants, a skirt, and a blazer. How many outfits can you make? Did you consider that the blazer is optional? How many outfits can you make if the manufacturer adds a sweater to the collection?

8. As a freshman, suppose you had to take two of four lab science courses, one of two literature courses, two of three math courses, and one of seven physical education courses. Disregarding possible time conflicts, how many different schedules do you have to choose from?

9. (a) Suppose each single character stored in a computer uses eight bits. Then each character is represented by a different sequence of eight 0's and 1's called a bit pattern. How many different bit patterns are there? (That is, how many different characters could be represented?)

    (b) How many bit patterns are palindromes (the same backwards as forwards)?

    (c) How many different bit patterns have an even number of 1's?

10. Automobile license plates in Massachusetts usually consist of three digits followed by three letters. The first digit is never zero. How many different plates of this type could be made?

11. (a) Let $A = \{1, 2, 3, 4\}$. Determine the number of different subsets of A.

    (b) Let $A = \{1, 2, 3, 4, 5\}$. Determine the number of proper subsets of A.

12. How many integers from 100 to 999 can be written in base ten without using the digit 7?

13. Consider three persons, A, B, and C, who are to be seated in a row of three chairs. Suppose A and B are identical twins. How many seating arrangements of these persons can there be

    (a) If you are a total stranger?   (b) If you are A and B's mother?

    This problem is designed to show you that different people can have different correct answers to the same problem.

14. How many ways can a student do a ten-question true-false exam if he or she can choose not to answer any number of questions?

15. Suppose you have a choice of fish, lamb, or beef for a main course, a choice of peas or carrots for a vegetable, and a choice of pie, cake, or ice cream for dessert. If you must order one item from each category, how many different dinners are possible?

16. Suppose you have a choice of vanilla, chocolate, or strawberry for ice cream, a choice of peanuts or walnuts for chopped nuts, and a choice of hot fudge or marshmallow for topping. If you must order one item from each category, how many different sundaes are possible?

17. A questionnaire contains six questions each having yes-no answers. For each yes response, there is a follow-up question with four possible responses.

    (a) Draw a tree diagram that illustrates how many ways a single question in the questionnaire can be answered.

    (b) How many ways can the questionnaire be answered?

18. Ten people are invited to a dinner party. How many ways are there of seating them at a round table? If the ten people consist of five men and five women, how many ways are there of seating them if each man must be surrounded by two women around the table?

19. How many ways can you separate a set with n elements into two nonempty subsets if the order of the subsets is immaterial? What if the order of the subsets is important?

20. A gardener has three flowering shrubs and four nonflowering shrubs, where all shrubs are distinguishable from one another. He must plant these shrubs in a row using an alternating pattern, that is, a shrub must be of a different type from that on either side. How many ways can he plant these shrubs? If he has to plant these shrubs in a circle using the same pattern, how many ways can he plant this circle? Note that one nonflowering shrub will be left out at the end.

2.2 Permutations

2.2.1 Ordering Things

A number of applications of the rule of products are of a specific type, and because of their frequent appearance they are given their own designation, permutations. Consider the following examples.

Example 2.2.1 Ordering the elements of a set. How many different ways can we order the three different elements of the set $A = \{a, b, c\}$? Since we have three choices for position one, two choices for position two, and one choice for the third position, we have, by the rule of products, $3 \cdot 2 \cdot 1 = 6$ different ways of ordering the three letters. We illustrate through a tree diagram.

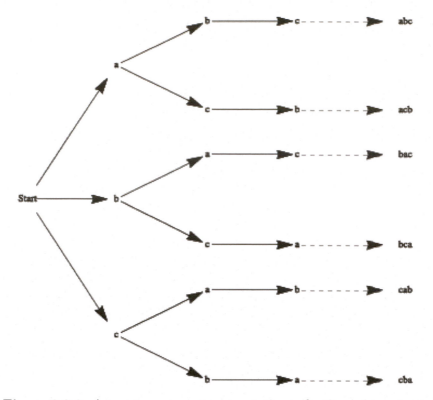

Figure 2.2.2: A tree to enumerate permutations of a three element set.

Each of the six orderings is called a permutation of the set A. □

Example 2.2.3 Ordering a schedule. A student is taking five courses in the fall semester. How many different ways can the five courses be listed? There are $5 \cdot 4 \cdot 3 \cdot 2 \cdot 1 = 120$ different permutations of the set of courses. □

In each of the above examples of the rule of products we observe that:

(a) We are asked to order or arrange elements from a single set.

(b) Each element is listed exactly once in each list (permutation). So if there are n choices for position one in a list, there are $n-1$ choices for position two, $n-2$ choices for position three, etc.

Example 2.2.4 Some orderings of a baseball team. The alphabetical ordering of the players of a baseball team is one permutation of the set of players. Other orderings of the players' names might be done by batting average, age, or height. The information that determines the ordering is called the key. We would expect that each key would give a different permutation of the names. If there are twenty-five players on the team, there are $25 \cdot 24 \cdot 23 \cdots \cdot 3 \cdot 2 \cdot 1$ different permutations of the players.

This number of permutations is huge. In fact it is 15511210043330985984000000, but writing it like this isn't all that instructive, while leaving it as a product as we originally had makes it easier to see where the number comes from. We just need to find a more compact way of writing these products. □

We now develop notation that will be useful for permutation problems.

Definition 2.2.5 Factorial. If n is a positive integer then n factorial is the product of the first n positive integers and is denoted $n!$. Additionally, we define zero factorial, $0!$, to be 1. ◊

The first few factorials are

$$
\begin{array}{c|cccccccc}
n & 0 & 1 & 2 & 3 & 4 & 5 & 6 & 7 \\
\hline
n! & 1 & 1 & 2 & 6 & 24 & 120 & 720 & 5040
\end{array} \cdot
$$

Note that 4! is 4 times 3!, or 24, and 5! is 5 times 4!, or 120. In addition, note that as n grows in size, $n!$ grows extremely quickly. For example, $11! = 39916800$. If the answer to a problem happens to be 25!, as in the previous example, you would never be expected to write that number out completely. However, a problem with an answer of $\frac{25!}{23!}$ can be reduced to $25 \cdot 24$, or 600.

If $|A| = n$, there are $n!$ ways of permuting all n elements of A . We next consider the more general situation where we would like to permute k elements out of a set of n objects, where $k \leq n$.

Example 2.2.6 Choosing Club Officers. A club of twenty-five members will hold an election for president, secretary, and treasurer in that order. Assume a person can hold only one position. How many ways are there of choosing these three officers? By the rule of products there are $25 \cdot 24 \cdot 23$ ways of making a selection. $\qquad \square$

Definition 2.2.7 Permutation. An ordered arrangement of k elements selected from a set of n elements, $0 \leq k \leq n$, where no two elements of the arrangement are the same, is called a permutation of n objects taken k at a time. The total number of such permutations is denoted by $P(n, k)$. $\qquad \Diamond$

Theorem 2.2.8 Permutation Counting Formula. *The number of possible permutations of k elements taken from a set of n elements is*

$$
P(n,k) = n \cdot (n-1) \cdot (n-2) \cdots \cdots (n-k+1) = \prod_{j=0}^{k-1}(n-j) = \frac{n!}{(n-k)!}.
$$

Proof. Case I: If $k = n$ we have $P(n,n) = n! = \frac{n!}{(n-n)!}$.

Case II: If $0 \leq k < n$, then we have k positions to fill using n elements and

(a) Position 1 can be filled by any one of $n - 0 = n$ elements

(b) Position 2 can be filled by any one of $n - 1$ elements

(c) $\cdots$

(d) Position k can be filled by any one of $n - (k-1) = n - k + 1$ elements

Hence, by the rule of products,

$$
P(n,k) = n \cdot (n-1) \cdot (n-2) \cdots \cdots (n-k+1) = \frac{n!}{(n-k)!}.
$$

$\blacksquare$

It is important to note that the derivation of the permutation formula given above was done solely through the rule of products. This serves to reiterate our introductory remarks in this section that permutation problems are really rule-of-products problems. We close this section with several examples.

Example 2.2.9 Another example of choosing officers. A club has eight members eligible to serve as president, vice-president, and treasurer. How many ways are there of choosing these officers?

Solution 1: Using the rule of products. There are eight possible choices for the presidency, seven for the vice-presidency, and six for the office of treasurer.

By the rule of products there are $8 \cdot 7 \cdot 6 = 336$ ways of choosing these officers.

Solution 2: Using the permutation formula. We want the total number of permutations of eight objects taken three at a time:

$$P(8,3) = \frac{8!}{(8-3)!} = 8 \cdot 7 \cdot 6 = 336$$

$\square$

Example 2.2.10 Course ordering, revisited. To count the number of ways to order five courses, we can use the permutation formula. We want the number of permutations of five courses taken five at a time:

$$P(5,5) = \frac{5!}{(5-5)!} = 5! = 120.$$

$\square$

Example 2.2.11 Ordering of digits under different conditions. Consider only the digits 1, 2, 3, 4, and 5.

a How many three-digit numbers can be formed if no repetition of digits can occur?

b How many three-digit numbers can be formed if repetition of digits is allowed?

c How many three-digit numbers can be formed if only non-consecutive repetition of digits are allowed?

Solutions to (a): Solution 1: Using the rule of products. We have any one of five choices for digit one, any one of four choices for digit two, and three choices for digit three. Hence, $5 \cdot 4 \cdot 3 = 60$ different three-digit numbers can be formed.

Solution 2; Using the permutation formula. We want the total number of permutations of five digits taken three at a time:

$$P(5,3) = \frac{5!}{(5-3)!} = 5 \cdot 4 \cdot 3 = 60.$$

Solution to (b): The definition of permutation indicates "...no two elements in each list are the same." Hence the permutation formula cannot be used. However, the rule of products still applies. We have any one of five choices for the first digit, five choices for the second, and five for the third. So there are $5 \cdot 5 \cdot 5 = 125$ possible different three-digit numbers if repetition is allowed.

Solution to (c): Again, the rule of products applies here. We have any one of five choices for the first digit, but then for the next two digits we have four choices since we are not allowed to repeat the previous digit So there are $5 \cdot 4 \cdot 4 = 80$ possible different three-digit numbers if only non-consecutive repetitions are allowed. $\square$

2.2.2 Exercises

1. If a raffle has three different prizes and there are 1,000 raffle tickets sold, how many different ways can the prizes be distributed?

2.

    (a) How many three-digit numbers can be formed from the digits 1, 2, 3 if no repetition of digits is allowed? List the three-digit numbers.

    (b) How many two-digit numbers can be formed if no repetition of digits is allowed? List them.

    (c) How many two-digit numbers can be obtained if repetition is allowed?

3. How many eight-letter words can be formed from the 26 letters in the alphabet? Even without concerning ourselves about whether the words make sense, there are two interpretations of this problem. Answer both.

4. Let A be a set with $|A| = n$. Determine

    (a) $|A^3|$

    (b) $|\{(a,b,c) \mid a,b,c \in A \text{ and each coordinate is different}\}|$

5. The state finals of a high school track meet involves fifteen schools. How many ways can these schools be listed in the program?

6. Consider the three-digit numbers that can be formed from the digits 1, 2, 3, 4, and 5 with no repetition of digits allowed.
    a. How many of these are even numbers?
    b. How many are greater than 250?

7. All 15 players on the Tall U. basketball team are capable of playing any position.

    (a) How many ways can the coach at Tall U. fill the five starting positions in a game?

    (b) What is the answer if the center must be one of two players?

8.

    (a) How many ways can a gardener plant five different species of shrubs in a circle?

    (b) What is the answer if two of the shrubs are the same?

    (c) What is the answer if all the shrubs are identical?

9. The president of the Math and Computer Club would like to arrange a meeting with six attendees, the president included. There will be three computer science majors and three math majors at the meeting. How many ways can the six people be seated at a circular table if the president does not want people with the same majors to sit next to one other?

10. Six people apply for three identical jobs and all are qualified for the positions. Two will work in New York and the other one will work in San Diego. How many ways can the positions be filled?

11. Let $A = \{1, 2, 3, 4\}$. Determine the cardinality of

    (a) $\{(a_1, a_2) \mid a_1 \neq a_2\}$

    (b) What is the answer to the previous part if $|A| = n$

    (c) If $|A| = n$, determine the number of m-tuples in A, $m \leq n$, where each coordinate is different from the other coordinates.

2.3 Partitions of Sets and the Law of Addition

2.3.1 Partitions

One way of counting the number of students in your class would be to count the number in each row and to add these totals. Of course this problem is simple because there are no duplications, no person is sitting in two different rows. The basic counting technique that you used involves an extremely important first step, namely that of partitioning a set. The concept of a partition must be clearly understood before we proceed further.

Definition 2.3.1 Partition. A partition of set A is a set of one or more nonempty subsets of A: $A_1, A_2, A_3, \cdots$, such that every element of A is in exactly one set. Symbolically,

(a) $A_1 \cup A_2 \cup A_3 \cup \cdots = A$

(b) If $i \neq j$ then $A_i \cap A_j = \emptyset$

$\Diamond$

The subsets in a partition are often referred to as blocks. Note how our definition allows us to partition infinite sets, and to partition a set into an infinite number of subsets. Of course, if A is finite the number of subsets can be no larger than $|A|$.

Example 2.3.2 Some partitions of a four element set. Let $A = \{a, b, c, d\}$. Examples of partitions of A are:

- $\{\{a\}, \{b\}, \{c, d\}\}$

- $\{\{a, b\}, \{c, d\}\}$

- $\{\{a\}, \{b\}, \{c\}, \{d\}\}$

How many others are there, do you suppose?

There are 15 different partitions. The most efficient way to count them all is to classify them by the size of blocks. For example, the partition $\{\{a\}, \{b\}, \{c, d\}\}$ has block sizes 1, 1, and 2. $\square$

Example 2.3.3 Some Integer Partitions. Two examples of partitions of set of integers $\mathbb{Z}$ are

- $\{\{n\} \mid n \in \mathbb{Z}\}$ and

- $\{\{n \in \mathbb{Z} \mid n < 0\}, \{0\}, \{n \in \mathbb{Z} \mid 0 < n\}\}$.

The set of subsets $\{\{n \in \mathbb{Z} \mid n \geq 0\}, \{n \in \mathbb{Z} \mid n \leq 0\}\}$ is not a partition because the two subsets have a nonempty intersection. A second example of a non-partition is $\{\{n \in \mathbb{Z} \mid |n| = k\} \mid k = -1, 0, 1, 2, \cdots\}$ because one of the blocks, when $k = -1$ is empty. $\square$

One could also think of the concept of partitioning a set as a "packaging problem." How can one "package" a carton of, say, twenty-four cans? We could use: four six-packs, three eight-packs, two twelve-packs, etc. In all cases: (a) the sum of all cans in all packs must be twenty-four, and (b) a can must be in one and only one pack.

2.3.2 Addition Laws

Theorem 2.3.4 The Basic Law Of Addition:. *If A is a finite set, and if $\{A_1, A_2, \ldots, A_n\}$ is a partition of A, then*

$$|A| = |A_1| + |A_2| + \cdots + |A_n| = \sum_{k=1}^{n} |A_k|$$

The basic law of addition can be rephrased as follows: If A is a finite set where $A_1 \cup A_2 \cup \cdots \cup A_n = A$ and where $A_i \cap A_j = \emptyset$ whenever $i \neq j$, then

$$|A| = |A_1 \cup A_2 \cup \cdots \cup A_n| = |A_1| + |A_2| + \cdots + |A_n|$$

Example 2.3.5 Counting All Students. The number of students in a class could be determined by adding the numbers of students who are freshmen, sophomores, juniors, and seniors, and those who belong to none of these categories. However, you probably couldn't add the students by major, since some students may have double majors. □

Example 2.3.6 Counting Students in Disjoint Classes. The sophomore computer science majors were told they must take one and only one of the following courses that are open only to them: Cryptography, Data Structures, or Javascript. The numbers in each course, respectively, for sophomore CS majors, were 75, 60, 55. How many sophomore CS majors are there? The Law of Addition applies here. There are exactly $75 + 60 + 55 = 190$ CS majors since the rosters of the three courses listed above would be a partition of the CS majors. □

Example 2.3.7 Counting Students in Non-disjoint Classes. It was determined that all junior computer science majors take at least one of the following courses: Algorithms, Logic Design, and Compiler Construction. Assume the number in each course was 75, 60 and 55, respectively for the three courses listed. Further investigation indicated ten juniors took all three courses, twenty-five took Algorithms and Logic Design, twelve took Algorithms and Compiler Construction, and fifteen took Logic Design and Compiler Construction. How many junior C.S. majors are there?

Example 2.3.6 was a simple application of the law of addition, however in this example some students are taking two or more courses, so a simple application of the law of addition would lead to double or triple counting. We rephrase information in the language of sets to describe the situation more explicitly.

A = the set of all junior computer science majors

A_1 = the set of all junior computer science majors who took Algorithms

A_2 = the set of all junior computer science majors who took Logic Design

A_3 = the set of all junior computer science majors who took Compiler Construction

Since all junior CS majors must take at least one of the courses, the number we want is:

$$|A| = |A_1 \cup A_2 \cup A_3| = |A_1| + |A_2| + |A_3| - \text{repeats}.$$

A Venn diagram is helpful to visualize the problem. In this case the universal set U can stand for all students in the university.

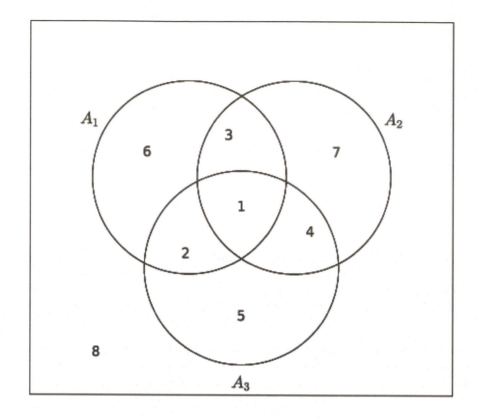

Figure 2.3.8: Venn Diagram

We see that the whole universal set is naturally partitioned into subsets that are labeled by the numbers 1 through 8, and the set A is partitioned into subsets labeled 1 through 7. The region labeled 8 represents all students who are not junior CS majors. Note also that students in the subsets labeled 2, 3, and 4 are double counted, and those in the subset labeled 1 are triple counted. To adjust, we must subtract the numbers in regions 2, 3 and 4. This can be done by subtracting the numbers in the intersections of each pair of sets. However, the individuals in region 1 will have been removed three times, just as they had been originally added three times. Therefore, we must finally add their number back in.

$$
\begin{aligned}
|A| &= |A_1 \cup A_2 \cup A_3| \\
&= |A_1| + |A_2| + |A_3| - \text{repeats} \\
&= |A_1| + |A_2| + |A_3| - \text{duplicates} + \text{triplicates} \\
&= |A_1| + |A_2| + |A_3| - (|A_1 \cap A_2| + |A_1 \cap A_3| + |A_2 \cap A_3|) + |A_1 \cap A_2 \cap A_3| \\
&= 75 + 60 + 55 - 25 - 12 - 15 + 10 = 148
\end{aligned}
$$

$\square$

The ideas used in this latest example gives rise to a basic counting technique:

Theorem 2.3.9 Laws of Inclusion-Exclusion. *Given finite sets* $A_1, A_2, A_3,$

then

(a) The Two Set Inclusion-Exclusion Law:

$$|A_1 \cup A_2| = |A_1| + |A_2| - |A_1 \cap A_2|$$

(b) The Three Set Inclusion-Exclusion Law:

$$\begin{aligned}
|A_1 \cup A_2 \cup A_3| = & \ |A_1| + |A_2| + |A_3| \\
& - (|A_1 \cap A_2| + |A_1 \cap A_3| + |A_2 \cap A_3|) \\
& + |A_1 \cap A_2 \cap A_3|
\end{aligned}$$

The inclusion-exclusion laws extend to more than three sets, as will be explored in the exercises.

In this section we saw that being able to partition a set into disjoint subsets gives rise to a handy counting technique. Given a set, there are many ways to partition depending on what one would wish to accomplish. One natural partitioning of sets is apparent when one draws a Venn diagram. This particular partitioning of a set will be discussed further in Chapters 4 and 13.

2.3.3 Exercises for Section 2.3

1. List all partitions of the set $A = \{a, b, c\}$.

2. Which of the following collections of subsets of the plane, $\mathbb{R}^2$, are partitions?

    (a) $\{\{(x, y) \mid x + y = c\} \mid c \in \mathbb{R}\}$

    (b) The set of all circles in $\mathbb{R}^2$

    (c) The set of all circles in $\mathbb{R}^2$ centered at the origin together with the set $\{(0, 0)\}$

    (d) $\{\{(x, y)\} \mid (x, y) \in \mathbb{R}^2\}$

3. A student, on an exam paper, defined the term partition the following way: "Let A be a set. A partition of A is any set of nonempty subsets $A_1, A_2, A_3, \ldots$ of A such that each element of A is in one of the subsets." Is this definition correct? Why?

4. Let A_1 and A_2 be subsets of a set U. Draw a Venn diagram of this situation and shade in the subsets $A_1 \cap A_2$, $A_1^c \cap A_2$, $A_1 \cap A_2^c$, and $A_1^c \cap A_2^c$
    . Use the resulting diagram and the definition of partition to convince yourself that the subset of these four subsets that are nonempty form a partition of U.

5. Show that $\{\{2n \mid n \in \mathbb{Z}\}, \{2n + 1 \mid n \in \mathbb{Z}\}\}$ is a partition of $\mathbb{Z}$. Describe this partition using only words.

6.

    (a) A group of 30 students were surveyed and it was found that 18 of them took Calculus and 12 took Physics. If all students took at least one course, how many took both Calculus and Physics? Illustrate using a Venn diagram.

    (b) What is the answer to the question in part (a) if five students did not take either of the two courses? Illustrate using a Venn diagram.

7. A survey of 90 people, 47 of them played tennis and 42 of them swam. If 17 of them participated in both activities, how many of them participated in neither?

8. A survey of 300 people found that 60 owned an iPhone, 75 owned a Blackberry, and 30 owned an Android phone. Furthermore, 40 owned both an iPhone and a Blackberry, 12 owned both an iPhone and an Android phone, and 8 owned a Blackberry and an Android phone. Finally, 3 owned all three phones.

    (a) How many people surveyed owned none of the three phones?

    (b) How many people owned a Blackberry but not an iPhone?

    (c) How many owned a Blackberry but not an Android?

9. Regarding the Theorem 2.3.9,

    (a) Use the two set inclusion-exclusion law to derive the three set inclusion-exclusion law. Note: A knowledge of basic set laws is needed for this exercise.

    (b) State and derive the inclusion-exclusion law for four sets.

10. To complete your spring schedule, you must add Calculus and Physics. At 9:30, there are three Calculus sections and two Physics sections; while at 11:30, there are two Calculus sections and three Physics sections. How many ways can you complete your schedule if your only open periods are 9:30 and 11:30?

11. The definition of $\mathbb{Q} = \{a/b \mid a, b \in \mathbb{Z}, b \neq 0\}$ given in Chapter 1 is awkward. If we use the definition to list elements in $\mathbb{Q}$, we will have duplications such as $\frac{1}{2}$, $\frac{-2}{-4}$ and $\frac{300}{600}$ Try to write a more precise definition of the rational numbers so that there is no duplication of elements.

2.4 Combinations and the Binomial Theorem

2.4.1 Combinations

In Section 2.1 we investigated the most basic concept in combinatorics, namely, the rule of products. It is of paramount importance to keep this fundamental rule in mind. In Section 2.2 we saw a subclass of rule-of-products problems, permutations, and we derived a formula as a computational aid to assist us. In this section we will investigate another counting formula, one that is used to count combinations, which are subsets of a certain size.

In many rule-of-products applications the ordering is important, such as the batting order of a baseball team. In other cases it is not important, as in placing coins in a vending machine or in the listing of the elements of a set. Order is important in permutations. Order is not important in combinations.

Example 2.4.1 Counting Permutations. How many different ways are there to permute three letters from the set $A = \{a, b, c, d\}$? From the Permutation Counting Formula there are $P(4,3) = \frac{4!}{(4-3)!} = 24$ different orderings of three letters from A □

Example 2.4.2 Counting with No Order. How many ways can we select a set of three letters from $A = \{a, b, c, d\}$? Note here that we are not concerned with the order of the three letters. By trial and error, abc, abd, acd, and bcd are the only listings possible. To repeat, we were looking for all three-element subsets of the set A. Order is not important in sets. The notation for choosing 3 elements from 4 is most commonly $\binom{4}{3}$ or occasionally $C(4,3)$, either of which is read "4 choose 3" or the number of combinations for four objects taken three at a time. □

Definition 2.4.3 Binomial Coefficient. Let n and k be nonnegative integers. The binomial coefficient $\binom{n}{k}$ represents the number of combinations of n objects taken k at a time, and is read "n choose k." ◊

We would now like to investigate the relationship between permutation and combination problems in order to derive a formula for $\binom{n}{k}$

Let us reconsider the Counting with No Order. There are $3! = 6$ different orderings for each of the three-element subsets. The table below lists each subset of A and all permutations of each subset on the same line.

subset	permutations
$\{a,b,c\}$	$abc, acb, bca, bac, cab, cba$
$\{a,b,d\}$	$abd, adb, bda, bad, dab, dba$
$\{a,c,d\}$	$acd, adc, cda, cad, dac, dca$
$\{b,c,d\}$	$bcd, bdc, cdb, cbd, dbc, dcb$

Hence, $\binom{4}{3} = \frac{P(4,3)}{3!} = \frac{4!}{(4-3)! \cdot 3!} = 4$

We generalize this result in the following theorem:

Theorem 2.4.4 Binomial Coefficient Formula. *If n and k are nonnegative integers with $0 \le k \le n$, then the number k-element subsets of an n element set is equal to*

$$\binom{n}{k} = \frac{n!}{(n-k)! \cdot k!}.$$

Proof. Proof 1: There are $k!$ ways of ordering the elements of any k element set. Therefore,

$$\binom{n}{k} = \frac{P(n,k)}{k!} = \frac{n!}{(n-k)!k!}..$$

Proof 2: To "construct" a permutation of k objects from a set of n elements, we can first choose one of the subsets of objects and second, choose one of the $k!$ permutations of those objects. By the rule of products,

$$P(n,k) = \binom{n}{k} \cdot k!$$

and solving for $\binom{n}{k}$ we get the desired formula. ∎

Example 2.4.5 Flipping Coins. Assume an evenly balanced coin is tossed five times. In how many ways can three heads be obtained? This is a combination problem, because the order in which the heads appear does not matter. We can think of this as a situation involving sets by considering the set of flips of the coin, 1 through 5, in which heads comes up. The number of ways to get three heads is $\binom{5}{3} = \frac{5 \cdot 4}{2 \cdot 1} = 10$. □

Example 2.4.6 Counting five ordered flips two ways. We determine the total number of ordered ways a fair coin can land if tossed five consecutive times. The five tosses can produce any one of the following mutually exclusive, disjoint events: 5 heads, 4 heads, 3 heads, 2 heads, 1 head, or 0 heads. For example, by the previous example, there are $\binom{5}{3} = 10$ sequences in which three heads appear. Counting the other possibilities in the same way, by the law of addition we have:

$$\binom{5}{5} + \binom{5}{4} + \binom{5}{3} + \binom{5}{2} + \binom{5}{1} + \binom{5}{0} = 1 + 5 + 10 + 10 + 5 + 1 = 32$$

ways to observe the five flips.

Of course, we could also have applied the extended rule of products, and since there are two possible outcomes for each of the five tosses, we have $2^5 = 32$ ways. $\square$

You might think that counting something two ways is a waste of time but solving a problem two different ways often is instructive and leads to valuable insights. In this case, it suggests a general formula for the sum $\sum_{k=0}^{n} \binom{n}{k}$. In the case of $n = 5$, we get 2^5 so it is reasonable to expect that the general sum is 2^n, and it is. A logical argument to prove the general statment simply involves generalizing the previous example to n coin flips.

Example 2.4.7 A Committee of Five. A committee usually starts as an unstructured set of people selected from a larger membership. Therefore, a committee can be thought of as a combination. If a club of 25 members has a five-member social committee, there are $\binom{25}{5} = \frac{25 \cdot 24 \cdot 23 \cdot 22 \cdot 21}{5!} = 53130$ different possible social committees. If any structure or restriction is placed on the way the social committee is to be selected, the number of possible committees will probably change. For example, if the club has a rule that the treasurer must be on the social committee, then the number of possibilities is reduced to $\binom{24}{4} = \frac{24 \cdot 23 \cdot 22 \cdot 21}{4!} = 10626$.

If we further require that a chairperson other than the treasurer be selected for the social committee, we have $\binom{24}{4} \cdot 4 = 42504$ different possible social committees. The choice of the four non-treasurers accounts for the factor $\binom{24}{4}$ while the need to choose a chairperson accounts for the 4. $\square$

Example 2.4.8 Binomial Coefficients - Extreme Cases. By simply applying the definition of a Binomial Coefficient as a number of subsets we see that there is $\binom{n}{0} = 1$ way of choosing a combination of zero elements from a set of n. In addition, we see that there is $\binom{n}{n} = 1$ way of choosing a combination of n elements from a set of n.

We could compute these values using the formula we have developed, but no arithmetic is really needed here. Other properties of binomial coefficients that can be derived using the subset definition will be seen in the exercises $\square$

2.4.2 The Binomial Theorem

The binomial theorem gives us a formula for expanding $(x + y)^n$, where n is a nonnegative integer. The coefficients of this expansion are precisely the binomial coefficients that we have used to count combinations. Using high school algebra we can expand the expression for integers from 0 to 5:

n	$(x+y)^n$
0	1
1	$x + y$
2	$x^2 + 2xy + y^2$
3	$x^3 + 3x^2y + 3xy^2 + y^3$
4	$x^4 + 4x^3y + 6x^2y^2 + 4xy^3 + y^4$
5	$x^5 + 5x^4y + 10x^3y^2 + 10x^2y^3 + 5xy^4 + y^5$

In the expansion of $(x+y)^5$ we note that the coefficient of the third term is

$\binom{5}{3} = 10$, and that of the sixth term is $\binom{5}{5} = 1$. We can rewrite the expansion as

$$\binom{5}{0}x^5 + \binom{5}{1}x^4y + \binom{5}{2}x^3y^2 + \binom{5}{3}x^2y^3 + \binom{5}{4}xy^4 + \binom{5}{5}y^5.$$

In summary, in the expansion of $(x+y)^n$ we note:

(a) The first term is x^n and the last term is y^n.

(b) With each successive term, exponents of x decrease by 1 as those of y increase by 1. For any term the sum of the exponents is n.

(c) The coefficient of $x^{n-k}y^k$ is $\binom{n}{k}$.

(d) The triangular array of binomial coefficients is called Pascal's triangle after the seventeenth-century French mathematician Blaise Pascal. Note that each number in the triangle other than the 1's at the ends of each row is the sum of the two numbers to the right and left of it in the row above.

Theorem 2.4.9 The Binomial Theorem. *If $n \geq 0$, and x and y are numbers, then*

$$(x+y)^n = \sum_{k=0}^{n} \binom{n}{k} x^{n-k} y^k.$$

Proof. This theorem will be proven using a logical procedure called mathematical induction, which will be introduced in Chapter 3. ■

Example 2.4.10 Identifying a term in an expansion. Find the third term in the expansion of $(x-y)^4 = (x+(-y))^4$. The third term, when $k=2$, is $\binom{4}{2}x^{4-2}(-y)^2 = 6x^2y^2$. □

Example 2.4.11 A Binomial Expansion. Expand $(3x-2)^3$. If we replace x and y in the Binomial Theorem with $3x$ and -2, respectively, we get

$$\sum_{k=0}^{3} \binom{3}{k}(3x)^{n-k}(-2)^k = \binom{3}{0}(3x)^3(-2)^0 + \binom{3}{1}(3x)^2(-2)^1 + \binom{3}{2}(3x)^1(-2)^2 + \binom{3}{3}(3x)^0(-2)^3.$$

$$= 27x^3 - 54x^2 + 36x - 8$$

□

2.4.3 SageMath Note

A bridge hand is a 13 element subset of a standard 52 card deck. The order in which the cards come to the player doesn't matter. From the point of view of a single player, the number of possible bridge hands is $\binom{52}{13}$, which can be easily computed with *Sage*.

```
binomial(52,13)
```

635013559600

In bridge, the location of a hand in relation to the dealer has some bearing on the game. An even truer indication of the number of possible hands takes into account *each* player's possible hand. It is customary to refer to bridge positions as West, North, East and South. We can apply the rule of product to get the total number of bridge hands with the following logic. West can get any of the $\binom{52}{13}$ hands identified above. Then North get 13 of the remaining 39 cards and so has $\binom{39}{13}$ possible hands. East then gets 13 of the 26 remaining cards, which has $\binom{26}{13}$ possibilities. South gets the remaining cards. Therefore the number of bridge hands is computed using the Product Rule.

```
binomial(52,13)*binomial(39,13)*binomial(26,13)
```

53644737765488792839237440000

2.4.4 Exercises

1. The judiciary committee at a college is made up of three faculty members and four students. If ten faculty members and 25 students have been nominated for the committee, how many judiciary committees could be formed at this point?

2. Suppose that a single character is stored in a computer using eight bits.
    a. How many bit patterns have exactly three 1's?
    b. How many bit patterns have at least two 1's?

    Hint. Think of the set of positions that contain a 1 to turn this is into a question about sets.

3. How many subsets of $\{1, 2, 3, \ldots, 10\}$ contain at least seven elements?

4. The congressional committees on mathematics and computer science are made up of five representatives each, and a congressional rule is that the two committees must be disjoint. If there are 385 members of congress, how many ways could the committees be selected?

5. The image below shows a 6 by 6 grid and an example of a **lattice path** that could be taken from $(0,0)$ to $(6,6)$, which is a path taken by traveling along grid lines going only to the right and up. How many different lattice paths are there of this type? Generalize to the case of lattice paths from $(0,0)$ to (m,n) for any nonnegative integers m and n.

Figure 2.4.12: A lattice path

    Hint. Think of each path as a sequence of instructions to go right (R) and up (U).

6. How many of the lattice paths from $(0,0)$ to $(6,6)$ pass through $(3,3)$ as the one in Figure 12 does?

7. A poker game is played with 52 cards. At the start of a game, each player gets five of the cards. The order in which cards are dealt doesn't matter.

    (a) How many "hands" of five cards are possible?

    (b) If there are four people playing, how many initial five-card "hands" are possible, taking into account all players and their positions at the table? Position with respect to the dealer does matter.

8. A flush in a five-card poker hand is five cards of the same suit. The suits are spades, clubs, diamonds and hearts. How many spade flushes are possible in a 52-card deck? How many flushes are possible in any suit?

9. How many five-card poker hands using 52 cards contain exactly two aces?

10. In poker, a full house is three-of-a-kind and a pair in one hand; for example, three fives and two queens. How many full houses are possible from a 52-card deck? You can use the sage cell in the SageMath Note to do this calculation, but also write your answer in terms of binomial coefficients.

11. A class of twelve computer science students are to be divided into three groups of 3, 4, and 5 students to work on a project. How many ways can this be done if every student is to be in exactly one group?

12. Explain in words why the following equalities are true based on number of subsets, and then verify the equalities using the formula for binomial coefficients.

    (a) $\binom{n}{1} = n$

    (b) $\binom{n}{k} = \binom{n}{n-k}$, $0 \le k \le n$

13. There are ten points, $P_1, P_2, \ldots, P_{10}$ on a plane, no three on the same line.

    (a) How many lines are determined by the points?

    (b) How many triangles are determined by the points?

14. How many ways can n persons be grouped into pairs when n is even? Assume the order of the pairs matters, but not the order within the pairs. For example, if $n = 4$, the six different groupings would be

$$\begin{array}{cc} \{1,2\} & \{3,4\} \\ \{1,3\} & \{2,4\} \\ \{1,4\} & \{2,3\} \\ \{2,3\} & \{1,4\} \\ \{2,4\} & \{1,3\} \\ \{3,4\} & \{1,2\} \end{array}$$

15. Use the binomial theorem to prove that if A is a finite set, then $|P(A)| = 2^{|A|}$

16.

    (a) A state's lottery involves choosing six different numbers out of a possible 36. How many ways can a person choose six numbers?

    (b) What is the probability of a person winning with one bet?

17. Use the binomial theorem to calculate 9998^3.

    Hint. $9998 = 10000 - 2$

18. In the card game Blackjack, there are one or more players and a dealer. Initially, each player is dealt two cards and the dealer is dealt one card down and one facing up. As in bridge, the order of the hands, but not the order of the cards in the hands, matters. Starting with a single 52 card deck, and three players, how many ways can the first two cards be dealt out? You can use the sage cell in the SageMath Note to do this calculation.

Chapter 3

Logic

In this chapter, we will introduce some of the basic concepts of mathematical logic. In order to fully understand some of the later concepts in this book, you must be able to recognize valid logical arguments. Although these arguments will usually be applied to mathematics, they employ the same techniques that are used by a lawyer in a courtroom or a physician examining a patient. An added reason for the importance of this chapter is that the circuits that make up digital computers are designed using the same algebra of propositions that we will be discussing.

3.1 Propositions and Logical Operators

3.1.1 Propositions

Definition 3.1.1 Proposition. A proposition is a sentence to which one and only one of the terms *true* or *false* can be meaningfully applied. $\Diamond$

Example 3.1.2 Some Propositions. "Four is even,", "$4 \in \{1,3,5\}$" and "$43 > 21$" are propositions. $\Box$

In traditional logic, a declarative statement with a definite truth value is considered a proposition. Although our ultimate aim is to discuss mathematical logic, we won't separate ourselves completely from the traditional setting. This is natural because the basic assumptions, or postulates, of mathematical logic are modeled after the logic we use in everyday life. Since compound sentences are frequently used in everyday speech, we expect that logical propositions contain connectives like the word "and." The statement "Europa supports life or Mars supports life" is a proposition and, hence, must have a definite truth value. Whatever that truth value is, it should be the same as the truth value of "Mars supports life or Europa supports life."

3.1.2 Logical Operations

There are several ways in which we commonly combine simple statements into compound ones. The words/phrases *and, or, not, if ... then...,* and *...if and only if ...* can be added to one or more propositions to create a new proposition. To avoid any confusion, we will precisely define each one's meaning and introduce its standard symbol. With the exception of negation (*not*), all of the operations act on pairs of propositions. Since each proposition has two possible truth values, there are four ways that truth can be assigned to two propositions. In defining the effect that a logical operation has on two propositions, the result must be specified for all four cases. The most convenient way of doing this is with a truth table, which we will illustrate by defining the word *and*.

Definition 3.1.3 Logical Conjunction. If p and q are propositions, their conjunction, p and q (denoted $p \wedge q$), is defined by the truth table

p	q	$p \wedge q$
0	0	0
0	1	0
1	0	0
1	1	1

$\Diamond$

Notes:

(a) To read this truth table, you must realize that any one line represents a case: one possible set of values for p and q.

(b) The numbers 0 and 1 are used to denote false and true, respectively. This is consistent with the way that many programming languages treat logical, or Boolean, variables since a single bit, 0 or 1, can represent a truth value.

(c) For each case, the symbol under p represents the truth value of p. The same is true for q. The symbol under $p \wedge q$ represents its truth value for that case. For example, the second row of the truth table represents the case in which p is false, q is true, and the resulting truth value for $p \wedge q$ is false. As in everyday speech, $p \wedge q$ is true only when both propositions are true.

(d) Just as the letters x, y and z are frequently used in algebra to represent numeric variables, p, q and r seem to be the most commonly used symbols for logical variables. When we say that p is a logical variable, we mean that any proposition can take the place of p.

(e) One final comment: The order in which we list the cases in a truth table is standardized in this book. If the truth table involves two simple propositions, the numbers under the simple propositions can be interpreted as the two-digit binary integers in increasing order, 00, 01, 10, and 11, for 0, 1, 2, and 3, respectively.

Definition 3.1.4 Logical Disjunction. If p and q are propositions, their disjunction, p or q (denoted $p \vee q$), is defined by the truth table

p	q	$p \vee q$
0	0	0
0	1	1
1	0	1
1	1	1

$\Diamond$

Definition 3.1.5 Logical Negation. If p is a proposition, its negation, not p, denoted $\neg p$, and is defined by the truth table

p	$\neg p$
0	1
1	0

$\Diamond$

Note: Negation is the only standard operator that acts on a single proposition; hence only two cases are needed.

Consider the following propositions from everyday speech:

(a) I'm going to quit if I don't get a raise.

(b) If I pass the final, then I'll graduate.

(c) I'll be going to the movies provided that my car starts.

All three propositions are conditional, they can all be restated to fit into the form "If *Condition*, then *Conclusion*." For example, the first statement can be rewritten as "If I don't get a raise, then I'm going to quit."

A conditional statement is meant to be interpreted as a guarantee; if the condition is true, then the conclusion is expected to be true. It says no more and no less.

Definition 3.1.6 Conditional Statement. The conditional statement "If p then q," denoted $p \rightarrow q$, is defined by the truth table

p	q	$p \rightarrow q$
0	0	1
0	1	1
1	0	0
1	1	1

Table 3.1.7: Truth Table for $p \rightarrow q$

$\Diamond$

Example 3.1.8 Analysis of a Conditional Proposition. Assume your instructor told you "If you receive a grade of 95 or better in the final examination, then you will receive an A in this course." Your instructor has made a promise to you. If you fulfill his condition, you expect the conclusion (getting an A) to be forthcoming. Suppose your graded final has been returned to you. Has your instructor told the truth or is your instructor guilty of a falsehood?

Case I: Your final exam score was less than 95 (the condition is false) and you did not receive an A (the conclusion is false). The instructor told the truth.

Case II: Your final exam score was less than 95, yet you received an A for the course. The instructor told the truth. (Perhaps your overall course average was excellent.)

Case III: Your final exam score was greater than 95, but you did not receive an A. The instructor lied.

Case IV: Your final exam score was greater than 95, and you received an A. The instructor told the truth.

To sum up, the only case in which a conditional proposition is false is when the condition is true and the conclusion is false. □

The order of the condition and conclusion in a conditional proposition is important. If the condition and conclusion are exchanged, a different proposition is produced.

Definition 3.1.9 Converse. The converse of the proposition $p \rightarrow q$ is the proposition $q \rightarrow p$.   ◊

The converse of "If you receive a grade of 95 or better in the final exam, then you will receive an A in this course," is "If you receive an A in this course, then you received a grade of 95 or better in the final exam." It should be clear that these two statements say different things.

There *is* a proposition related to $p \rightarrow q$ that does have the same logical meaning. This is the contrapositive.

Definition 3.1.10 Contrapositive. The contrapositive of the proposition $p \rightarrow q$ is the proposition $\neg q \rightarrow \neg p$.   ◊

As we will see when we discuss logical proofs, we can prove a conditional proposition by proving its contrapositive, which may be somewhat easier.

Definition 3.1.11 Biconditional Proposition. If p and q are propositions, the biconditional statement "p if and only if q," denoted $p \leftrightarrow q$, is defined by the truth table

p	q	$p \leftrightarrow q$
0	0	1
0	1	0
1	0	0
1	1	1

◊

Note that $p \leftrightarrow q$ is true when p and q have the same truth values. It is common to abbreviate "if and only if" to "iff."

Although "if ... then..." and "...if and only if ..." are frequently used in everyday speech, there are several alternate forms that you should be aware of. They are summarized in the following lists.

All of the following are equivalent to "If p then q":

- p implies q.

- q follows from p.

- p, only if q.

- q, if p.

- p is sufficient for q.

- q is necessary for p.

All of the following are equivalent to "p if and only if q":

- p is necessary and sufficient for q.

- p is equivalent to q.

- If p, then q, and if q, then p.

- If p, then q and conversely.

3.1.3 Exercises for Section 3.1

1. Let $d = $ "I like discrete structures", $c = $ "I will pass this course" and $s = $ "I will do my assignments." Express each of the following propositions in symbolic form:

    (a) I like discrete structures and I will pass this course.

    (b) I will do my assignments or I will not pass this course.

    (c) It is not true that I both like discrete structures, and will do my assignments.

    (d) I will not do my assignment and I will not pass this course.

2. For each of the following propositions, identify simple propositions, express the compound proposition in symbolic form, and determine whether it is true or false:

    (a) The world is flat or zero is an even integer.

    (b) If 432,802 is a multiple of 4, then 432,802 is even.

    (c) 5 is a prime number and 6 is not divisible by 4.

    (d) $3 \in \mathbb{Z}$ and $3 \in \mathbb{Q}$.

    (e) $2/3 \in \mathbb{Z}$ and $2/3 \in \mathbb{Q}$.

    (f) The sum of two even integers is even and the sum of two odd integers is odd.

3. Let $p = $ "$2 \leq 5$", $q = $ "8 is an even integer," and $r = $ "11 is a prime number." Express the following as a statement in English and determine whether the statement is true or false:

    (a) $\neg p \wedge q$ (d) $p \rightarrow (q \vee (\neg r))$

    (b) $p \rightarrow q$ (e) $p \rightarrow ((\neg q) \vee (\neg r))$

    (c) $(p \wedge q) \rightarrow r$ (f) $(\neg q) \rightarrow (\neg p)$

4. Rewrite each of the following statements using the other conditional forms:

    (a) If an integer is a multiple of 4, then it is even.

    (b) The fact that a polygon is a square is a sufficient condition that it is a rectangle.

    (c) If $x = 5$, then $x^2 = 25$.

    (d) If $x^2 - 5x + 6 = 0$, then $x = 2$ or $x = 3$.

    (e) $x^2 = y^2$ is a necessary condition for $x = y$.

5. Write the converse of the propositions in exercise 4. Compare the truth of each proposition and its converse.

3.2 Truth Tables and Propositions Generated by a Set

3.2.1 Truth Tables

Consider the compound proposition $c = (p \wedge q) \vee (\neg q \wedge r)$, where p, q, and r are propositions. This is an example of a proposition generated by p, q, and r. We will define this terminology later in the section. Since each of the three simple propositions has two possible truth values, it follows that there are eight different combinations of truth values that determine a value for c. These values can be obtained from a truth table for c. To construct the truth table, we build c from p, q, and r and from the logical operators. The result is the truth table below. Strictly speaking, the first three columns and the last column make up the truth table for c. The other columns are work space needed to build up to c.

p	q	r	$p \wedge q$	$\neg q$	$\neg q \wedge r$	$(p \wedge q) \vee (\neg q \wedge r)$
0	0	0	0	1	0	0
0	0	1	0	1	1	1
0	1	0	0	0	0	0
0	1	1	0	0	0	0
1	0	0	0	1	0	0
1	0	1	0	1	1	1
1	1	0	1	0	0	1
1	1	1	1	0	0	1

Table 3.2.1: Truth Table for $c = (p \wedge q) \vee (\neg q \wedge r)$

Note that the first three columns of the truth table are an enumeration of the eight three-digit binary integers. This standardizes the order in which the cases are listed. In general, if c is generated by n simple propositions, then the truth table for c will have 2^n rows with the first n columns being an enumeration of the n digit binary integers. In our example, we can see at a glance that for exactly four of the eight cases, c will be true. For example, if p and r are true and q is false (the sixth case), then c is true.

Let S be any set of propositions. We will give two definitions of a proposition generated by S. The first is a bit imprecise, but should be clear. The second definition is called a *recursive definition*. If you find it confusing, use the first definition and return to the second later.

3.2.2 Propositions Generated by a Set

Definition 3.2.2 Proposition Generated by a Set. Let S be any set of propositions. A proposition generated by S is any valid combination of propositions in S with conjunction, disjunction, and negation. Or, to be more precise,

(a) If $p \in S$, then p is a proposition generated by S, and

(b) If x and y are propositions generated by S, then so are (x), $\neg x$, $x \vee y$, and $x \wedge y$.

$\Diamond$

Note: We have not included the conditional and biconditional in the definition because they can both be generated from conjunction, disjunction, and negation, as we will see later.

If S is a finite set, then we may use slightly different terminology. For example, if $S = \{p, q, r\}$, we might say that a proposition is generated by p, q, and r instead of from $\{p, q, r\}$.

It is customary to use the following hierarchy for interpreting propositions, with parentheses overriding this order:

- First: Negation

- Second: Conjunction

- Third: Disjunction

- Fourth: The conditional operation

- Fifth: The biconditional operation

Within any level of the hierarchy, work from left to right. Using these rules, $p \wedge q \vee r$ is taken to mean $(p \wedge q) \vee r$. These precedence rules are universal, and are exactly those used by computer languages to interpret logical expressions.

Example 3.2.3 Examples of the Hierarchy of Logical Operations. A few shortened expressions and their fully parenthesized versions:

(a) $p \wedge q \wedge r$ is $(p \wedge q) \wedge r$.

(b) $\neg p \vee \neg r$ is $(\neg p) \vee (\neg r)$.

(c) $\neg \neg p$ is $\neg(\neg p)$.

(d) $p \leftrightarrow q \wedge r \to s$ is $p \leftrightarrow ((q \wedge r) \to s)$.

□

A proposition generated by a set S need not include each element of S in its expression. For example, $\neg q \wedge r$ is a proposition generated by p, q, and r.

3.2.3 Exercises for Section 3.2

1. Construct the truth tables of:
   (a) $p \vee p$ (c) $p \vee (\neg p)$

   (b) $p \wedge (\neg p)$ (d) $p \wedge p$

2. Construct the truth tables of:
   (a) $\neg(p \wedge q)$ (d) $(p \wedge q) \vee (q \wedge r) \vee (r \wedge p)$

   (b) $p \wedge (\neg q)$ (e) $\neg p \vee \neg q$

   (c) $(p \wedge q) \wedge r$ (f) $p \vee q \vee r \vee s$

3. Rewrite the following with as few extraneous parentheses as possible:
   (a) $(\neg((p) \wedge (r))) \vee (s)$ (b) $((p) \vee (q)) \wedge ((r) \vee (q))$

4. In what order are the operations in the following propositions performed?
   (a) $p \vee \neg q \vee r \wedge \neg p$ (b) $p \wedge \neg q \wedge r \wedge \neg p$

5. Determine the number of rows in the truth table of a proposition containing four variables p, q, r, and s.

6. If there are 45 lines on a sheet of paper, and you want to reserve one line for each line in a truth table, how large could $|S|$ be if you can write truth tables of propositions generated by S on the sheet of paper?

3.3 Equivalence and Implication

Consider two propositions generated by p and q: $\neg(p \wedge q)$ and $\neg p \vee \neg q$. At first glance, they are different propositions. In form, they are different, but they have the same meaning. One way to see this is to substitute actual propositions for p and q; such as p: I've been to Toronto; and q: I've been to Chicago.

Then $\neg(p \wedge q)$ translates to "I haven't been to both Toronto and Chicago," while $\neg p \vee \neg q$ is "I haven't been to Toronto or I haven't been to Chicago." Determine the truth values of these propositions. Naturally, they will be true for some people and false for others. What is important is that no matter what truth values they have, $\neg(p \wedge q)$ and $\neg p \vee \neg q$ will have the same truth value. The easiest way to see this is by examining the truth tables of these propositions.

p	q	$\neg(p \wedge q)$	$\neg p \vee \neg q$
0	0	1	1
0	1	1	1
1	0	1	1
1	1	0	0

Table 3.3.1: Truth Tables for $\neg(p \wedge q)$ and $\neg p \vee \neg q$

In all four cases, $\neg(p \wedge q)$ and $\neg p \vee \neg q$ have the same truth value. Furthermore, when the biconditional operator is applied to them, the result is a value of true in all cases. A proposition such as this is called a tautology.

3.3.1 Tautologies and Contradictions

Definition 3.3.2 Tautology. An expression involving logical variables that is true in all cases is a tautology. The number 1 is used to symbolize a tautology.
◊

Example 3.3.3 Some Tautologies. All of the following are tautologies because their truth tables consist of a column of 1's.

(a) $(\neg(p \wedge q)) \leftrightarrow (\neg p \vee \neg q)$.

(b) $p \vee \neg p$

(c) $(p \wedge q) \to p$

(d) $q \to (p \vee q)$

(e) $(p \vee q) \leftrightarrow (q \vee p)$

□

Definition 3.3.4 Contradiction. An expression involving logical variables that is false for all cases is called a contradiction. The number 0 is used to symbolize a contradiction. ◊

Example 3.3.5 Some Contradictions. $p \wedge \neg p$ and $(p \vee q) \wedge (\neg p) \wedge (\neg q)$ are contradictions.   □

3.3.2 Equivalence

Definition 3.3.6 Equivalence. Let S be a set of propositions and let r and s be propositions generated by S. r and s are equivalent if and only if $r \leftrightarrow s$ is a tautology. The equivalence of r and s is denoted $r \iff s$.   ◇

Equivalence is to logic as equality is to algebra. Just as there are many ways of writing an algebraic expression, the same logical meaning can be expressed in many different ways.

Example 3.3.7 Some Equivalences. The following are all equivalences:

(a) $(p \wedge q) \vee (\neg p \wedge q) \iff q$.

(b) $p \rightarrow q \iff \neg q \rightarrow \neg p$

(c) $p \vee q \iff q \vee p$.

□

All tautologies are equivalent to one another.

Example 3.3.8 An equivalence to 1. $p \vee \neg p \iff 1$.   □

All contradictions are equivalent to one another.

Example 3.3.9 An equivalence to 0. $p \wedge \neg p \iff 0$.   □

3.3.3 Implication

Consider the two propositions:

$$x: \text{The money is behind Door A; and}$$
$$y: \text{The money is behind Door A or Door B.}$$

Table 3.3.10

Imagine that you were told that there is a large sum of money behind one of two doors marked A and B, and that one of the two propositions x and y is true and the other is false. Which door would you choose? All that you need to realize is that if x is true, then y will also be true. Since we know that this can't be the case, y must be the true proposition and the money is behind Door B.

This is an example of a situation in which the truth of one proposition leads to the truth of another. Certainly, y can be true when x is false; but x can't be true when y is false. In this case, we say that x implies y.

Consider the truth table of $p \rightarrow q$, Table 3.1.7. If p implies q, then the third case can be ruled out, since it is the case that makes a conditional proposition false.

Definition 3.3.11 Implication. Let S be a set of propositions and let r and s be propositions generated by S. We say that r implies s if $r \rightarrow s$ is a tautology. We write $r \Rightarrow s$ to indicate this implication.   ◇

Example 3.3.12 Disjunctive Addition. A commonly used implication called "disjunctive addition" is $p \Rightarrow (p \vee q)$, which is verified by truth table

Table 3.3.13.   □

p	q	$p \vee q$	$p \to p \vee q$
0	0	0	1
0	1	1	1
1	0	1	1
1	1	1	1

Table 3.3.13: Truth Table to verify that $p \Rightarrow (p \vee q)$

If we let p represent "The money is behind Door A" and q represent "The money is behind Door B," $p \Rightarrow (p \vee q)$ is a formalized version of the reasoning used in Example 3.3.12. A common name for this implication is disjunctive addition. In the next section we will consider some of the most commonly used implications and equivalences.

When we defined what we mean by a Proposition Generated by a Set, we didn't include the conditional and biconditional operators. This was because of the two equivalences $p \to q \Leftrightarrow \neg p \vee q$ and $p \leftrightarrow q \Leftrightarrow (p \wedge q) \vee (\neg p \wedge \neg q)$. Therefore, any proposition that includes the conditional or biconditional operators can be written in an equivalent way using only conjunction, disjunction, and negation. We could even dispense with disjunction since $p \vee q$ is equivalent to a proposition that uses only conjunction and negation.

3.3.4 A Universal Operation

We close this section with a final logical operation, the Sheffer Stroke, that has the interesting property that all other logical operations can be created from it. You can explore this operation in Exercise 3.3.5.8

Definition 3.3.14 The Sheffer Stroke. The Sheffer Stroke is the logical operator defined by the following truth table:

p	q	$p \mid q$
0	0	1
0	1	1
1	0	1
1	1	0

Table 3.3.15: Truth Table for the Sheffer Stroke

◇

3.3.5 Exercises for Section 3.3

1. Given the following propositions generated by p, q, and r, which are equivalent to one another?

    (a) $(p \wedge r) \vee q$ (e) $(p \vee q) \wedge (r \vee q)$

    (b) $p \vee (r \vee q)$ (f) $r \to p$

    (c) $r \wedge p$ (g) $r \vee \neg p$

    (d) $\neg r \vee p$ (h) $p \to r$

2.

    (a) Construct the truth table for $x = (p \wedge \neg q) \vee (r \wedge p)$.

(b) Give an example other than x itself of a proposition generated by p, q, and r that is equivalent to x.

(c) Give an example of a proposition other than x that implies x.

(d) Give an example of a proposition other than x that is implied by x.

3. Is an implication equivalent to its converse? Verify your answer using a truth table.

4. Suppose that x is a proposition generated by p, q, and r that is equivalent to $p \vee \neg q$. Write out the truth table for x.

5. How large is the largest set of propositions generated by p and q with the property that no two elements are equivalent?

6. Find a proposition that is equivalent to $p \vee q$ and uses only conjunction and negation.

7. Explain why a contradiction implies any proposition and any proposition implies a tautology.

8. The significance of the Sheffer Stroke is that it is a "universal" operation in that all other logical operations can be built from it.

(a) Prove that $p|q$ is equivalent to $\neg(p \wedge q)$.

(b) Prove that $\neg p \Leftrightarrow p|p$.

(c) Build $\wedge$ using only the Sheffer Stroke.

(d) Build $\vee$ using only the Sheffer Stroke.

3.4 The Laws of Logic

3.4.1

In this section, we will list the most basic equivalences and implications of logic. Most of the equivalences listed in Table Table 3.4.3 should be obvious to the reader. Remember, 0 stands for contradiction, 1 for tautology. Many logical laws are similar to algebraic laws. For example, there is a logical law corresponding to the associative law of addition, $a + (b + c) = (a + b) + c$. In fact, associativity of both conjunction and disjunction are among the laws of logic. Notice that with one exception, the laws are paired in such a way that exchanging the symbols $\wedge$, $\vee$, 1 and 0 for $\vee$, $\wedge$, 0, and 1, respectively, in any law gives you a second law. For example, $p \vee 0 \Leftrightarrow p$ results in $p \wedge 1 \Leftrightarrow p$. This is called a *duality principle*. For now, think of it as a way of remembering two laws for the price of one. We will leave it to the reader to verify a few of these laws with truth tables. However, the reader should be careful in applying duality to the conditional operator and implication since the dual involves taking the converse. For example, the dual of $p \wedge q \Rightarrow p$ is $p \vee q \Leftarrow p$, which is usually written $p \Rightarrow p \vee q$.

Example 3.4.1 Verification of an Identity Law. The Identity Law can be verified with this truth table. The fact that $(p \wedge 1) \leftrightarrow p$ is a tautology serves as a valid proof.

$$
\begin{array}{cccc}
p & 1 & p \wedge 1 & (p \wedge 1) \leftrightarrow p \\
0 & 1 & 0 & 1 \\
1 & 1 & 1 & 1
\end{array}
$$

Table 3.4.2: Truth table to demonstrate the identity law for conjunction.

$\square$

Some of the logical laws in Table Table 3.4.4 might be less obvious to you. For any that you are not comfortable with, substitute actual propositions for the logical variables. For example, if p is "John owns a pet store" and q is "John likes pets," the detachment law should make sense.

Commutative Laws	
$p \vee q \Leftrightarrow q \vee p$	$p \wedge q \Leftrightarrow q \wedge p$
Associative Laws	
$(p \vee q) \vee r \Leftrightarrow p \vee (q \vee r)$	$(p \wedge q) \wedge r \Leftrightarrow p \wedge (q \wedge r)$
Distributive Laws	
$p \wedge (q \vee r) \Leftrightarrow (p \wedge q) \vee (p \wedge r)$	$p \vee (q \wedge r) \Leftrightarrow (p \vee q) \wedge (p \vee r)$
Identity Laws	
$p \vee 0 \Leftrightarrow p$	$p \wedge 1 \Leftrightarrow p$
Negation Laws	
$p \wedge \neg p \Leftrightarrow 0$	$p \vee \neg p \Leftrightarrow 1$
Idempotent Laws	
$p \vee p \Leftrightarrow p$	$p \wedge p \Leftrightarrow p$
Null Laws	
$p \wedge 0 \Leftrightarrow 0$	$p \vee 1 \Leftrightarrow 1$
Absorption Laws	
$p \wedge (p \vee q) \Leftrightarrow p$	$p \vee (p \wedge q) \Leftrightarrow p$
DeMorgan's Laws	
$\neg(p \vee q) \Leftrightarrow (\neg p) \wedge (\neg q)$	$\neg(p \wedge q) \Leftrightarrow (\neg p) \vee (\neg q)$
Involution Law	
$\neg(\neg p) \Leftrightarrow p$	

Table 3.4.3: Basic Logical Laws - Equivalences

Detachment	$(p \rightarrow q) \wedge p \Rightarrow q$
Indirect Reasoning	$(p \rightarrow q) \wedge \neg q \Rightarrow \neg p$
Disjunctive Addition	$p \Rightarrow (p \vee q)$
Conjunctive Simplification	$(p \wedge q) \Rightarrow p$ and $(p \wedge q) \Rightarrow q$
Disjunctive Simplification	$(p \vee q) \wedge \neg p \Rightarrow q$ and $(p \vee q) \wedge \neg q \Rightarrow p$
Chain Rule	$(p \rightarrow q) \wedge (q \rightarrow r) \Rightarrow (p \rightarrow r)$
Conditional Equivalence	$p \rightarrow q \Leftrightarrow \neg p \vee q$
Biconditional Equivalences	$(p \leftrightarrow q) \Leftrightarrow (p \rightarrow q) \wedge (q \rightarrow p) \Leftrightarrow (p \wedge q) \vee (\neg p \wedge \neg q)$
Contrapositive	$(p \rightarrow q) \Leftrightarrow (\neg q \rightarrow \neg p)$

Table 3.4.4: Basic Logical Laws - Common Implications and Equivalences

3.4.2 Exercises for Section 3.4

1. Write the following in symbolic notation and determine whether it is a tautology: "If I study then I will learn. I will not learn. Therefore, I do not study."

2. Show that the common fallacy $(p \rightarrow q) \land \neg p \Rightarrow \neg q$ is not a law of logic.

3. Describe, in general, how duality can be applied to implications if we introduce the relation $\Leftarrow$, read "is implied by." We define this relation by

$$(p \Leftarrow q) \Leftrightarrow (q \Rightarrow p).$$

4. Write the dual of the following statements:

   (a) $(p \land q) \Rightarrow p$

   (b) $(p \lor q) \land \neg q \Rightarrow p$

3.5 Mathematical Systems and Proofs

3.5.1 Mathematical Systems

In this section, we present an overview of what a mathematical system is and how logic plays an important role in one. The axiomatic method that we will use here will not be duplicated with as much formality anywhere else in the book, but we hope an emphasis on how mathematical facts are developed and organized will help to unify the concepts we will present. The system of propositions and logical operators we have developed will serve as a model for our discussion. Roughly, a mathematical system can be defined as follows.

Definition 3.5.1 Mathematical System. A mathematical system consists of:

(1) A set or universe, U.

(2) Definitions: sentences that explain the meaning of concepts that relate to the universe. Any term used in describing the universe itself is said to be undefined. All definitions are given in terms of these undefined concepts of objects.

(3) Axioms: assertions about the properties of the universe and rules for creating and justifying more assertions. These rules always include the system of logic that we have developed to this point.

(4) Theorems: the additional assertions mentioned above.

$\Diamond$

Example 3.5.2 Euclidean Geometry. In Euclidean geometry the universe consists of points and lines (two undefined terms). Among the definitions is a definition of parallel lines and among the axioms is the axiom that two distinct parallel lines never meet. □

Example 3.5.3 Propositional Calculus. Propositional calculus is a formal name for the logical system that we've been discussing. The universe consists of propositions. The axioms are the truth tables for the logical operators and the key definitions are those of equivalence and implication. We use propositions to describe any other mathematical system; therefore, this is the minimum amount of structure that a mathematical system can have. □

Definition 3.5.4 Theorem. A true proposition derived from the axioms of a mathematical system is called a theorem. ◊

Theorems are normally expressed in terms of a finite number of propositions, $p_1, p_2, ..., p_n$, called the *premises*, and a proposition,C, called the *conclusion*. These theorems take the form

$$p_1 \wedge p_2 \wedge \cdots \wedge p_n \Rightarrow C$$

or more informally,

$$p_1, p_2, ..., \text{ and } p_n \text{ imply } C$$

For a theorem of this type, we say that the premises imply the conclusion. When a theorem is stated, it is assumed that the axioms of the system are true. In addition, any previously proven theorem can be considered an extension of the axioms and can be used in demonstrating that the new theorem is true. When the proof is complete, the new theorem can be used to prove subsequent theorems. A mathematical system can be visualized as an inverted pyramid with the axioms at the base and the theorems expanding out in various directions.

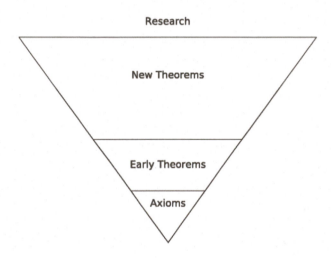

Figure 3.5.5: The body of knowledge in a mathematical system

Definition 3.5.6 Proof. A proof of a theorem is a finite sequence of logically valid steps that demonstrate that the premises of a theorem imply its conclusion. ◊

Exactly what constitutes a proof is not always clear. For example, a research mathematician might require only a few steps to prove a theorem to a colleague, but might take an hour to give an effective proof to a class of students. Therefore, what constitutes a proof often depends on the audience. But the audience is not the only factor. One of the most famous theorems in graph theory, The Four-Color Theorem, was proven in 1976, after over a century of effort by many mathematicians. Part of the proof consisted of having a computer check many different graphs for a certain property. Without the aid of the computer, this checking would have taken years. In the eyes of some mathematicians, this proof was considered questionable. Shorter proofs have been developed since 1976 and there is no controversy associated with The Four Color Theorem at this time.

3.5.2 Direct Proof

Theoretically, you can prove anything in propositional calculus with truth tables. In fact, the laws of logic stated in Section 3.4 are all theorems. Propositional calculus is one of the few mathematical systems for which any valid sentence can be determined true or false by mechanical means. A program to write truth tables is not too difficult to write; however, what can be done theoretically is not always practical. For example,

$$a, a \rightarrow b, b \rightarrow c, ..., y \rightarrow z \Rightarrow z$$

is a theorem in propositional calculus. However, suppose that you wrote such a program and you had it write the truth table for

$$(a \wedge (a \rightarrow b) \wedge (b \rightarrow c) \wedge \cdots \wedge (y \rightarrow z)) \rightarrow z$$

The truth table will have 2^{26} cases. At one million cases per second, it would take approximately one hour to verify the theorem. Now if you decided to check a similar theorem,

$$p_1, p_1 \rightarrow p_2, ..., p_{99} \rightarrow p_{100} \Rightarrow p_{100}$$

you would really have time trouble. There would be $2^{100} \approx 1.26765 \times 10^{30}$ cases to check in the truth table. At one million cases per second it would take approximately 1.46719×10^{19} days to check all cases. For most of the remainder of this section, we will discuss an alternate method for proving theorems in propositional calculus. It is the same method that we will use in a less formal way for proofs in other systems. Formal axiomatic methods would be too unwieldy to actually use in later sections. However, none of the theorems in later chapters would be stated if they couldn't be proven by the axiomatic method.

We will introduce two types of proof here, direct and indirect.

Example 3.5.7 A typical direct proof. This is a theorem: $p \rightarrow r, q \rightarrow s, p \vee q \Rightarrow s \vee r$. A direct proof of this theorem is:

Step	Proposition	Justification
1.	$p \vee q$	Premise
2.	$\neg p \rightarrow q$	(1), conditional rule
3.	$q \rightarrow s$	Premise
4.	$\neg p \rightarrow s$	(2), (3), chain rule
5.	$\neg s \rightarrow p$	(4), contrapositive
6.	$p \rightarrow r$	Premise
7.	$\neg s \rightarrow r$	(5), (6), chain rule
8.	$s \vee r$	(7), conditional rule $\square$

Table 3.5.8: Direct proof of $p \rightarrow r, q \rightarrow s, p \vee q \Rightarrow s \vee r$

$\square$

Note that $\square$ marks the end of a proof.

Example 3.5.7 illustrates the usual method of formal proof in a formal mathematical system. The rules governing these proofs are:

(1) A proof must end in a finite number of steps.

(2) Each step must be either a premise or a proposition that is implied from previous steps using any valid equivalence or implication.

(3) For a direct proof, the last step must be the conclusion of the theorem. For an indirect proof (see below), the last step must be a contradiction.

(4) Justification Column. The column labeled "justification" is analogous to the comments that appear in most good computer programs. They simply make the proof more readable.

Example 3.5.9 Two proofs of the same theorem. Here are two direct proofs of $\neg p \vee q, s \vee p, \neg q \Rightarrow s$:

1.	$\neg p \vee q$	Premise
2.	$\neg q$	Premise
3.	$\neg p$	Disjunctive simplification, (1), (2)
4.	$s \vee p$	Premise
5.	s	Disjunctive simplification, (3), (4). $\square$

Table 3.5.10: Direct proof of $\neg p \vee q, s \vee p, \neg q \Rightarrow s$

You are invited to justify the steps in this second proof:

1.	$\neg p \vee q$
2.	$\neg q \rightarrow \neg p$
3.	$s \vee p$
4.	$p \vee s$
5.	$\neg p \rightarrow s$
6.	$\neg q \rightarrow s$
7.	$\neg q$
8.	s $\square$

Table 3.5.11: Alternate proof of $\neg p \vee q, s \vee p, \neg q \Rightarrow s$

$\square$

The conclusion of a theorem is often a conditional proposition. The condition of the conclusion can be included as a premise in the proof of the theorem. The object of the proof is then to prove the consequence of the conclusion. This rule is justified by the logical law

$$p \rightarrow (h \rightarrow c) \Leftrightarrow (p \wedge h) \rightarrow c$$

Example 3.5.12 Example of a proof with a conditional conclusion. The following proof of $p \rightarrow (q \rightarrow s), \neg r \vee p, q \Rightarrow r \rightarrow s$ includes r as a fourth premise. Inference of truth of s completes the proof.

1.	$\neg r \vee p$	Premise
2.	r	Added premise
3.	p	(1), (2), disjunction simplification
4.	$p \rightarrow (q \rightarrow s)$	Premise
5.	$q \rightarrow s$	(3), (4), detachment
6.	q	Premise
7.	s	(5), (6), detachment. $\square$

Table 3.5.13: Proof of a theorem with a conditional conclusion.

3.5.3 Indirect Proof

Consider a theorem $P \Rightarrow C$, where P represents $p_1, p_2, ...,$ and p_n, the premises. The method of **indirect proof** is based on the equivalence $P \rightarrow C \Leftrightarrow \neg(P \wedge \neg C)$. In words, this logical law states that if $P \Rightarrow C$, then $P \wedge \neg C$ is always false; that is, $P \wedge \neg C$ is a contradiction. This means that a valid method of proof is to negate the conclusion of a theorem and add this negation to the premises. If a contradiction can be implied from this set of propositions, the proof is complete. For the proofs in this section, a contradiction will often take the form $t \wedge \neg t$.

For proofs involving numbers, a contradiction might be $1 = 0$ or $0 < 0$. Indirect proofs involving sets might conclude with $x \in \emptyset$ or $(x \in A$ and $x \in A^c)$. Indirect proofs are often more convenient than direct proofs in certain situations. Indirect proofs are often called *proofs by contradiction*.

Example 3.5.14 An Indirect Proof. Here is an example of an indirect proof of the theorem in Example 3.5.7.

1.	$\neg(s \vee r)$	Negated conclusion
2.	$\neg s \wedge \neg r$	DeMorgan's Law, (1)
3.	$\neg s$	Conjunctive simplification, (2)
4.	$q \rightarrow s$	Premise
5.	$\neg q$	Indirect reasoning, (3), (4)
6.	$\neg r$	Conjunctive simplification, (2)
7.	$p \rightarrow r$	Premise
8.	$\neg p$	Indirect reasoning, (6), (7)
9.	$(\neg p) \wedge (\neg q)$	Conjunctive, (5), (8)
10.	$\neg(p \vee q)$	DeMorgan's Law, (9)
11.	$p \vee q$	Premise
12.	0	(10), (11) $\square$

Table 3.5.15: An Indirect proof of $p \rightarrow r, q \rightarrow s, p \vee q \Rightarrow s \vee r$

Note 3.5.16 Proof Style. The rules allow you to list the premises of a theorem immediately; however, a proof is much easier to follow if the premises are only listed when they are needed.

Example 3.5.17 Yet Another Indirect Proof. Here is an indirect proof of $a \rightarrow b, \neg(b \vee c) \Rightarrow \neg a$.

1.	a	Negation of the conclusion
2.	$a \rightarrow b$	Premise
3.	b	(1), (2), detachment
4.	$b \vee c$	(3), disjunctive addition
5.	$\neg(b \vee c)$	Premise
6.	0	(4), (5) $\square$

Table 3.5.18: Indirect proof of $a \rightarrow b, \neg(b \vee c) \Rightarrow \neg a$

As we mentioned at the outset of this section, we are only presenting an overview of what a mathematical system is. For greater detail on axiomatic theories, see Stoll (1961). An excellent description of how propositional calculus plays a part in artificial intelligence is contained in Hofstadter (1980). If you enjoy the challenge of constructing proofs in propositional calculus, you should enjoy the game WFF'N PROOF (1962), by L.E. Allen.

3.5.4 Exercises for Section 3.5

1.   Prove with truth tables:

   (a) $p \vee q, \neg q \Rightarrow p$

   (b) $p \rightarrow q, \neg q \Rightarrow \neg p$

2.   Prove with truth tables:

   (a) $q, \neg q \Rightarrow p$

   (b) $p \rightarrow q \Rightarrow \neg p \vee q$

3.   Give direct and indirect proofs of:

   (a) $a \rightarrow b, c \rightarrow b, d \rightarrow (a \vee c), d \Rightarrow b.$

   (b) $(p \rightarrow q) \wedge (r \rightarrow s), (q \rightarrow t) \wedge (s \rightarrow u), \neg(t \wedge u), p \rightarrow r \Rightarrow \neg p.$

   (c) $p \rightarrow (q \rightarrow r), \neg s \vee p, q \Rightarrow s \rightarrow r.$

   (d) $p \rightarrow q, q \rightarrow r, \neg(p \wedge r), p \vee r \Rightarrow r.$

   (e) $\neg q, p \rightarrow q, p \vee t \Rightarrow t$

4.   Give direct and indirect proofs of:

   (a) $p \rightarrow q, \neg r \rightarrow \neg q, \neg r \Rightarrow \neg p.$

   (b) $p \rightarrow \neg q, \neg r \rightarrow q, p \Rightarrow r.$

   (c) $a \vee b, c \wedge d, a \rightarrow \neg c \Rightarrow b.$

5.   Are the following arguments valid? If they are valid, construct formal proofs; if they aren't valid, explain why not.

   (a) If wages increase, then there will be inflation. The cost of living will not increase if there is no inflation. Wages will increase. Therefore, the cost of living will increase.

   (b) If the races are fixed or the casinos are crooked, then the tourist trade will decline. If the tourist trade decreases, then the police will be happy. The police force is never happy. Therefore, the races are not fixed.

6.   Determine the validity of the following argument: For students to do well in a discrete mathematics course, it is necessary that they study hard. Students who do well in courses do not skip classes. Students who study hard do well in courses. Therefore students who do well in a discrete

mathematics course do not skip class.

7.   Describe how $p_1, p_1 \rightarrow p_2, \ldots, p_{99} \rightarrow p_{100} \Rightarrow p_{100}$ could be proved in 199 steps.

3.6 Propositions over a Universe

3.6.1 Propositions over a Universe

Consider the sentence "He was a member of the Boston Red Sox." There is no way that we can assign a truth value to this sentence unless "he" is specified. For that reason, we would not consider it a proposition. However, "he" can be considered a variable that holds a place for any name. We might want to restrict the value of "he" to all names in the major-league baseball record books. If that is the case, we say that the sentence is a proposition over the set of major-league baseball players, past and present.

Definition 3.6.1 Proposition over a Universe. Let U be a nonempty set. A proposition over U is a sentence that contains a variable that can take on any value in U and that has a definite truth value as a result of any such substitution. ◇

Example 3.6.2 Some propositions over a variety of universes.

(a) A few propositions over the integers are $4x^2 - 3x = 0$, $0 \le n \le 5$, and "k is a multiple of 3."

(b) A few propositions over the rational numbers are $4x^2 - 3x = 0$, $y^2 = 2$, and $(s-1)(s+1) = s^2 - 1$.

(c) A few propositions over the subsets of $\mathbb{P}$ are $(A = \emptyset) \vee (A = \mathbb{P})$, $3 \in A$, and $A \cap \{1, 2, 3\} \ne \emptyset$.

□

All of the laws of logic that we listed in Section 3.4 are valid for propositions over a universe. For example, if p and q are propositions over the integers, we can be certain that $p \wedge q \Rightarrow p$, because $(p \wedge q) \rightarrow p$ is a tautology and is true no matter what values the variables in p and q are given. If we specify p and q to be $p(n) : n < 4$ and $q(n) : n < 8$, we can also say that p implies $p \wedge q$. This is not a usual implication, but for the propositions under discussion, it is true. One way of describing this situation in general is with truth sets.

3.6.2 Truth Sets

Definition 3.6.3 Truth Set. If p is a proposition over U, the truth set of p is $T_p = \{a \in U \mid p(a) \text{ is true}\}$. ◇

Example 3.6.4 Truth Set Example. The truth set of the proposition $\{1, 2\} \cap A = \emptyset$, taken as a proposition over the power set of $\{1, 2, 3, 4\}$ is $\{\emptyset, \{3\}, \{4\}, \{3, 4\}\}$. □

Example 3.6.5 Truth sets depend on the universe. Over the universe $\mathbb{Z}$ (the integers), the truth set of $4x^2 - 3x = 0$ is $\{0\}$. If the universe is expanded to the rational numbers, the truth set becomes $\{0, 3/4\}$. The term *solution set* is often used for the truth set of an equation such as the one in this example. □

Definition 3.6.6 Tautologies and Contradictions over a Universe. A proposition over U is a tautology if its truth set is U. It is a contradiction if its truth set is empty. ◊

Example 3.6.7 Tautology, Contradiction over $\mathbb{Q}$. $(s-1)(s+1) = s^2 - 1$ is a tautology over the rational numbers. $x^2 - 2 = 0$ is a contradiction over the rationals. □

The truth sets of compound propositions can be expressed in terms of the truth sets of simple propositions. For example, if $a \in T_{p \wedge q}$ if and only if a makes $p \wedge q$ true. This is true if and only if a makes both p and q true, which, in turn, is true if and only if $a \in T_p \cap T_q$. This explains why the truth set of the conjunction of two propositions equals the intersection of the truth sets of the two propositions. The following list summarizes the connection between compound and simple truth sets

$$T_{p \wedge q} = T_p \cap T_q$$
$$T_{p \vee q} = T_p \cup T_q$$
$$T_{\neg p} = T_p{}^c$$
$$T_{p \leftrightarrow q} = (T_p \cap T_q) \cup (T_p{}^c \cap T_q{}^c)$$
$$T_{p \rightarrow q} = T_p{}^c \cup T_q$$

Table 3.6.8: Truth Sets of Compound Statements

Definition 3.6.9 Equivalence of propositions over a universe. Two propositions, p and q, are equivalent if $p \leftrightarrow q$ is a tautology. In terms of truth sets, this means that p and q are equivalent if $T_p = T_q$. ◊

Example 3.6.10 Some pairs of equivalent propositions.

(a) $n + 4 = 9$ and $n = 5$ are equivalent propositions over the integers.

(b) $A \cap \{4\} \neq \emptyset$ and $4 \in A$ are equivalent propositions over the power set of the natural numbers.

□

Definition 3.6.11 Implication for propositions over a universe. If p and q are propositions over U, p implies q if $p \rightarrow q$ is a tautology. ◊

Since the truth set of $p \rightarrow q$ is $T_p{}^c \cup T_q$, the Venn diagram for $T_{p \rightarrow q}$ in Figure 12 shows that $p \Rightarrow q$ when $T_p \subseteq T_q$.

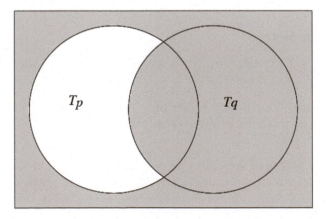

Figure 3.6.12: Venn Diagram for $T_{p \rightarrow q}$

Example 3.6.13 Examples of Implications.

(a) Over the natural numbers: $n \leq 4 \Rightarrow n \leq 8$ since $\{0, 1, 2, 3, 4\} \subseteq \{0, 1, 2, 3, 4, 5, 6, 7, 8\}$

(b) Over the power set of the integers: $|A^c| = 1$ implies $A \cap \{0, 1\} \neq \emptyset$

(c) Over the power set of the integers, $A \subseteq$ even integers $\Rightarrow A \cap$ odd integers $= \emptyset$

$\square$

3.6.3 Exercises for Section 3.6

1. If $U = \mathcal{P}(\{1, 2, 3, 4\})$, what are the truth sets of the following propositions?

(a) $A \cap \{2, 4\} = \emptyset$.

(b) $3 \in A$ and $1 \notin A$.

(c) $A \cup \{1\} = A$.

(d) A is a proper subset of $\{2, 3, 4\}$.

(e) $|A| = |A^c|$.

2. Over the universe of positive integers, define

$$
\begin{aligned}
p(n): &\quad n \text{ is prime and } n < 32. \\
q(n): &\quad n \text{ is a power of } 3. \\
r(n): &\quad n \text{ is a divisor of } 27.
\end{aligned}
$$

Table 3.6.14

(a) What are the truth sets of these propositions?

(b) Which of the three propositions implies one of the others?

3. If $U = \{0, 1, 2\}$, how many propositions over U could you list without listing two that are equivalent?

4. Given the propositions over the natural numbers:

$$p : n < 4, \quad q : 2n > 17, \text{ and } \quad r : n \text{ is a divisor of } 18$$

Table 3.6.15

What are the truth sets of:

(a) q (c) r

(b) $p \wedge q$ (d) $q \rightarrow r$

5. Suppose that s is a proposition over $\{1, 2, \ldots, 8\}$. If $T_s = \{1, 3, 5, 7\}$, give two examples of propositions that are equivalent to s.

6.

(a) Determine the truth sets of the following propositions over the positive integers:

$$p(n) : n \text{ is a perfect square and } n < 100$$

$$q(n) : n = |\mathcal{P}(A)| \text{ for some set } A.$$

(b) Determine $T_{p \wedge q}$ for p and q above.

7. Let the universe be $\mathbb{Z}$, the set of integers. Which of the following propositions are equivalent over $\mathbb{Z}$?

$$
\begin{array}{ll}
a: & 0 < n^2 < 9 \\
b: & 0 < n^3 < 27 \\
c: & 0 < n < 3
\end{array}
$$

Table 3.6.16

3.7 Mathematical Induction

3.7.1 Introduction, First Example

In this section, we will examine mathematical induction, a technique for proving propositions over the positive integers. Mathematical induction reduces the proof that all of the positive integers belong to a truth set to a finite number of steps.

Example 3.7.1 Formula for Triangular Numbers. Consider the following proposition over the positive integers, which we will label $p(n)$: The sum of the positive integers from 1 to n is $\frac{n(n+1)}{2}$. This is a well-known formula that is quite simple to verify for a given value of n. For example, $p(5)$ is: The sum of the positive integers from 1 to 5 is $\frac{5(5+1)}{2}$. Indeed, $1+2+3+4+5 = 15 = \frac{5(5+1)}{2}$. However, this doesn't serve as a proof that $p(n)$ is a tautology. All that we've established is that 5 is in the truth set of p. Since the positive integers are infinite, we certainly can't use this approach to prove the formula. $\square$

An Analogy: A proof by mathematical induction is similar to knocking over a row of closely spaced dominos that are standing on end. To knock over the dominos in Figure 3.7.2, all you need to do is push the first domino over. To be assured that they all will be knocked over, some work must be done ahead of time. The dominos must be positioned so that if any domino is pushed is knocked over, it will push the next domino in the line.

Figure 3.7.2: An analogy for Mathematical Induction, Creative Commons photo by Ranveig Thattai

Returning to Example 3.7.1 imagine the propositions $p(1), p(2), p(3), \ldots$ to be an infinite line of dominos. Let's see if these propositions are in the same formation as the dominos were. First, we will focus on one specific point of the line: $p(99)$ and $p(100)$. We are not going to prove that either of these propositions is true, just that the truth of $p(99)$ implies the truth of $p(100)$. In terms of our analogy, if $p(99)$ is knocked over, it will knock over $p(100)$.

In proving $p(99) \Rightarrow p(100)$, we will use $p(99)$ as our premise. We must prove: The sum of the positive integers from 1 to 100 is $\frac{100(100+1)}{2}$. We start by observing that the sum of the positive integers from 1 to 100 is $(1 + 2 + \cdots + 99) + 100$. That is, the sum of the positive integers from 1 to 100 equals the sum of the first ninety-nine plus the final number, 100. We can now apply our premise, $p(99)$, to the sum $1 + 2 + \cdots + 99$. After rearranging our numbers, we obtain the desired expression for $1 + 2 + \cdots + 100$:

$$
\begin{aligned}
1 + 2 + \cdots + 99 + 100 &= (1 + 2 + \cdots + 99) + 100 \\
&= \frac{99 \cdot (99 + 1)}{2} + 100 \text{ by our assumption of } p(99) \\
&= \frac{99 \cdot 100}{2} + \frac{2 \cdot 100}{2} \\
&= \frac{100 \cdot 101}{2} \\
&= \frac{100 \cdot (100 + 1)}{2}
\end{aligned}
$$

What we've just done is analogous to checking two dominos in a line and finding that they are properly positioned. Since we are dealing with an infinite line, we must check all pairs at once. This is accomplished by proving that $p(n) \Rightarrow p(n + 1)$ for all $n \geq 1$:

$$
\begin{aligned}
1 + 2 + \cdots + n + (n + 1) &= (1 + 2 + \cdots + n) + (n + 1) \\
&= \frac{n(n + 1)}{2} + (n + 1) \text{ by } p(n) \\
&= \frac{n(n + 1)}{2} + \frac{2(n + 1)}{2} \\
&= \frac{(n + 1)(n + 2)}{2} \\
&= \frac{(n + 1)((n + 1) + 1)}{2}
\end{aligned}
$$

They are all lined up! Now look at $p(1)$: The sum of the positive integers from 1 to 1 is $\frac{1+1}{2}$. Clearly, $p(1)$ is true. This sets off a chain reaction. Since $p(1) \Rightarrow p(2)$, $p(2)$ is true. Since $p(2) \Rightarrow p(3)$, $p(3)$ is true; and so on. ∎

Theorem 3.7.3 The Principle of Mathematical Induction. *Let $p(n)$ be a proposition over the positive integers. If*

(1) $p(1)$ is true, and

(2) for all $n \geq 1$, $p(n) \Rightarrow p(n + 1)$,

then $p(n)$ is a tautology.

Note: The truth of $p(1)$ is called the *basis* for the induction proof. The premise that $p(n)$ is true in the second part is called the *induction hypothesis*. The proof that $p(n)$ implies $p(n + 1)$ is called the *induction step* of the proof. Despite our analogy, the basis is usually done first in an induction proof. However, order doesn't really matter.

3.7.2 More Examples

Example 3.7.4 Generalized Detachment. Consider the implication over the positive integers.

$$
p(n) : q_0 \to q_1, q_1 \to q_2, \ldots, q_{n-1} \to q_n, q_0 \Rightarrow q_n
$$

A proof that $p(n)$ is a tautology follows. Basis: $p(1)$ is $q_0 \to q_1, q_0 \Rightarrow q_1$. This is the logical law of detachment which we know is true. If you haven't done so yet, write out the truth table of $((q_0 \to q_1) \land q_0) \to q_1$ to verify this step.

Induction: Assume that $p(n)$ is true for some $n \geq 1$. We want to prove that $p(n+1)$ must be true. That is:

$$q_0 \to q_1, q_1 \to q_2, \ldots, q_{n-1} \to q_n, q_n \to q_{n+1}, q_0 \Rightarrow q_{n+1}$$

Here is a direct proof of $p(n+1)$:

Step	Proposition	Justification
$1 - (n+1)$	$q_0 \to q_1, q_1 \to q_2, \ldots, q_{n-1} \to q_n, q_0$	Premises
$n+2$	q_n	$(1) - (n+1), p(n)$
$n+3$	$q_n \to q_{n+1}$	Premise
$n+4$	q_{n+1}	$(n+2), (n+3)$, detachment $\square$

Table 3.7.5

$\square$

Example 3.7.6 An example from Number Theory. For all $n \geq 1$, $n^3 + 2n$ is a multiple of 3. An inductive proof follows:

Basis: $1^3 + 2(1) = 3$ is a multiple of 3. The basis is almost always this easy!

Induction: Assume that $n \geq 1$ and $n^3 + 2n$ is a multiple of 3. Consider $(n+1)^3 + 2(n+1)$. Is it a multiple of 3?

$$\begin{aligned}
(n+1)^3 + 2(n+1) &= n^3 + 3n^2 + 3n + 1 + (2n + 2) \\
&= n^3 + 2n + 3n^2 + 3n + 3 \qquad . \\
&= (n^3 + 2n) + 3(n^2 + n + 1)
\end{aligned}$$

Yes, $(n+1)^3 + 2(n+1)$ is the sum of two multiples of 3; therefore, it is also a multiple of 3. $\square$ $\square$

Now we will discuss some of the variations of the principle of mathematical induction. The first simply allows for universes that are similar to $\mathbb{P}$ such as $\{-2, -1, 0, 1, \ldots\}$ or $\{5, 6, 7, 8, \ldots\}$.

Theorem 3.7.7 Principle of Mathematical Induction (Generalized). *If $p(n)$ is a proposition over $\{k_0, k_0 + 1, k_0 + 2, \ldots\}$, where k_0 is any integer, then $p(n)$ is a tautology if*

(1) $p(k_0)$ is true, and

(2) for all $n \geq k_0$, $p(n) \Rightarrow p(n+1)$.

Example 3.7.8 A proof of the permutations formula. In Chapter 2, we stated that the number of different permutations of k elements taken from an n element set, $P(n; k)$, can be computed with the formula $\frac{n!}{(n-k)!}$. We can prove this statement by induction on n. For $n \geq 0$, let $q(n)$ be the proposition

$$P(n; k) = \frac{n!}{(n-k)!} \text{ for all } k, 0 \leq k \leq n.$$

Basis: $q(0)$ states that $P(0; 0)$ if is the number of ways that 0 elements can be selected from the empty set and arranged in order, then $P(0; 0) = \frac{0!}{0!} = 1$. This is true. A general law in combinatorics is that there is exactly one way of doing nothing.

Induction: Assume that $q(n)$ is true for some natural number n. It is left for us to prove that this assumption implies that $q(n+1)$ is true. Suppose that we have a set of cardinality $n+1$ and want to select and arrange k of its elements. There are two cases to consider, the first of which is easy. If $k = 0$, then there is one way of selecting zero elements from the set; hence

$$P(n+1;0) = 1 = \frac{(n+1)!}{(n+1+0)!}$$

and the formula works in this case.

The more challenging case is to verify the formula when k is positive and less than or equal to $n+1$. Here we count the value of $P(n+1;k)$ by counting the number of ways that the first element in the arrangement can be filled and then counting the number of ways that the remaining $k-1$ elements can be filled in using the induction hypothesis.

There are $n+1$ possible choices for the first element. Since that leaves n elements to fill in the remaining $k-1$ positions, there are $P(n;k-1)$ ways of completing the arrangement. By the rule of products,

$$P(n+1;k) = (n+1)P(n;k-1)$$
$$= (n+1)\frac{n!}{(n-(k-1))!}$$
$$= \frac{(n+1)n!}{(n-k+1)!}$$
$$= \frac{(n+1)!}{((n+1)-k)!}$$

∎   □

3.7.3 Course of Values Induction

A second variation allows for the expansion of the induction hypothesis. The course-of-values principle includes the previous generalization. It is also sometimes called *strong induction*.

Theorem 3.7.9 The Course-of-Values Principle of Mathematical Induction. *If $p(n)$ is a proposition over $\{k_0, k_0 + 1, k_0 + 2, \ldots\}$, where k_0 is any integer, then $p(n)$ is a tautology if*

(1) $p(k_0)$ is true, and

(2) for all $n \geq k_0$, $p(k_0), p(k_0 + 1), \ldots, p(n) \Rightarrow p(n+1)$.

A prime number is defined as a positive integer that has exactly two positive divisors, 1 and itself. There are an infinite number of primes. The list of primes starts with $2, 3, 5, 7, 11, \ldots$. The proposition over $\{2, 3, 4, \ldots\}$ that we will prove here is $p(n)$: n can be written as the product of one or more primes. In most texts, the assertion that $p(n)$ is a tautology would appear as

Theorem 3.7.10 Existence of Prime Factorizations. *Every positive integer greater than or equal to 2 has a prime decomposition.*

Proof. If you were to encounter this theorem outside the context of a discussion of mathematical induction, it might not be obvious that the proof can be done by induction. Recognizing when an induction proof is appropriate is mostly a matter of experience. Now on to the proof!

Basis: Since 2 is a prime, it is already decomposed into primes (one of them).

Induction: Suppose that for some $n \geq 2$ all of the integers $2, 3, ..., n$ have a prime decomposition. Notice the course-of-value hypothesis. Consider $n + 1$. Either $n + 1$ is prime or it isn't. If $n + 1$ is prime, it is already decomposed into primes. If not, then $n + 1$ has a divisor, d, other than 1 and $n + 1$. Hence, $n + 1 = cd$ where both c and d are between 2 and n. By the induction hypothesis, c and d have prime decompositions, $c_1 c_2 \cdots c_s$ and $d_1 d_2 \cdots d_t$, respectively. Therefore, $n+1$ has the prime decomposition $c_1 c_2 \cdots c_s d_1 d_2 \cdots d_t$.

∎

Peano Postulates and Induction. Mathematical induction originated in the late nineteenth century. Two mathematicians who were prominent in its development were Richard Dedekind and Giuseppe Peano. Dedekind developed a set of axioms that describe the positive integers. Peano refined these axioms and gave a logical interpretation to them. The axioms are usually called the Peano Postulates.

Axiom 3.7.11 Peano Postulates. *The system of positive integers consists of a nonempty set, $\mathbb{P}$; a least element of $\mathbb{P}$, denoted 1; and a "successor function," s, with the properties*

(1) *If $k \in \mathbb{P}$, then there is an element of $\mathbb{P}$ called the successor of k, denoted $s(k)$.*

(2) *No two elements of $\mathbb{P}$ have the same successor.*

(3) *No element of $\mathbb{P}$ has 1 as its successor.*

(4) *If $S \subseteq \mathbb{P}$, $1 \in S$, and $k \in S \Rightarrow s(k) \in S$, then $S = \mathbb{P}$.*

Notes:

- You might recognize $s(k)$ as simply being $k + 1$.

- Axiom 4 is the one that makes mathematical induction possible. In an induction proof, we simply apply that axiom to the truth set of a proposition.

3.7.4 Exercises for Section 3.7

1. Prove that the sum of the first n odd integers equals n^2 .

2. Prove that if $n \geq 1$, then $1(1!) + 2(2!) + \cdots + n(n!) = (n + 1)! - 1$.

3. Prove that for $n \geq 1$: $\sum_{k=1}^{n} k^2 = \frac{1}{6}n(n + 1)(2n + 1)$.

4. Prove that for $n \geq 1$: $\sum_{k=0}^{n} 2^k = 2^{n+1} - 1$.

5. Use mathematical induction to show that for $n \geq 1$,

$$\frac{1}{1 \cdot 2} + \frac{1}{2 \cdot 3} + \cdots + \frac{1}{n(n + 1)} = \frac{n}{n + 1}.$$

6. Prove that if $n \geq 2$, the generalized DeMorgan's Law is true:

$$\neg(p_1 \wedge p_2 \wedge ... \wedge p_n) \Leftrightarrow (\neg p_1) \vee (\neg p_2) \vee \cdots \vee (\neg p_n).$$

7. The number of strings of n zeros and ones that contain an even number of ones is 2^{n-1}. Prove this fact by induction for $n \geq 1$.

8. Let $p(n)$ be $8^n - 3^n$ is a multiple of 5. Prove that $p(n)$ is a tautology over $\mathbb{N}$.

9. Suppose that there are n people in a room, $n \geq 1$, and that they all shake hands with one another. Prove that $\frac{n(n-1)}{2}$ handshakes will have occurred.

10. Prove that it is possible to make up any postage of eight cents or more using only three- and five-cent stamps.

11. Generalized associativity. It is well known that if a_1, a_2, and a_3 are numbers, then no matter what order the sums in the expression $a_1 + a_2 + a_3$ are taken in, the result is always the same. Call this fact $p(3)$ and assume it is true. Prove using course-of-values induction that if a_1, a_2, ..., and a_n are numbers, then no matter what order the sums in the expression $a_1 + a_2 + \cdots + a_n$ are taken in, the result is always the same.

12. Let S be the set of all numbers that can be produced by applying any of the rules below in any order a finite number of times.

    * Rule 1: $\frac{1}{2} \in S$

    * Rule 2: $1 \in S$

    * Rule 3: If a and b have been produced by the rules, then $ab \in S$.

    * Rule 4: If a and b have been produced by the rules, then $\frac{a+b}{2} \in S$.

    Prove that $a \in S \Rightarrow 0 \leq a \leq 1$.

    Hint. The number of times the rules are applied should be the integer that you do the induction on.

13. Proofs involving objects that are defined recursively are often inductive. A recursive definition is similar to an inductive proof. It consists of a basis, usually the simple part of the definition, and the recursion, which defines complex objects in terms of simpler ones. For example, if x is a real number and n is a positive integer, we can define x^n as follows:

    * Basis: $x^1 = x$.

    * Recursion: if $n \geq 2$, $x^n = x^{n-1}x$.

    For example, $x^3 = x^2 x = (x^1 x)x = (xx)x$.
        Prove that if $n, m \in \mathbb{P}$, $x^{m+n} = x^m x^n$. There is much more on recursion in Chapter 8.

    Hint. Let $p(m)$ be the proposition that $x^{m+n} = x^m x^n$ for all $n \geq 1$.

14. Let S be a finite set and let P_n be defined recursively by $P_1 = S$ and $P_n = S \times P_{n-1}$ for $n \geq 2$.

    * List the elements of P_3 for the case $S = \{a, b\}$.

    * Determine the formula for $|P_n|$, given that $|S| = k$, and prove your formula by induction.

3.8 Quantifiers

As we saw in Section 3.6, if $p(n)$ is a proposition over a universe U, its truth set T_p is equal to a subset of U. In many cases, such as when $p(n)$ is an equation, we are most concerned with whether T_p is empty or not. In other cases, we might be interested in whether $T_p = U$; that is, whether $p(n)$ is a tautology. Since the conditions $T_p \neq \emptyset$ and $T_p = U$ are so often an issue, we have a special system of notation for them.

3.8.1 The Existential Quantifier

Definition 3.8.1   The Existential Quantifier. If $p(n)$ is a proposition over U with $T_p \neq \emptyset$, we commonly say "There exists an n in U such that $p(n)$ (is true)." We abbreviate this with the symbols $(\exists n)_U(p(n))$. The symbol $\exists$ is called the existential quantifier. If the context is clear, the mention of U is dropped: $(\exists n)(p(n))$.   ◊

Example 3.8.2   Some examples of existential quantifiers.

(a) $(\exists k)_{\mathbb{Z}}(k^2 - k - 12 = 0)$ is another way of saying that there is an integer that solves the equation $k^2 - k - 12 = 0$. The fact that two such integers exist doesn't affect the truth of this proposition in any way.

(b) $(\exists k)_{\mathbb{Z}}(3k = 102)$ simply states that 102 is a multiple of 3, which is true. On the other hand, $(\exists k)_{\mathbb{Z}}(3k = 100)$ states that 100 is a multiple of 3, which is false.

(c) $(\exists x)_{\mathbb{R}}(x^2 + 1 = 0)$ is false since the solution set of the equation $x^2 + 1 = 0$ in the real numbers is empty. It is common to write $(\nexists x)_{\mathbb{R}}(x^2 + 1 = 0)$ in this case.

□

There are a wide variety of ways that you can write a proposition with an existential quantifier. Table 3.8.5 contains a list of different variations that could be used for both the existential and universal quantifiers.

3.8.2 The Universal Quantifier

Definition 3.8.3   The Universal Quantifier. If $p(n)$ is a proposition over U with $T_p = U$, we commonly say "For all n in U, $p(n)$ (is true)." We abbreviate this with the symbols $(\forall n)_U(p(n))$. The symbol $\forall$ is called the universal quantifier. If the context is clear, the mention of U is dropped: $(\forall n)(p(n))$.   ◊

Example 3.8.4   Some Universal Quantifiers.

(a) We can say that the square of every real number is non-negative symbolically with a universal quantifier: $(\forall x)_{\mathbb{R}}(x^2 \geq 0)$.

(b) $(\forall n)_{\mathbb{Z}}(n + 0 = 0 + n = n)$ says that the sum of zero and any integer n is n. This fact is called the identity property of zero for addition.

□

Universal Quantifier	Existential Quantifier
$(\forall n)_U(p(n))$	$(\exists n)_U(p(n))$
$(\forall n \in U)(p(n))$	$(\exists n \in U)(p(n))$
$\forall n \in U, p(n)$	$\exists n \in U$ such that $p(n)$
$p(n), \forall n \in U$	$p(n)$ is true for some $n \in U$
$p(n)$ is true for all $n \in U$	

Table 3.8.5: Notational Variations with Quantified Expressions

3.8.3 The Negation of Quantified Propositions

When you negate a quantified proposition, the existential and universal quantifiers complement one another.

Example 3.8.6 Negation of an Existential Quantifier. Over the universe of animals, define $F(x)$: x is a fish and $W(x)$: x lives in the water. We know that the proposition $W(x) \to F(x)$ is not always true. In other words, $(\forall x)(W(x) \to F(x))$ is false. Another way of stating this fact is that there exists an animal that lives in the water and is not a fish; that is,

$$\neg(\forall x)(W(x) \to F(x)) \Leftrightarrow (\exists x)(\neg(W(x) \to F(x)))$$
$$\Leftrightarrow (\exists x)(W(x) \wedge \neg F(x)) \quad .$$

$\square$

Note that the negation of a universally quantified proposition is an existentially quantified proposition. In addition, when you negate an existentially quantified proposition, you get a universally quantified proposition. Symbolically,

$$\neg((\forall n)_U(p(n))) \Leftrightarrow (\exists n)_U(\neg p(n))$$
$$\neg((\exists n)_U(p(n))) \Leftrightarrow (\forall n)_U(\neg p(n))$$

Table 3.8.7: Negation of Quantified Expressions

Example 3.8.8 More Negations of Quantified Expressions.

(a) The ancient Greeks first discovered that $\sqrt{2}$ is an irrational number; that is, $\sqrt{2}$ is not a rational number. $\neg((\exists r)_{\mathbb{Q}}(r^2 = 2))$ and $(\forall r)_{\mathbb{Q}}(r^2 \neq 2)$ both state this fact symbolically.

(b) $\neg((\forall n)_{\mathbb{P}}(n^2 - n + 41 \text{ is prime}))$ is equivalent to $(\exists n)_{\mathbb{P}}(n^2 - n + 41 \text{ is composite})$. They are either both true or both false.

$\square$

3.8.4 Multiple Quantifiers

If a proposition has more than one variable, then you can quantify it more than once. For example, $p(x, y) : x^2 - y^2 = (x + y)(x - y)$ is a tautology over the set of all pairs of real numbers because it is true for each pair (x, y) in $\mathbb{R} \times \mathbb{R}$. Another way to look at this proposition is as a proposition with two variables. The assertion that $p(x, y)$ is a tautology could be quantified as $(\forall x)_{\mathbb{R}}((\forall y)_{\mathbb{R}}(p(x, y)))$ or $(\forall y)_{\mathbb{R}}((\forall x)_{\mathbb{R}}(p(x, y)))$

In general, multiple universal quantifiers can be arranged in any order without logically changing the meaning of the resulting proposition. The same is true for multiple existential quantifiers. For example, $p(x, y) : x + y = 4$ and $x-$

$y = 2$ is a proposition over $\mathbb{R} \times \mathbb{R}$. $(\exists x)_{\mathbb{R}}((\exists y)_{\mathbb{R}}(x + y = 4 \text{ and } x - y = 2))$ and $(\exists y)_{\mathbb{R}}((\exists x)_{\mathbb{R}}(x + y = 4 \text{ and } x - y = 2))$ are equivalent. A proposition with multiple existential quantifiers such as this one says that there are simultaneous values for the quantified variables that make the proposition true. A similar example is $q(x, y) : 2x - y = 2$ and $4x - 2y = 5$, which is always false; and the following are all equivalent:

$$\neg((\exists x)_{\mathbb{R}}((\exists y)_{\mathbb{R}}(q(x, y)))) \Leftrightarrow \neg(\exists y)_{\mathbb{R}}((\exists x)_{\mathbb{R}}(q(x, y))))$$
$$\Leftrightarrow (\forall y)_{\mathbb{R}}(\neg((\exists x)_{\mathbb{R}}(q(x, y))))$$
$$\Leftrightarrow ((\forall y)_{\mathbb{R}}((\forall x)_{\mathbb{R}}(\neg q(x, y))))$$
$$\Leftrightarrow ((\forall x)_{\mathbb{R}}((\forall y)_{\mathbb{R}}(\neg q(x, y))))$$

When existential and universal quantifiers are mixed, the order cannot be exchanged without possibly changing the meaning of the proposition. For example, let $\mathbb{R}^{+}$ be the positive real numbers, $x : (\forall a)_{\mathbb{R}+}((\exists b)_{\mathbb{R}+}(ab = 1))$ and $y : (\exists b)_{\mathbb{R}+}((\forall a)_{\mathbb{R}+}(ab = 1))$ have different logical values; x is true, while y is false.

Tips on Reading Multiply-Quantified Propositions. It is understandable that you would find propositions such as x difficult to read. The trick to deciphering these expressions is to "peel" one quantifier off the proposition just as you would peel off the layers of an onion (but quantifiers shouldn't make you cry!). Since the outermost quantifier in x is universal, x says that $z(a) : (\exists b)_{\mathbb{R}+}(ab = 1)$ is true for each value that a can take on. Now take the time to select a value for a, like 6. For the value that we selected, we get $z(6) : (\exists b)_{\mathbb{R}+}(6b = 1)$, which is obviously true since $6b = 1$ has a solution in the positive real numbers. We will get that same truth value no matter which positive real number we choose for a; therefore, $z(a)$ is a tautology over $\mathbb{R}^{+}$ and we are justified in saying that x is true. The key to understanding propositions like x on your own is to experiment with actual values for the outermost variables as we did above.

Now consider y. To see that y is false, we peel off the outer quantifier. Since it is an existential quantifier, all that y says is that some positive real number makes $w(b) : (\forall a)_{\mathbb{R}+}(ab = 1)$ true. Choose a few values of b to see if you can find one that makes $w(b)$ true. For example, if we pick $b = 2$, we get $(\forall a)_{\mathbb{R}+}(2a = 1)$, which is false, since $2a$ is almost always different from 1. You should be able to convince yourself that no value of b will make $w(b)$ true. Therefore, y is false.

Another way of convincing yourself that y is false is to convince yourself that $\neg y$ is true:

$$\neg((\exists b)_{\mathbb{R}+}((\forall a)_{\mathbb{R}+}(ab = 1))) \Leftrightarrow (\forall b)_{\mathbb{R}+}\neg((\forall a)_{\mathbb{R}+}(ab = 1))$$
$$\Leftrightarrow (\forall b)_{\mathbb{R}+}((\exists a)_{\mathbb{R}+}(ab \neq 1))$$

In words, for each value of b, there is a value for a that makes $ab \neq 1$. One such value is $a = \frac{1}{b} + 1$. Therefore, $\neg y$ is true.

3.8.5 Exercises for Section 3.8

1. Let $C(x)$ be "x is cold-blooded," let $F(x)$ be "x is a fish," and let $S(x)$ be "x lives in the sea."

    (a) Translate into a formula: Every fish is cold-blooded.

    (b) Translate into English: $(\exists x)(S(x) \land \neg F(x))$.

(c) Translate into English: $(\forall x)(F(x) \to S(x))$.

2. Let $M(x)$ be "x is a mammal," let $A(x)$ be "x is an animal," and let $W(x)$ be "x is warm-blooded."

    (a) Translate into a formula: Every mammal is warm-blooded.

    (b) Translate into English: $(\exists x)(A(x) \wedge (\neg M(x)))$.

3. Over the universe of books, define the propositions $B(x)$: x has a blue cover, $M(x)$: x is a mathematics book, $U(x)$: x is published in the United States, and $R(x, y)$: The bibliography of x includes y.
    Translate into words:

    (a) $(\exists x)(\neg B(x))$.

    (b) $(\forall x)(M(x) \wedge U(x) \to B(x))$.

    (c) $(\exists x)(M(x) \wedge \neg B(x))$.

    (d) $(\exists y)((\forall x)(M(x) \to R(x, y)))$.

    (e) Express using quantifiers: Every book with a blue cover is a mathematics book.

    (f) Express using quantifiers: There are mathematics books that are published outside the United States.

    (g) Express using quantifiers: Not all books have bibliographies.

4. Let the universe of discourse, U, be the set of all people, and let $M(x, y)$ be "x is the mother of y."
    Which of the following is a true statement? Translate it into English.

    (a) $(\exists x)_U((\forall y)_U(M(x, y)))$

    (b) $(\forall y)_U((\exists x)_U(M(x, y)))$

    (c) Translate the following statement into logical notation using quantifiers and the proposition $M(x, y)$: "Everyone has a maternal grandmother."

5. Translate into your own words and indicate whether it is true or false that $(\exists u)_{\mathbb{Z}}(4u^2 - 9 = 0)$.

6. Use quantifiers to say that $\sqrt{3}$ is an irrational number.

    Hint.   Your answer will depend on your choice of a universe

7. What do the following propositions say, where U is the power set of $\{1, 2, \ldots, 9\}$? Which of these propositions are true?

    (a) $(\forall A)_U |A| \neq |A^c|$.

    (b) $(\exists A)_U (\exists B)_U (|A| = 5, |B| = 5, \text{ and } A \cap B = \emptyset)$.

    (c) $(\forall A)_U (\forall B)_U (A - B = B^c - A^c)$.

8. Use quantifiers to state that for every positive integer, there is a larger positive integer.

9. Use quantifiers to state that the sum of any two rational numbers is rational.

10. Over the universe of real numbers, use quantifiers to say that the equation $a + x = b$ has a solution for all values of a and b.

    Hint. You will need three quantifiers.

11. Let n be a positive integer. Describe using quantifiers:

    (a) $x \in \bigcup\limits_{k=1}^{n} A_k$

    (b) $x \in \bigcap\limits_{k=1}^{n} A_k$

12. Prove that $(\exists x)(\forall y)(p(x, y)) \Rightarrow (\forall y)(\exists x)(p(x, y))$, but that converse is not true.

3.9 A Review of Methods of Proof

One of the major goals of this chapter is to acquaint the reader with the key concepts in the nature of proof in logic, which of course carries over into all areas of mathematics and its applications. In this section we will stop, reflect, and "smell the roses," so that these key ideas are not lost in the many concepts covered in logic. In Chapter 4 we will use set theory as a vehicle for further practice and insights into methods of proof.

3.9.1 Key Concepts in Proof

All theorems in mathematics can be expressed in form "If P then C" ($P \Rightarrow C$), or in the form "C_1 if and only if C_2" ($C_1 \Leftrightarrow C_2$). The latter is equivalent to "If C_1 then C_2," and "If C_2 then C_1."

In "If P then C," P is the premise (or hypothesis) and C is the conclusion. It is important to realize that a theorem makes a statement that is dependent on the premise being true.

There are two basic methods for proving $P \Rightarrow C$:

- *Directly:* Assume P is true and prove C is true.

- *Indirectly (or by contradiction):* Assume P is true and C is false and prove that this leads to a contradiction of some premise, theorem, or basic truth.

The method of proof for "If and only if" theorems is found in the law $(P \leftrightarrow C) \Leftrightarrow ((P \rightarrow C) \wedge (C \rightarrow P))$. Hence to prove an "If and only if" statement one must prove an "if . . . then ..." statement and its converse.

The initial response of most people when confronted with the task of being told they must be able to read and do proofs is often "Why?" or "I can't do proofs." To answer the first question, doing proofs or problem solving, even on the most trivial level, involves being able to read statements. First we must understand the problem and know the hypothesis; second, we must realize when we are done and we must understand the conclusion. To apply theorems or algorithms we must be able to read theorems and their proofs intelligently.

To be able to do the actual proofs of theorems we are forced to learn:

- the actual meaning of the theorems, and

- the basic definitions and concepts of the topic discussed.

For example, when we discuss rational numbers and refer to a number x as being rational, this means we can substitute a fraction $\frac{p}{q}$ in place of x, with the understanding that p and q are integers and $q \neq 0$. Therefore, to prove a theorem about rational numbers it is absolutely necessary that you know what a rational number "looks like."

It's easy to comment on the response, "I cannot do proofs." Have you tried? As elementary school students we may have been awe of anyone who could handle algebraic expressions, especially complicated ones. We learned by trying and applying ourselves. Maybe we cannot solve all problems in algebra or calculus, but we are comfortable enough with these subjects to know that we can solve many and can express ourselves intelligently in these areas. The same remarks hold true for proofs.

3.9.2 The Art of Proving $P \Rightarrow C$

First one must completely realize what is given, the hypothesis. The importance of this is usually overlooked by beginners. It makes sense, whenever you begin any task, to spend considerable time thinking about the tools at your disposal. Write down the premise in precise language. Similarly, you have to know when the task is finished. Write down the conclusion in precise language. Then you usually start with P and attempt to show that C follows logically. How do you begin? Basically you attack the proof the same way you solve a complicated equation in elementary algebra. You may not know exactly what each and every step is but you must try something. If we are lucky, C follows naturally; if it doesn't, try something else. Often what is helpful is to work backward from C. Finally, we have all learned, possibly the hard way, that mathematics is a participating sport, not a spectator sport. One learns proofs by doing them, not by watching others do them. We give several illustrations of how to set up the proofs of several examples. Our aim here is not to prove the statements given, but to concentrate on the logical procedure.

Example 3.9.1 The Sum of Odd Integers. We will outline a proof that the sum of any two odd integers is even. Our first step will be to write the theorem in the familiar conditional form: If j and k are odd integers, then $j + k$ is even. The premise and conclusion of this theorem should be clear now. Notice that if j and k are not both odd, then the conclusion may or may not be true. Our only objective is to show that the truth of the premise forces the conclusion to be true. Therefore, we can express the integers j and k in the form that all odd integers take; that is:

$$n \in \mathbb{Z} \text{ is odd implies that } (\exists m \in \mathbb{Z})(n = 2m + 1)$$

This observation allows us to examine the sum $j + k$ and to verify that it must be even. $\square$

Example 3.9.2 The Square of an Even Integer. Let $n \in \mathbb{Z}$. We will outline a proof that n^2 is even if and only if n is even.

Outline of a proof: Since this is an "If and only if" theorem we must prove two things:

(i) ($\Rightarrow$) If n^2 is even, then n is even. To do this directly, assume that n^2 is even and prove that n is even. To do this indirectly, assume n^2 is even and that n is odd, and reach a contradiction. It turns out that the latter of the two approaches is easiest here.

(ii) ($\Leftarrow$) If n is even, then n^2 is even. To do this directly, assume that n is even and prove that n^2 is even.

Now that we have broken the theorem down into two parts and know what to prove, we proceed to prove the two implications. The final ingredient that we need is a convenient way of describing even integers. When we refer to an integer n (or m, or k, . . .) as even, we can always replace it with a product of the form $2q$, where q is an integer (more precisely, $(\exists q)_\mathbb{Z}(n = 2q)$). In other words, for an integer to be even it must have a factor of two in its prime decomposition. $\square$

Example 3.9.3 $\sqrt{2}$ **is irrational.** Our final example will be an outline of the proof that the square root of 2 is irrational (not an element of $\mathbb{Q}$). This is an example of the theorem that does not appear to be in the standard $P \Rightarrow C$ form. One way to rephrase the theorem is: If x is a rational number, then $x^2 \neq 2$. A direct proof of this theorem would require that we verify that the square of every rational number is not equal to 2. There is no convenient way of doing this, so we must turn to the indirect method of proof. In such a proof, we assume that x is a rational number and that $x^2 = 2$. This will lead to a contradiction. In order to reach this contradiction, we need to use the following facts:

- A rational number is a quotient of two integers.

- Every fraction can be reduced to lowest terms, so that the numerator and denominator have no common factor greater than 1.

- If n is an integer, n^2 is even if and only if n is even.

$\square$

3.9.3 Exercises for Section 3.9

1. Prove that the sum of two odd positive integers is even positive integer.

2. Write out a complete proof that if n is an integer, n^2 is even if and only if n is even.

3. Write out a complete proof that $\sqrt{2}$ is irrational.

4. Prove that $\sqrt[3]{2}$ is an irrational number.

5. Prove that if x and y are real numbers such that $x + y \leq 1$, then either $x \leq \frac{1}{2}$ or $y \leq \frac{1}{2}$.

6. Use the following definition of absolute value to prove the given statements: If x is a real number, then the absolute value of x, $|x|$, is defined by:

$$|x| = \begin{cases} x & \text{if } x \geq 0 \\ -x & \text{if } x < 0 \end{cases}$$

(a) For any real number x, $|x| \geq 0$. Moreover, $|x| = 0$ implies $x = 0$.

(b) For any two real numbers x and y, $|x| \cdot |y| = |xy|$.

(c) For any two real numbers x and y, $|x + y| \leq |x| + |y|$.

Chapter 4

More on Sets

In this chapter we shall look more closely at some basic facts about sets. One question we could ask ourselves is: Can we manipulate sets similarly to the way we manipulated expressions in basic algebra, or to the way we manipulated propositions in logic? In basic algebra we are aware that $a \cdot (b+c) = a \cdot b + a \cdot c$ for all real numbers a, b, and c. In logic we verified an analogue of this statement, namely, $p \wedge (q \vee r) \Leftrightarrow (p \wedge q) \vee (p \wedge r))$, where p, q, and r were arbitrary propositions. If A, B, and C are arbitrary sets, is $A \cap (B \cup C) = (A \cap B) \cup (A \cap C)$? How do we convince ourselves of it is truth, or discover that it is false? Let us consider some approaches to this problem, look at their pros and cons, and determine their validity. Later in this chapter, we introduce partitions of sets and minsets.

4.1 Methods of Proof for Sets

If A, B, and C are arbitrary sets, is it always true that $A \cap (B \cup C) = (A \cap B) \cup (A \cap C)$? There are a variety of ways that we could attempt to prove that this distributive law for intersection over union is indeed true. We start with a common "non-proof" and then work toward more acceptable methods.

4.1.1 Examples and Counterexamples

We could, for example, let $A = \{1, 2\}$, $B = \{5, 8, 10\}$, and $C = \{3, 2, 5\}$, and determine whether the distributive law is true for these values of A, B, and C. In doing this we will have only determined that the distributive law is true for this one example. It does not prove the distributive law for all possible sets A, B, and C and hence is an invalid method of proof. However, trying a few examples has considerable merit insofar as it makes us more comfortable with the statement in question. Indeed, if the statement is not true for the example, we have disproved the statement.

Definition 4.1.1 Counterexample. An example that disproves a statement is called a counterexample. ◇

Example 4.1.2 Disproving distributivity of addition over multiplication. From basic algebra we learned that multiplication is distributive over addition. Is addition distributive over multiplication? That is, is $a + (b \cdot c) = (a + b) \cdot (a + c)$ always true? If we choose the values $a = 3$, $b = 4$, and $c = 1$, we find that $3 + (4 \cdot 1) \neq (3 + 4) \cdot (3 + 1)$. Therefore, this set of values serves as a counterexample to a distributive law of addition over multiplication. □

4.1.2 Proof Using Venn Diagrams

In this method, we illustrate both sides of the statement via a Venn diagram and determine whether both Venn diagrams give us the same "picture," For example, the left side of the distributive law is developed in Figure Figure 4.1.3 and the right side in Figure Figure 4.1.4. Note that the final results give you the same shaded area.

The advantage of this method is that it is relatively quick and mechanical. The disadvantage is that it is workable only if there are a small number of sets under consideration. In addition, it doesn't work very well in a static environment like a book or test paper. Venn diagrams tend to work well if you have a potentially dynamic environment like a blackboard or video.

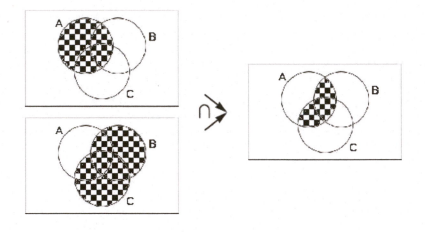

Figure 4.1.3: Development of the left side of the distributive law for sets

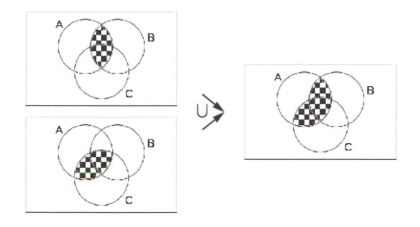

Figure 4.1.4: Development of the right side of the distributive law for sets

4.1.3 Proof using Set-membership Tables

Let A be a subset of a universal set U and let $u \in U$. To use this method we note that exactly one of the following is true: $u \in A$ or $u \notin A$. Denote the situation where $u \in A$ by 1 and that where $u \notin A$ by 0. Working with two sets, A and B, and if $u \in U$, there are four possible outcomes of "where u can be." What are they? The set-membership table for $A \cup B$ is:

A	B	$A \cup B$
0	0	0
0	1	1
1	0	1
1	1	1

Table 4.1.5: Membership Table for $A \cup B$

This table illustrates that $u \in A \cup B$ if and only if $u \in A$ or $u \in B$.

In order to prove the distributive law via a set-membership table, write out the table for each side of the set statement to be proved and note that if S and T are two columns in a table, then the set statement S is equal to the set statement T if and only if corresponding entries in each row are the same.

To prove $A \cap (B \cup C) = (A \cap B) \cup (A \cap C)$, first note that the statement involves three sets, A, B, and C, so there are $2^3 = 8$ possibilities for the membership of an element in the sets.

A	B	C	$B \cup C$	$A \cap B$	$A \cap C$	$A \cap (B \cup C)$	$(A \cap B) \cup (A \cap C)$
0	0	0	0	0	0	0	0
0	0	1	1	0	0	0	0
0	1	0	1	0	0	0	0
0	1	1	1	0	0	0	0
1	0	0	0	0	0	0	0
1	0	1	1	0	1	1	1
1	1	0	1	1	0	1	1
1	1	1	1	1	1	1	1

Table 4.1.6: Membership table to prove the distributive law of intersection over union

Since each entry in Column 7 is the same as the corresponding entry in Column 8, we have shown that $A \cap (B \cup C) = (A \cap B) \cup (A \cap C)$ for any sets A, B, and C. The main advantage of this method is that it is mechanical. The main disadvantage is that it is reasonable to use only for a relatively small number of sets. If we are trying to prove a statement involving five sets, there are $2^5 = 32$ rows, which would test anyone's patience doing the work by hand.

4.1.4 Proof Using Definitions

This method involves using definitions and basic concepts to prove the given statement. This procedure forces one to learn, relearn, and understand basic definitions and concepts. It helps individuals to focus their attention on the main ideas of each topic and therefore is the most useful method of proof. One does not learn a topic by memorizing or occasionally glancing at core topics, but by using them in a variety of contexts. The word proof panics most people; however, everyone can become comfortable with proofs. Do not expect to prove every statement immediately. In fact, it is not our purpose to prove every theorem or fact encountered, only those that illustrate methods and/or basic concepts. Throughout the text we will focus in on main techniques of proofs. Let's illustrate by proving the distributive law.

Proof Technique 1. State or restate the theorem so you understand what is given (the hypothesis) and what you are trying to prove (the conclusion).

Theorem 4.1.7 The Distributive Law of Intersection over Union. *If A, B, and C are sets, then $A \cap (B \cup C) = (A \cap B) \cup (A \cap C)$.*

Proof. What we can assume: A, B, and C are sets.

What we are to prove: $A \cap (B \cup C) = (A \cap B) \cup (A \cap C)$.

Commentary: What types of objects am I working with: sets? real numbers? propositions? The answer is sets: sets of elements that can be anything you care to imagine. The universe from which we draw our elements plays no part in the proof of this theorem.

We need to show that the two sets are equal. Let's call them the left-hand set (LHS) and the right-hand set (RHS). To prove that $LHS = RHS$, we must prove two things: (a) $LHS \subseteq RHS$, and (b) $RHS \subseteq LHS$.

To prove part a and, similarly, part b, we must show that each element of LHS is an element of RHS. Once we have diagnosed the problem we are ready to begin.

We must prove: (a) $A \cap (B \cup C) \subseteq (A \cap B) \cup (A \cap C)$.

Let $x \in A \cap (B \cup C)$:

$$x \in A \cap (B \cup C) \Rightarrow x \in A \text{ and } (x \in B \text{ or } x \in C)$$

def. of union and intersection

$$\Rightarrow (x \in A \text{ and } x \in B) \text{ or } (x \in A \text{ and } x \in C)$$

distributive law of logic

$$\Rightarrow (x \in A \cap B) \text{ or } (x \in A \cap C)$$

def. of intersection

$$\Rightarrow x \in (A \cap B) \cup (A \cap C)$$

def. of union

We must also prove (b) $(A \cap B) \cup (A \cap C) \subseteq A \cap (B \cup C)$.

$$x \in (A \cap B) \cup (A \cap C) \Rightarrow (x \in A \cap B) \text{or} (x \in A \cap C)$$

Why?

$$\Rightarrow (x \in A \text{ and } x \in B) \text{ or } (x \in A \text{ and } x \in C)$$

Why?

$$\Rightarrow x \in A \text{ and } (x \in B \text{ or } x \in C)$$

Why?

$$\Rightarrow x \in A \cap (B \cup C)$$

Why? $\square$

■

Proof Technique 2

(1) To prove that $A \subseteq B$, we must show that if $x \in A$, then $x \in B$.

(2) To prove that $A = B$, we must show:

    (a) $A \subseteq B$ and

    (b) $B \subseteq A$.

To further illustrate the Proof-by-Definition technique, let's prove the following theorem.

Theorem 4.1.8 Another Proof using Definitions. *If A, B, and C are any sets, then $A \times (B \cap C) = (A \times B) \cap (A \times C)$.*

Proof. Commentary; We again ask ourselves: What are we trying to prove? What types of objects are we dealing with? We realize that we wish to prove two facts: (a) $LHS \subseteq RHS$, and (b) $RHS \subseteq LHS$.

To prove part (a), and similarly part (b), we'll begin the same way. Let _____ $\in LHS$ to show _____ $\in RHS$. What should _____ be? What does a typical object in the *LHS* look like?

Now, on to the actual proof.

(a) $A \times (B \cap C) \subseteq (A \times B) \cap (A \times C)$.

Let $(x, y) \in A \times (B \cap C)$.

$$(x, y) \in A \times (B \cap C) \Rightarrow x \in A \text{ and } y \in (B \cap C)$$

Why?

$$\Rightarrow x \in A \text{ and } (y \in B \text{ and } y \in C)$$

Why?

$$\Rightarrow (x \in A \text{ and } y \in B) \text{ and } (x \in A \text{ and } y \in C)$$

Why?

$$\Rightarrow (x, y) \in (A \times B) \text{ and } (x, y) \in (A \times C)$$

Why?

$$\Rightarrow (x, y) \in (A \times B) \cap (A \times C)$$

Why?

(b) $(A \times B) \cap (A \times C) \subseteq A \times (B \cap C)$.

Let $(x, y) \in (A \times B) \cap (A \times C)$.

$$(x, y) \in (A \times B) \cap (A \times C) \Rightarrow (x, y) \in A \times B \text{ and } (x, y) \in A \times C$$

Why?

$$\Rightarrow (x \in A \text{ and } y \in B) \text{ and } (x \in A \text{ and } y \in C)$$

Why?

$$\Rightarrow x \in A \text{ and } (y \in B \text{ and } y \in C)$$

Why?

$$\Rightarrow x \in A \text{ and } y \in (B \cap C)$$

Why?

$$\Rightarrow (x, y) \in A \times (B \cap C)$$

Why?

■

4.1.5 Exercises for Section 4.1

1. Prove the following:

(a) Let A, B, and C be sets. If $A \subseteq B$ and $B \subseteq C$, then $A \subseteq C$.

(b) Let A and B be sets. Then $A - B = A \cap B^c$.

(c) Let A, B, and C be sets. If $(A \subseteq B \text{ and } A \subseteq C)$ then $A \subseteq B \cap C$.

(d) Let A and B be sets. $A \subseteq B$ if and only if $B^c \subseteq A^c$.

(e) Let A, B, and C be sets. If $A \subseteq B$ then $A \times C \subseteq B \times C$.

2. Write the converse of parts (a), (c), and (e) of Exercise 1 and prove or disprove them.

3. Disprove the following, assuming A, B, and C are sets:

(a) $A - B = B - A$.

(b) $A \times B = B \times A$.

(c) $A \cap B = A \cap C$ implies $B = C$.

(d) $A \oplus (B \cap C) = (A \oplus B) \cap (A \oplus C)$

4. Let A, B, and C be sets. Write the following in "if . . . then . . ." language and prove:

(a) $x \in B$ is a sufficient condition for $x \in A \cup B$.

(b) $A \cap B \cap C = \emptyset$ is a necessary condition for $A \cap B = \emptyset$.

(c) $A \cup B = B$ is a necessary and sufficient condition for $A \subseteq B$.

5. Prove by induction that if $A, B_1, B_2, \ldots, B_n$ are sets, $n \geq 2$, then $A \cap (B_1 \cup B_2 \cup \cdots \cup B_n) = (A \cap B_1) \cup (A \cap B_2) \cup \cdots \cup (A \cap B_n)$.

6. Let A, B and C be sets. Prove or disprove:

$$A \cap B \neq \emptyset, B \cap C \neq \emptyset \Rightarrow A \cap C \neq \emptyset$$

4.2 Laws of Set Theory

4.2.1 Tables of Laws

The following basic set laws can be derived using either the Basic Definition or the Set-Membership approach and can be illustrated by Venn diagrams.

Commutative Laws	
(1) $A \cup B = B \cup A$	(1') $A \cap B = B \cap A$
Associative Laws	
(2) $A \cup (B \cup C) = (A \cup B) \cup C$	(2') $A \cap (B \cap C) = (A \cap B) \cap C$
Distributive Laws	
(3) $A \cap (B \cup C) = (A \cap B) \cup (A \cap C)$	(3') $A \cup (B \cap C) = (A \cup B) \cap (A \cup C)$
Identity Laws	
(4) $A \cup \emptyset = \emptyset \cup A = A$	(4') $A \cap U = U \cap A = A$
Complement Laws	
(5) $A \cup A^c = U$	(5') $A \cap A^c = \emptyset$
Idempotent Laws	
(6) $A \cup A = A$	(6') $A \cap A = A$
Null Laws	
(7) $A \cup U = U$	(7') $A \cap \emptyset = \emptyset$
Absorption Laws	
(8) $A \cup (A \cap B) = A$	(8') $A \cap (A \cup B) = A$
DeMorgan's Laws	
(9) $(A \cup B)^c = A^c \cap B^c$	(9') $(A \cap B)^c = A^c \cup B^c$
Involution Law	
(10) $(A^c)^c = A$	

Table 4.2.1: Basic Laws of Set Theory

It is quite clear that most of these laws resemble or, in fact, are analogues of laws in basic algebra and the algebra of propositions.

4.2.2 Proof Using Previously Proven Theorems

Once a few basic laws or theorems have been established, we frequently use them to prove additional theorems. This method of proof is usually more efficient than that of proof by Definition. To illustrate, let us prove the following Corollary to the Distributive Law. The term "corollary" is used for theorems that can be proven with relative ease from previously proven theorems.

Corollary 4.2.2 A Corollary to the Distributive Law of Sets. *Let A and B be sets. Then* $(A \cap B) \cup (A \cap B^c) = A$.

Proof.

$$(A \cap B) \cup (A \cap B^c) = A \cap (B \cup B^c)$$
$$\text{Why?}$$
$$= A \cap U$$
$$\text{Why?}$$
$$= A$$
$$\text{Why?}$$

∎

4.2.3 Proof Using the Indirect Method/Contradiction

The procedure one most frequently uses to prove a theorem in mathematics is the Direct Method, as illustrated in Theorem 4.1.7 and Theorem 4.1.8. Occasionally there are situations where this method is not applicable. Consider the following:

Theorem 4.2.3 An Indirect Proof in Set Theory. *Let* A, B, C *be sets. If* $A \subseteq B$ *and* $B \cap C = \emptyset$, *then* $A \cap C = \emptyset$.

Proof. Commentary: The usual and first approach would be to assume $A \subseteq B$ and $B \cap C = \emptyset$ is true and to attempt to prove $A \cap C = \emptyset$ is true. To do this you would need to show that nothing is contained in the set $A \cap C$. Think about how you would show that something doesn't exist. It is very difficult to do directly.

The Indirect Method is much easier: If we assume the conclusion is false and we obtain a contradiction --- then the theorem must be true. This approach is on sound logical footing since it is exactly the same method of indirect proof that we discussed in Subsection 3.5.3.

Assume $A \subseteq B$ and $B \cap C = \emptyset$, and $A \cap C \neq \emptyset$. To prove that this cannot occur, let $x \in A \cap C$.

$$x \in A \cap C \Rightarrow x \in A \text{ and } x \in C$$
$$\Rightarrow x \in B \text{ and } x \in C.$$
$$\Rightarrow x \in B \cap C$$

But this contradicts the second premise. Hence, the theorem is proven. ∎

4.2.4 Exercises for Section 4.2

In the exercises that follow it is most important that you outline the logical procedures or methods you use.

1.

    (a) Prove the associative law for intersection (Law 2′) with a Venn diagram.

    (b) Prove DeMorgan's Law (Law 9) with a membership table.

    (c) Prove the Idempotent Law (Law 6) using basic definitions.

2.

    (a) Prove the Absorption Law (Law 8′) with a Venn diagram.

    (b) Prove the Identity Law (Law 4) with a membership table.

    (c) Prove the Involution Law (Law 10) using basic definitions.

3. Prove the following using the set theory laws, as well as any other theorems proved so far.

    (a) $A \cup (B - A) = A \cup B$

    (b) $A - B = B^c - A^c$

    (c) $A \subseteq B, A \cap C \neq \emptyset \Rightarrow B \cap C \neq \emptyset$

    (d) $A \cap (B - C) = (A \cap B) - (A \cap C)$

    (e) $A - (B \cup C) = (A - B) \cap (A - C)$

4. Use previously proven theorems to prove the following.

    (a) $A \cap (B \cap C)^c = (A \cap B^c) \cup (A \cap C^c)$

    (b) $A \cap (B \cap (A \cap B)^c) = \emptyset$

    (c) $(A \cap B) \cup B^c = A \cup B^c$

    (d) $A \cup (B - C) = (A \cup B) - (C - A)$.

5. **Hierarchy of Set Operations.** The rules that determine the order of evaluation in a set expression that involves more than one operation are similar to the rules for logic. In the absence of parentheses, complementations are done first, intersections second, and unions third. Parentheses are used to override this order. If the same operation appears two or more consecutive times, evaluate from left to right. In what order are the following expressions performed?

    (a) $A \cup B^c \cap C$.　　　　　　　(c) $A \cup B \cup C^c$

    (b) $A \cap B \cup C \cap B$.

6. There are several ways that we can use to format the proofs in this chapter. One that should be familiar to you from Chapter 3 is illustrated with the following alternate proof of part (a) in Theorem 4.1.7:

(1)	$x \in A \cap (B \cup C)$	Premise
(2)	$(x \in A) \wedge (x \in B \cup C)$	(1), definition of intersection
(3)	$(x \in A) \wedge ((x \in B) \vee (x \in C))$	(2), definition of union
(4)	$(x \in A) \wedge (x \in B) \vee (x \in A) \wedge (x \in C)$	(3), distribute $\wedge$ over $\vee$
(5)	$(x \in A \cap B) \vee (x \in A \cap C)$	(4), definition of intersection
(6)	$x \in (A \cap B) \cup (A \cap C)$	(5), definition of union ∎

Table 4.2.4: An alternate format for the proof of Theorem 4.1.7

Prove part (b) of Theorem 4.1.8 and Theorem 4.2.3 using this format.

4.3 Minsets

4.3.1 Definition of Minsets

Let B_1 and B_2 be subsets of a set A. Notice that the Venn diagram of Figure 4.3.1 is naturally partitioned into the subsets A_1, A_2, A_3, and A_4. Further we observe that A_1, A_2, A_3, and A_4 can be described in terms of B_1 and B_2 as follows:

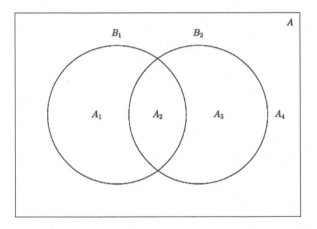

Figure 4.3.1: Venn Diagram of Minsets

$$A_1 = B_1 \cap B_2^c$$
$$A_2 = B_1 \cap B_2$$
$$A_3 = B_1^c \cap B_2$$
$$A_4 = B_1^c \cap B_2^c$$

Table 4.3.2: Minsets generated by two sets

Each A_i is called a minset generated by B_1 and B_2. We note that each minset is formed by taking the intersection of two sets where each may be either B_k or its complement, B_k^c. Note also, given two sets, there are $2^2 = 4$ minsets.

Minsets are occasionally called *minterms*.

The reader should note that if we apply all possible combinations of the operations intersection, union, and complementation to the sets B_1 and B_2 of

Figure 1, the smallest sets generated will be exactly the minsets, the minimum sets. Hence the derivation of the term minset.

Next, consider the Venn diagram containing three sets, B_1, B_2, and B_3. Draw it right now and count the regions! What are the minsets generated by B_1, B_2, and B_3? How many are there? Following the procedures outlined above, we note that the following are three of the $2^3 = 8$ minsets. What are the others?

$$B_1 \cap B_2 \cap B_3^c$$
$$B_1 \cap B_2^c \cap B_3$$
$$B_1 \cap B_2^c \cap B_3^c$$

Table 4.3.3: Three of the minsets generated by B_1, B_2, and B_3

Definition 4.3.4 Minset. Let $\{B_1, B_2, \ldots, B_n\}$ be a set of subsets of set A. Sets of the form $D_1 \cap D_2 \cap \cdots \cap D_n$, where each D_i may be either B_i or B_i^c, is called a minset generated by B_1, $B_2,\ldots$ and B_n. $\Diamond$

Example 4.3.5 A concrete example of some minsets. Consider the following concrete example. Let $A = \{1, 2, 3, 4, 5, 6\}$ with subsets $B_1 = \{1, 3, 5\}$ and $B_2 = \{1, 2, 3\}$. How can we use set operations applied to B_1 and B_2 and produce a list of sets that contain elements of A efficiently without duplication? As a first attempt, we might try these three sets:

$$B_1 \cap B_2 = \{1, 3\}$$
$$B_1^c = \{2, 4, 6\}$$
$$B_2^c = \{4, 5, 6\}.$$

Table 4.3.6

We have produced all elements of A but we have 4 and 6 repeated in two sets. In place of B_1^c and B_2^c, let's try $B_1^c \cap B_2$ and $B_1 \cap B_2^c$, respectively:

$$B_1^c \cap B_2 = \{2\} \text{ and}$$
$$B_1 \cap B_2^c = \{5\}.$$

Table 4.3.7

We have now produced the elements 1, 2, 3, and 5 using $B_1 \cap B_2$, $B_1^c \cap B_2$ and $B_1 \cap B_2^c$ yet we have not listed the elements 4 and 6. Most ways that we could combine B_1 and B_2 such as $B_1 \cup B_2$ or $B_1 \cup B_2^c$ will produce duplications of listed elements and will not produce both 4 and 6. However we note that $B_1^c \cap B_2^c = \{4, 6\}$, exactly the elements we need. Each element of A appears exactly once in one of the four minsets $B_1 \cap B_2$, $B_1^c \cap B_2$, $B_1 \cap B_2^c$ and $B_1^c \cap B_2^c$. Hence, we have a partition of A. $\square$

4.3.2 Properties of Minsets

Theorem 4.3.8 Minset Partition Theorem. *Let A be a set and let B_1, $B_2 \ldots$, B_n be subsets of A. The set of nonempty minsets generated by B_1, B_2 $\ldots$, B_n is a partition of A.*

Proof. The proof of this theorem is left to the reader.   ∎

One of the most significant facts about minsets is that any subset of A that can be obtained from B_1, B_2 ..., B_n, using the standard set operations can be obtained in a standard form by taking the union of selected minsets.

Definition 4.3.9 Minset Normal Form. A set is said to be in minset normal form when it is expressed as the union of zero or more distinct nonempty minsets.   ◊

Notes:

- The union of zero sets is the empty set, $\emptyset$.

- Minset normal form is also called **canonical form**.

Example 4.3.10   Another Concrete Example of Minsets. Let $U = \{-2, -1, 0, 1, 2\}$, $B_1 = \{0, 1, 2\}$, and $B_2 = \{0, 2\}$. Then

$$B_1 \cap B_2 = \{0, 2\}$$
$$B_1^c \cap B_2 = \emptyset$$
$$B_1 \cap B_2^c = \{1\}$$
$$B_1^c \cap B_2^c = \{-2, -1\}$$

Table 4.3.11

In this case, there are only three nonempty minsets, producing the partition $\{\{0, 2\}, \{1\}, \{-2, -1\}\}$. An example of a set that could not be produced from just B_1 and B_2 is the set of even elements of U, $\{-2, 0, 2\}$. This is because -2 and -1 cannot be separated. They are in the same minset and any union of minsets either includes or excludes them both. In general, there are $2^3 = 8$ different minset normal forms because there are three nonempty minsets. This means that only 8 of the $2^5 = 32$ subsets of U could be generated from any two sets B_1 and B_2.   □

4.3.3 Exercises for Section 4.3

1.   Consider the subsets $A = \{1, 7, 8\}$, $B = \{1, 6, 9, 10\}$, and $C = \{1, 9, 10\}$, where $U = \{1, 2, ..., 10\}$.

    (a) List the nonempty minsets generated by A, B, and C.

    (b) How many elements of the power set of U can be generated by A, B, and C? Compare this number with $|\, \mathcal{P}(U)\,|$. Give an example of one subset that cannot be generated by A, B, and C.

2.

    (a) Partition $\{1, 2, ...9\}$ into the minsets generated by $B_1 = \{5, 6, 7\}$, $B_2 = \{2, 4, 5, 9\}$, and $B_3 = \{3, 4, 5, 6, 8, 9\}$.

    (b) How many different subsets of $\{1, 2, ..., 9\}$ can you create using B_1, B_2, and B_3 with the standard set operations?

    (c) Do there exist subsets C_1, C_2, C_3 whose minsets will generate every subset of $\{1, 2, ..., 9\}$?

3.   Partition the set of strings of 0's and 1's of length two or less, using the minsets generated by $B_1 = \{s \mid s$ has length 2$\}$, and $B_2 = \{s \mid s$ starts with a 0$\}$.

4. Let $B_1, B_2,$ and B_3 be subsets of a universal set U,

    (a) Symbolically list all minsets generated by $B_1, B_2,$ and B_3.

    (b) Illustrate with a Venn diagram all minsets obtained in part (a).

    (c) Express the following sets in minset normal form: B_1^c, $B_1 \cap B_2$, $B_1 \cup B_2^c$.

5.

    (a) Partition $A = \{0, 1, 2, 3, 4, 5\}$ with the minsets generated by $B_1 = \{0, 2, 4\}$ and $B_2 = \{1, 5\}$.

    (b) How many different subsets of A can you generate from B_1 and B_2?

6. If $\{B_1, B_2, \ldots, B_n\}$ is a partition of A, how many minsets are generated by $B_1, B_2, \ldots, B_n$?

7. Prove Theorem 4.3.8

8. Let S be a finite set of n elements. Let B_i „ $i = 1, 2, \ldots, k$ be nonempty subsets of S. There are 2^{2^k} minset normal forms generated by the k subsets. The number of subsets of S is 2^n. Since we can make $2^{2^k} > 2^n$ by choosing $k \geq \log_2 n$, it is clear that two distinct minset normal-form expressions do not always equal distinct subsets of S. Even for $k < \log_2 n$, it may happen that two distinct minset normal-form expressions equal the same subset of S. Determine necessary and sufficient conditions for distinct normal-form expressions to equal distinct subsets of S.

4.4 The Duality Principle

4.4.1

In Section 4.2, we observed that each of the Table 4.2.1 labeled 1 through 9 had an analogue $1'$ through $9'$. We notice that each of the laws in one column can be obtained from the corresponding law in the other column by replacing $\cup$ by $\cap$, $\cap$ by $\cup$, $\emptyset$ by U, U by $\emptyset$, and leaving the complement unchanged.

Definition 4.4.1 Duality Principle for Sets. Let S be any identity involving sets and the operations complement, intersection and union. If $S*$ is obtained from S by making the substitutions $\cup \to \cap$, $\cap \to \cup$, $\emptyset \to U$, and $U \to \emptyset$, then the statement $S*$ is also true and it is called the dual of the statement S. $\diamond$

Example 4.4.2 Example of a dual. The dual of $(A \cap B) \cup (A \cap B^c) = A$ is $(A \cup B) \cap (A \cup B^c) = A$. $\square$

One should not underestimate the importance of this concept. It gives us a whole second set of identities, theorems, and concepts. For example, we can consider the dual of *minsets* and *minset normal form* to obtain what is called *maxsets* and *maxset normal form*.

4.4.2 Exercises for Section 4.4

1.   State the dual of each of the following:

   (a) $A \cup (B \cap A) = A$.

   (b) $A \cup ((B^c \cup A) \cap B)^c = U$

   (c) $(A \cup B^c)^c \cap B = A^c \cap B$

2.   Examine Table 3.4.3 and then write a description of the principle of duality for logic.

3.   Write the dual of each of the following:

   (a) $p \vee \neg ((\neg q \vee p) \wedge q) \Leftrightarrow 1$

   (b) $(\neg (p \wedge (\neg q))) \vee q \Leftrightarrow (\neg p \vee q)$.

4.   Use the principle of duality and the definition of minset to write the definition of maxset.

5.   Let $A = \{1, 2, 3, 4, 5, 6\}$ and let $B_1 = \{1, 3, 5\}$ and $B_2 = \{1, 2, 3\}$.

   (a) Find the maxsets generated by B_1 and B_2. Note the set of maxsets does not constitute a partition of A. Can you explain why?

   (b) Write out the definition of maxset normal form.

   (c) Repeat Exercise 4.3.3.4 for maxsets.

6.   What is the dual of the expression in Exercise 4.1.5.5 ?

Chapter 5

Introduction to Matrix Algebra

The purpose of this chapter is to introduce you to matrix algebra, which has many applications. You are already familiar with several algebras: elementary algebra, the algebra of logic, the algebra of sets. We hope that as you studied the algebra of logic and the algebra of sets, you compared them with elementary algebra and noted that the basic laws of each are similar. We will see that matrix algebra is also similar. As in previous discussions, we begin by defining the objects in question and the basic operations.

5.1 Basic Definitions and Operations

5.1.1 Matrix Order and Equality

Definition 5.1.1 matrix. A matrix is a rectangular array of elements of the form

$$A = \begin{pmatrix} a_{11} & a_{12} & a_{13} & \cdots & a_{1n} \\ a_{21} & a_{22} & a_{23} & \cdots & a_{2n} \\ a_{31} & a_{32} & a_{33} & \cdots & a_{3n} \\ \vdots & \vdots & \vdots & \ddots & \vdots \\ a_{m1} & a_{m2} & a_{m3} & \cdots & a_{mn} \end{pmatrix}$$

$\Diamond$

A convenient way of describing a matrix in general is to designate each entry via its position in the array. That is, the entry a_{34} is the entry in the third row and fourth column of the matrix A. Depending on the situation, we will decide in advance to which set the entries in a matrix will belong. For example, we might assume that each entry a_{ij} ($1 \leq i \leq m$, $1 \leq j \leq n$) is a real number. In that case we would use $M_{m \times n}(\mathbb{R})$ to stand for the set of all m by n matrices whose entries are real numbers. If we decide that the entries in a matrix must come from a set S, we use $M_{m \times n}(S)$ to denote all such matrices.

Definition 5.1.2 The Order of a Matrix. A matrix A that has m rows and n columns is called an $m \times n$ (read "m by n") matrix, and is said to have order $m \times n$. $\Diamond$

Since it is rather cumbersome to write out the large rectangular array above each time we wish to discuss the generalized form of a matrix, it is common practice to replace the above by $A = (a_{ij})$. In general, matrices are often given names that are capital letters and the corresponding lower case letter is used for individual entries. For example the entry in the third row, second column of a matrix called C would be c_{32}.

Example 5.1.3 Orders of Some Matrices. $A = \begin{pmatrix} 2 & 3 \\ 0 & -5 \end{pmatrix}$, $B = \begin{pmatrix} 0 \\ \frac{1}{2} \\ 15 \end{pmatrix}$, and $D = \begin{pmatrix} 1 & 2 & 5 \\ 6 & -2 & 3 \\ 4 & 2 & 8 \end{pmatrix}$ are 2×2, 3×1, and 3×3 matrices, respectively. $\square$

Since we now understand what a matrix looks like, we are in a position to investigate the operations of matrix algebra for which users have found the most applications.

First we ask ourselves: Is the matrix $A = \begin{pmatrix} 1 & 2 \\ 3 & 4 \end{pmatrix}$ equal to the matrix $B = \begin{pmatrix} 1 & 2 \\ 3 & 5 \end{pmatrix}$? No, they are not because the corresponding entries in the second row, second column of the two matrices are not equal.

Next, is $A = \begin{pmatrix} 1 & 2 & 3 \\ 4 & 5 & 6 \end{pmatrix}$ equal to $B = \begin{pmatrix} 1 & 2 \\ 4 & 5 \end{pmatrix}$? No, although the corresponding entries in the first two columns are identical, B doesn't have a third column to compare to that of A. We formalize these observations in the following definition.

Definition 5.1.4 Equality of Matrices. A matrix A is said to be equal to matrix B (written $A = B$) if and only if:

(1) A and B have the same order, and

(2) all corresponding entries are equal: that is, $a_{ij} = b_{ij}$ for all appropriate i and j.

$\Diamond$

5.1.2 Matrix Addition and Scalar Multiplication

The first two operations we introduce are very natural and are not likely cause much confusion. The first is matrix addition. It seems natural that if $A = \begin{pmatrix} 1 & 0 \\ 2 & -1 \end{pmatrix}$ and $B = \begin{pmatrix} 3 & 4 \\ -5 & 2 \end{pmatrix}$, then

$$A + B = \begin{pmatrix} 1+3 & 0+4 \\ 2-5 & -1+2 \end{pmatrix} = \begin{pmatrix} 4 & 4 \\ -3 & 1 \end{pmatrix}.$$

However, if $A = \begin{pmatrix} 1 & 2 & 3 \\ 0 & 1 & 2 \end{pmatrix}$ and $B = \begin{pmatrix} 3 & 0 \\ 2 & 8 \end{pmatrix}$, is there a natural way to add them to give us $A + B$? No, the orders of the two matrices must be identical.

Definition 5.1.5 Matrix Addition. Let A and B be $m \times n$ matrices. Then $A + B$ is an $m \times n$ matrix where $(A+B)_{ij} = a_{ij} + b_{ij}$ (read "The ith jth entry of the matrix $A+B$ is obtained by adding the ith jth entry of A to the ith jth entry of B"). If the orders of A and B are not identical, $A + B$ is not defined.

$\Diamond$

In short, $A + B$ is defined if and only if A and B are of the same order.

Another frequently used operation is that of multiplying a matrix by a number, commonly called a scalar in this context. Scalars normally come from the same set as the entries in a matrix. For example, if $A \in M_{m \times n}(\mathbb{R})$, a scalar can be any real number.

Example 5.1.6 A Scalar Product. If $c = 3$ and if $A = \begin{pmatrix} 1 & -2 \\ 3 & 5 \end{pmatrix}$ and we wish to find cA, it seems natural to multiply each entry of A by 3 so that $3A = \begin{pmatrix} 3 & -6 \\ 9 & 15 \end{pmatrix}$, and this is precisely the way scalar multiplication is defined.

$\square$

Definition 5.1.7 Scalar Multiplication. Let A be an $m \times n$ matrix and c a scalar. Then cA is the $m \times n$ matrix obtained by multiplying c times each entry of A; that is $(cA)_{ij} = ca_{ij}$. $\diamondsuit$

5.1.3 Matrix Multiplication

A definition that is more awkward to motivate is the product of two matrices. See Exercise 5.1.4.8 for an attempt to do so. In time, the reader will see that the following definition of the product of matrices will be very useful, and will provide an algebraic system that is quite similar to elementary algebra.

Here is a video introduction to matrix multiplication.

Definition 5.1.8 Matrix Multiplication. Let A be an $m \times n$ matrix and let B be an $n \times p$ matrix. The product of A and B, denoted by AB, is an $m \times p$ matrix whose ith row jth column entry is

$$(AB)_{ij} = a_{i1}b_{1j} + a_{i2}b_{2j} + \cdots + a_{in}b_{nj}$$

$$= \sum_{k=1}^{n} a_{ik}b_{kj}$$

for $1 \leq i \leq m$ and $1 \leq j \leq p$. $\diamondsuit$

The mechanics of computing one entry in the product of two matrices is illustrated in Figure 5.1.9.

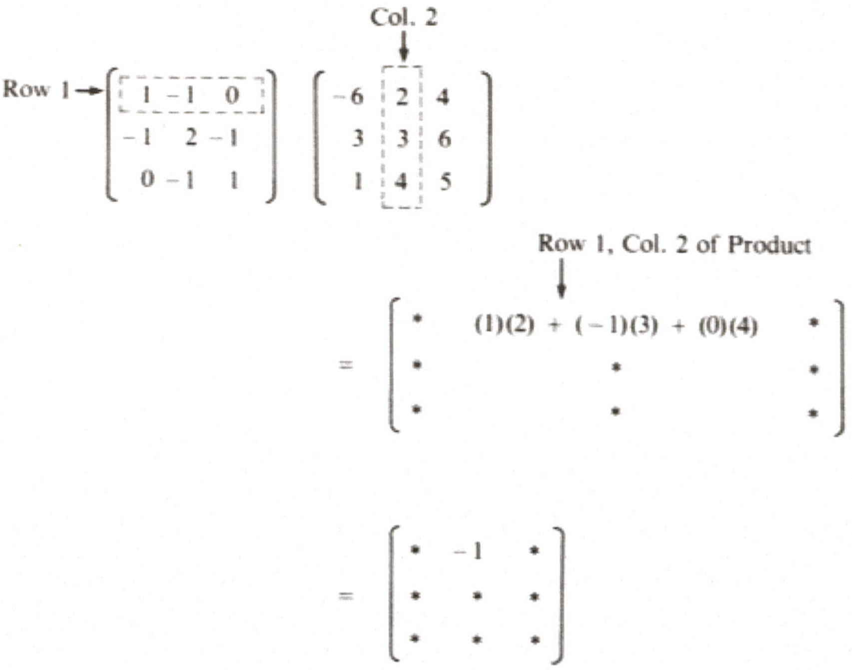

$$\text{Row 1} \rightarrow \begin{bmatrix} 1 & -1 & 0 \\ -1 & 2 & -1 \\ 0 & -1 & 1 \end{bmatrix} \quad \begin{bmatrix} -6 & 2 & 4 \\ 3 & 3 & 6 \\ 1 & 4 & 5 \end{bmatrix}$$

Col. 2

Row 1, Col. 2 of Product

$$= \begin{bmatrix} \bullet & (1)(2) + (-1)(3) + (0)(4) & \bullet \\ \bullet & \bullet & \bullet \\ \bullet & \bullet & \bullet \end{bmatrix}$$

$$= \begin{bmatrix} \bullet & -1 & \bullet \\ \bullet & \bullet & \bullet \\ \bullet & \bullet & \bullet \end{bmatrix}$$

Figure 5.1.9: Computation of one entry in the product of two 3 by 3 matrices

The computation of a product can take a considerable amount of time in comparison to the time required to add two matrices. Suppose that A and B are $n \times n$ matrices; then $(AB)_{ij}$ is determined performing n multiplications and $n - 1$ additions. The full product takes n^3 multiplications and $n^3 - n^2$ additions. This compares with n^2 additions for the sum of two $n \times n$ matrices. The product of two 10 by 10 matrices will require 1,000 multiplications and 900 additions, clearly a job that you would assign to a computer. The sum of two matrices requires a more modest 100 additions. This analysis is based on the assumption that matrix multiplication will be done using the formula that is given in the definition. There are more advanced methods that, in theory, reduce operation counts. For example, Strassen's algorithm (https:/ /en.wikipedia.org/wiki/Strassen_algorithm) computes the product of two n by n matrices in $7 \cdot 7^{\log_2 n} - 6 \cdot 4^{\log_2 n} \approx 7n^{2.808}$ operations. There are practical issues involved in actually using the algorithm in many situations. For example, round-off error can be more of a problem than with the standard formula.

Example 5.1.10 A Matrix Product. Let $A = \begin{pmatrix} 1 & 0 \\ 3 & 2 \\ -5 & 1 \end{pmatrix}$, a 3×2 matrix,

and let $B = \begin{pmatrix} 6 \\ 1 \end{pmatrix}$, a 2×1 matrix. Then AB is a 3×1 matrix:

$$AB = \begin{pmatrix} 1 & 0 \\ 3 & 2 \\ -5 & 1 \end{pmatrix} \begin{pmatrix} 6 \\ 1 \end{pmatrix} = \begin{pmatrix} 1 \cdot 6 + 0 \cdot 1 \\ 3 \cdot 6 + 2 \cdot 1 \\ -5 \cdot 6 + 1 \cdot 1 \end{pmatrix} = \begin{pmatrix} 6 \\ 20 \\ -29 \end{pmatrix}$$

□

Remarks:

(1) The product AB is defined only if A is an $m \times n$ matrix and B is an $n \times p$ matrix; that is, the two "inner" numbers must be equal. Furthermore, the order of the product matrix AB is the "outer" numbers, in this case $m \times p$.

(2) It is wise to first determine the order of a product matrix. For example, if A is a 3×2 matrix and B is a 2×2 matrix, then AB is a 3×2 matrix of the form

$$AB = \begin{pmatrix} c_{11} & c_{12} \\ c_{21} & c_{22} \\ c_{31} & c_{32} \end{pmatrix}$$

Then to obtain, for example, c_{31}, we multiply corresponding entries in the third row of A times the first column of B and add the results.

Example 5.1.11 Multiplication with a diagonal matrix. Let $A = \begin{pmatrix} -1 & 0 \\ 0 & 3 \end{pmatrix}$ and $B = \begin{pmatrix} 3 & 10 \\ 2 & 1 \end{pmatrix}$. Then $AB = \begin{pmatrix} -1 \cdot 3 + 0 \cdot 2 & -1 \cdot 10 + 0 \cdot 1 \\ 0 \cdot 3 + 3 \cdot 2 & 0 \cdot 10 + 3 \cdot 1 \end{pmatrix} = \begin{pmatrix} -3 & -10 \\ 6 & 3 \end{pmatrix}$

The net effect is to multiply the first row of B by -1 and the second row of B by 3.

Note: $BA = \begin{pmatrix} -3 & 30 \\ -2 & 3 \end{pmatrix} \neq AB$. The columns of B are multiplied by -1 and 3 when the order is switched. □

Remarks:

- An $n \times n$ matrix is called a *square matrix*.

- If A is a square matrix, AA is defined and is denoted by A^2, and $AAA = A^3$, etc.

- The $m \times n$ matrices whose entries are all 0 are denoted by $\mathbf{0}_{m \times n}$, or simply $\mathbf{0}$, when no confusion arises regarding the order.

5.1.4 Exercises for Section 5.1

1. Let $A = \begin{pmatrix} 1 & -1 \\ 2 & 3 \end{pmatrix}$, $B = \begin{pmatrix} 0 & 1 \\ 3 & -5 \end{pmatrix}$, and $C = \begin{pmatrix} 0 & 1 & -1 \\ 3 & -2 & 2 \end{pmatrix}$

    (a) Compute AB and BA.

    (b) Compute $A + B$ and $B + A$.

    (c) If $c = 3$, show that $c(A + B) = cA + cB$.

    (d) Show that $(AB)C = A(BC)$.

    (e) Compute $A^2 C$.

    (f) Compute $B + \mathbf{0}$.

    (g) Compute $A\mathbf{0}_{2 \times 2}$ and $\mathbf{0}_{2 \times 2} A$, where $\mathbf{0}_{2 \times 2}$ is the 2×2 zero matrix.

    (h) Compute $0A$, where 0 is the real number (scalar) zero.

    (i) Let $c = 2$ and $d = 3$. Show that $(c + d)A = cA + dA$.

2. Let $A = \begin{pmatrix} 1 & 0 & 2 \\ 2 & -1 & 5 \\ 3 & 2 & 1 \end{pmatrix}$, $B = \begin{pmatrix} 0 & 2 & 3 \\ 1 & 1 & 2 \\ -1 & 3 & -2 \end{pmatrix}$, and $C = \begin{pmatrix} 2 & 1 & 2 & 3 \\ 4 & 0 & 1 & 1 \\ 3 & -1 & 4 & 1 \end{pmatrix}$

   Compute, if possible;

   (a) $A - B$

   (b) AB

   (c) $AC - BC$

   (d) $A(BC)$

   (e) $CA - CB$

   (f) $C\begin{pmatrix} x \\ y \\ z \\ w \end{pmatrix}$

3. Let $A = \begin{pmatrix} 2 & 0 \\ 0 & 3 \end{pmatrix}$. Find a matrix B such that $AB = I$ and $BA = I$, where $I = \begin{pmatrix} 1 & 0 \\ 0 & 1 \end{pmatrix}$.

4. Find AI and BI where I is as in Exercise 3, where $A = \begin{pmatrix} 1 & 8 \\ 9 & 5 \end{pmatrix}$ and $B = \begin{pmatrix} -2 & 3 \\ 5 & -7 \end{pmatrix}$. What do you notice?

5. Find A^3 if $A = \begin{pmatrix} 1 & 0 & 0 \\ 0 & 2 & 0 \\ 0 & 0 & 3 \end{pmatrix}$. What is A^{15} equal to?

6.

   (a) Determine I^2 and I^3 if $I = \begin{pmatrix} 1 & 0 & 0 \\ 0 & 1 & 0 \\ 0 & 0 & 1 \end{pmatrix}$.

   (b) What is I^n equal to for any $n \geq 1$?

   (c) Prove your answer to part (b) by induction.

7.

   (a) If $A = \begin{pmatrix} 2 & 1 \\ 1 & -1 \end{pmatrix}$, $X = \begin{pmatrix} x_1 \\ x_2 \end{pmatrix}$, and $B = \begin{pmatrix} 3 \\ 1 \end{pmatrix}$, show that $AX = B$ is a way of expressing the system $\begin{array}{c} 2x_1 + x_2 = 3 \\ x_1 - x_2 = 1 \end{array}$ using matrices.

   (b) Express the following systems of equations using matrices:

   (i) $\begin{array}{c} 2x_1 - x_2 = 4 \\ x_1 + x_2 = 0 \end{array}$

   (ii) $\begin{array}{c} x_1 + x_2 + 2x_3 = 1 \\ x_1 + 2x_2 - x_3 = -1 \\ x_1 + 3x_2 + x_3 = 5 \end{array}$

   (iii) $\begin{array}{c} x_1 + x_2 = 3 \\ x_2 = 5 \\ x_1 + 3x_3 = 6 \end{array}$

8. In this exercise, we propose to show how matrix multiplication is a natural operation. Suppose a bakery produces bread, cakes and pies every weekday, Monday through Friday. Based on past sales history, the bakery produces various numbers of each product each day, summarized in the 5×3 matrix D. It should be noted that the order could be described as "number of days by number of products." For example, on Wednesday

(the third day) the number of cakes (second product in our list) that are produced is $d_{3,2} = 4$.

$$D = \begin{pmatrix} 25 & 5 & 5 \\ 14 & 5 & 8 \\ 20 & 4 & 15 \\ 18 & 5 & 7 \\ 35 & 10 & 9 \end{pmatrix}$$

The main ingredients of these products are flour, sugar and eggs. We assume that other ingredients are always in ample supply, but we need to be sure to have the three main ones available. For each of the three products, The amount of each ingredient that is needed is summarized in the 3×3, or "number of products by number of ingredients" matrix P. For example, to bake a cake (second product) we need $P_{2,1} = 1.5$ cups of flour (first ingredient). Regarding units: flour and sugar are given in cups per unit of each product, while eggs are given in individual eggs per unit of each product.

$$P = \begin{pmatrix} 2 & 0.5 & 0 \\ 1.5 & 1 & 2 \\ 1 & 1 & 1 \end{pmatrix}$$

These amounts are "made up", so don't used them to do your own baking!

(a) How many cups of flour will the bakery need every Monday? Pay close attention to how you compute your answer and the units of each number.

(b) How many eggs will the bakery need every Wednesday?

(c) Compute the matrix product DP. What do you notice?

(d) Suppose the costs of ingredients are \$0.12 for a cup of flour, \$0.15 for a cup of sugar and \$0.19 for one egg. How can this information be put into a matrix that can meaningfully be multiplied by one of the other matrices in this problem?

5.2 Special Types of Matrices

5.2.1 Diagonal Matrices

We have already investigated, in exercises in the previous section, one special type of matrix. That was the zero matrix, and found that it behaves in matrix algebra in an analogous fashion to the real number 0; that is, as the additive identity. We will now investigate the properties of a few other special matrices.

Definition 5.2.1 Diagonal Matrix. A square matrix D is called a diagonal matrix if $d_{ij} = 0$ whenever $i \neq j$. ◇

Example 5.2.2 Some diagonal matrices. $A = \begin{pmatrix} 1 & 0 & 0 \\ 0 & 2 & 0 \\ 0 & 0 & 5 \end{pmatrix}$, $B =$

$$\begin{pmatrix} 3 & 0 & 0 \\ 0 & 0 & 0 \\ 0 & 0 & -5 \end{pmatrix}, \text{ and } I = \begin{pmatrix} 1 & 0 & 0 \\ 0 & 1 & 0 \\ 0 & 0 & 1 \end{pmatrix} \text{ are all diagonal matrices.} \qquad \square$$

5.2.2 The Identity Matrix and Matrix Inverses

In the example above, the 3×3 diagonal matrix I whose diagonal entries are all 1's has the distinctive property that for any other 3×3 matrix A we have $AI = IA = A$. For example:

Example 5.2.3 Multiplying by the Identity Matrix. If $A = \begin{pmatrix} 1 & 2 & 5 \\ 6 & 7 & -2 \\ 3 & -3 & 0 \end{pmatrix}$,

then $AI = \begin{pmatrix} 1 & 2 & 5 \\ 6 & 7 & -2 \\ 3 & -3 & 0 \end{pmatrix}$ and $IA = \begin{pmatrix} 1 & 2 & 5 \\ 6 & 7 & -2 \\ 3 & -3 & 0 \end{pmatrix}$. $\qquad \square$

In other words, the matrix I behaves in matrix algebra like the real number 1; that is, as a multiplicative identity. In matrix algebra, the matrix I is called simply the identity matrix. Convince yourself that if A is any $n \times n$ matrix $AI = IA = A$.

Definition 5.2.4 Identity Matrix. The $n \times n$ diagonal matrix I_n whose diagonal components are all 1's is called the identity matrix. If the context is clear, we simply use I. $\qquad \diamond$

In the set of real numbers we recall that, given a nonzero real number x, there exists a real number y such that $xy = yx = 1$. We know that real numbers commute under multiplication so that the two equations can be summarized as $xy = 1$. Further we know that $y = x^{-1} = \frac{1}{x}$. Do we have an analogous situation in $M_{n \times n}(\mathbb{R})$? Can we define the multiplicative inverse of an $n \times n$ matrix A? It seems natural to imitate the definition of multiplicative inverse in the real numbers.

Definition 5.2.5 Matrix Inverse. Let A be an $n \times n$ matrix. If there exists an $n \times n$ matrix B such that $AB = BA = I$, then B is a multiplicative inverse of A (called simply an inverse of A) and is denoted by A^{-1} $\qquad \diamond$

When we are doing computations involving matrices, it would be helpful to know that when we find A^{-1}, the answer we obtain is the only inverse of the given matrix. This would let us refer to *the* inverse of a matrix. We refrained from saying that in the definition, but the theorem below justifies it.

Remark: Those unfamiliar with the laws of matrix algebra should return to the following proof after they have familiarized themselves with the Laws of Matrix Algebra in Section 5.5.

Theorem 5.2.6 Inverses are unique. *The inverse of an $n \times n$ matrix A, when it exists, is unique.*

Proof. Let A be an $n \times n$ matrix. Assume to the contrary, that A has two (different) inverses, say B and C. Then

$$\begin{aligned} B &= BI & &\text{Identity property of } I \\ &= B(AC) & &\text{Assumption that } C \text{ is an inverse of } A \\ &= (BA)C & &\text{Associativity of matrix multiplication} \\ &= IC & &\text{Assumption that } B \text{ is an inverse of } A \\ &= C & &\text{Identity property of } I \end{aligned}$$

Let $A = \begin{pmatrix} 2 & 0 \\ 0 & 3 \end{pmatrix}$. What is A^{-1}? Without too much difficulty, by trial and error, we determine that $A^{-1} = \begin{pmatrix} \frac{1}{2} & 0 \\ 0 & \frac{1}{3} \end{pmatrix}$. This might lead us to guess that the inverse is found by taking the reciprocal of all nonzero entries of a matrix. Alas, it isn't that easy!

If $A = \begin{pmatrix} 1 & 2 \\ -3 & 5 \end{pmatrix}$, the "reciprocal rule" would tell us that the inverse of A is $B = \begin{pmatrix} 1 & \frac{1}{2} \\ \frac{-1}{3} & \frac{1}{5} \end{pmatrix}$. Try computing AB and you will see that you don't get the identity matrix. So, what *is* A^{-1}? In order to understand more completely the notion of the inverse of a matrix, it would be beneficial to have a formula that would enable us to compute the inverse of at least a 2×2 matrix. To do this, we introduce the definition of the determinant of a 2×2 matrix.

Definition 5.2.7 Determinant of a 2 by 2 matrix. Let $A = \begin{pmatrix} a & b \\ c & d \end{pmatrix}$. The determinant of A is the number $\det A = ad - bc$. $\Diamond$

In addition to $\det A$, common notation for the determinant of matrix A is $|A|$. This is particularly common when writing out the whole matrix, which case we would write $\begin{vmatrix} a & b \\ c & d \end{vmatrix}$ for the determinant of the general 2×2 matrix.

Example 5.2.8 Some determinants of two by two matrices. If $A = \begin{pmatrix} 1 & 2 \\ -3 & 5 \end{pmatrix}$ then $\det A = 1 \cdot 5 - 2 \cdot (-3) = 11$. If $B = \begin{pmatrix} 1 & 2 \\ 2 & 4 \end{pmatrix}$ then $\det B = 1 \cdot 4 - 2 \cdot 2 = 0$. $\square$

Theorem 5.2.9 Inverse of 2 by 2 matrix. *Let* $A = \begin{pmatrix} a & b \\ c & d \end{pmatrix}$. *If* $\det A \neq 0$, *then* $A^{-1} = \frac{1}{\det A} \begin{pmatrix} d & -b \\ -c & a \end{pmatrix}$.

Proof. See Exercise 4 at the end of this section. ∎

Example 5.2.10 Finding Inverses. Can we find the inverses of the matrices in Example 5.2.8? If $A = \begin{pmatrix} 1 & 2 \\ -3 & 5 \end{pmatrix}$ then

$$A^{-1} = \frac{1}{11} \begin{pmatrix} 5 & -2 \\ 3 & 1 \end{pmatrix} = \begin{pmatrix} \frac{5}{11} & -\frac{2}{11} \\ \frac{3}{11} & \frac{1}{11} \end{pmatrix}$$

The reader should verify that $AA^{-1} = A^{-1}A = I$.

The second matrix, B, has a determinant equal to zero. If we tried to apply the formula in Theorem 5.2.9, we would be dividing by zero. For this reason, the formula can't be applied and in fact B^{-1} does not exist. $\square$

Remarks:

- In general, if A is a 2×2 matrix and if $\det A = 0$, then A^{-1} does not exist.

- A formula for the inverse of $n \times n$ matrices $n \geq 3$ can be derived that also involves $\det A$. Hence, in general, if the determinant of a matrix is zero, the matrix does not have an inverse. However the formula for even a 3×3 matrix is very long and is not the most efficient way to compute the inverse of a matrix.

- In Chapter 12 we will develop a technique to compute the inverse of a higher-order matrix, if it exists.

- Matrix inversion comes first in the hierarchy of matrix operations; therefore, AB^{-1} is $A(B^{-1})$.

5.2.3 Exercises for Section 5.2

1. For the given matrices A find A^{-1} if it exists and verify that $AA^{-1} = A^{-1}A = I$. If A^{-1} does not exist explain why.

    (a) $A = \begin{pmatrix} 1 & 3 \\ 2 & 1 \end{pmatrix}$

    (b) $A = \begin{pmatrix} 6 & -3 \\ 8 & -4 \end{pmatrix}$

    (c) $A = \begin{pmatrix} 1 & -3 \\ 0 & 1 \end{pmatrix}$

    (d) $A = \begin{pmatrix} 1 & 0 \\ 0 & 1 \end{pmatrix}$

    (e) Use the definition of the inverse of a matrix to find A^{-1}: $A = \begin{pmatrix} 3 & 0 & 0 \\ 0 & \frac{1}{2} & 0 \\ 0 & 0 & -5 \end{pmatrix}$

2. For the given matrices A find A^{-1} if it exists and verify that $AA^{-1} = A^{-1}A = I$. If A^{-1} does not exist explain why.

    (a) $A = \begin{pmatrix} 2 & -1 \\ -1 & 2 \end{pmatrix}$

    (b) $A = \begin{pmatrix} 0 & 1 \\ 0 & 2 \end{pmatrix}$

    (c) $A = \begin{pmatrix} 1 & c \\ 0 & 1 \end{pmatrix}$

    (d) $A = \begin{pmatrix} a & b \\ b & a \end{pmatrix}$, where $|a| \neq |b|$.

3.

    (a) Let $A = \begin{pmatrix} 2 & 3 \\ 1 & 4 \end{pmatrix}$ and $B = \begin{pmatrix} 3 & -3 \\ 2 & 1 \end{pmatrix}$. Verify that $(AB)^{-1} = B^{-1}A^{-1}$.

    (b) Let A and B be $n \times n$ invertible matrices. Prove that $(AB)^{-1} = B^{-1}A^{-1}$. Why is the right side of the above statement written "backwards"? Is this necessary? Hint: Use Theorem 5.2.6

4. Let $A = \begin{pmatrix} a & b \\ c & d \end{pmatrix}$. Derive the formula for A^{-1}.

5. **Linearity of Determinants.**

    (a) Let A and B be 2-by-2 matrices. Show that $\det(AB) = (\det A)(\det B)$.

    (b) It can be shown that the statement in part (a) is true for all $n \times n$ matrices. Let A be any invertible $n \times n$ matrix. Prove that $\det\left(A^{-1}\right) = (\det A)^{-1}$. Note: The determinant of the identity matrix I_n is 1 for all n.

    (c) Verify that the equation in part (b) is true for the matrix in exercise 1(a) of this section.

6. Prove by induction that for $n \geq 1$, $\begin{pmatrix} a & 0 \\ 0 & b \end{pmatrix}^n = \begin{pmatrix} a^n & 0 \\ 0 & b^n \end{pmatrix}$.

7. Use the assumptions in Exercise 5.2.3.5 to prove by induction that if $n \geq 1$, $\det\left(A^n\right) = (\det A)^n$.

8. Prove: If the determinant of a matrix A is zero, then A does not have an inverse. Hint: Use the indirect method of proof and exercise 5.

9.

    (a) Let $A, B,$ and D be $n \times n$ matrices. Assume that B is invertible. If $A = BDB^{-1}$, prove by induction that $A^m = BD^mB^{-1}$ is true for $m \geq 1$.

    (b) Given that $A = \begin{pmatrix} -8 & 15 \\ -6 & 11 \end{pmatrix} = B\begin{pmatrix} 1 & 0 \\ 0 & 2 \end{pmatrix}B^{-1}$ where $B = \begin{pmatrix} 5 & 3 \\ 3 & 2 \end{pmatrix}$ what is A^{10}?

5.3 Laws of Matrix Algebra

5.3.1 The Laws

The following is a summary of the basic laws of matrix operations. Assume that the indicated operations are defined; that is, that the orders of the matrices A, B and C are such that the operations make sense.

(1) Commutative Law of Addition	$A + B = B + A$
(2) Associative Law of Addition	$A + (B + C) = (A + B) + C$
(3) Distributive Law of a Scalar over Matrices	$c(A + B) = cA + cB$, where $c \in \mathbb{R}$.
(4) Distributive Law of Scalars over a Matrix	$(c_1 + c_2)\, A = c_1 A + c_2 A$, where $c_1, c_2 \in \mathbb{R}$.
(5) Associative Law of Scalar Multiplication	$c_1 (c_2 A) = (c_1 \cdot c_2)\, A$, where $c_1, c_2 \in \mathbb{R}$.
(6) Zero Matrix Annihilates all Products	$\mathbf{0}A = \mathbf{0}$, where $\mathbf{0}$ is the zero matrix.
(7) Zero Scalar Annihilates all Products	$0A = \mathbf{0}$, where 0 on the left is the scalar zero.
(8) Zero Matrix is an identity for Addition	$A + \mathbf{0} = A$.
(9) Negation produces additive inverses	$A + (-1)A = \mathbf{0}$.
(10) Right Distributive Law of Matrix Multiplication	$A(B + C) = AB + AC$.
(11) Left Distributive Law of Matrix Multiplication	$(B + C)A = BA + CA$.
(12) Associative Law of Multiplication	$A(BC) = (AB)C$.
(13) Identity Matrix is a Multiplicative Identity	$IA = A$ and $AI = A$.
(14) Involution Property of Inverses	If A^{-1} exists, $\left(A^{-1}\right)^{-1} = A$.
(15) Inverse of Product Rule	If A^{-1} and B^{-1} exist, $(AB)^{-1} = B^{-1}A^{-1}$

Table 5.3.1: Laws of Matrix Algebra

5.3.2 Commentary

Example 5.3.2 More Precise Statement of one Law. If we wished to write out each of the above laws more completely, we would specify the orders of the matrices. For example, Law 10 should read:

> Let A, B, and C be $m \times n$, $n \times p$, and $n \times p$ matrices, respectively, then $A(B + C) = AB + AC$

$\square$

Remarks:

- Notice the absence of the "law" $AB = BA$. Why?

- Is it really necessary to have both a right (No. 11) and a left (No. 10) distributive law? Why?

5.3.3 Exercises for Section 5.3

1. Rewrite the above laws specifying as in Example 5.3.2 the orders of the matrices.

2. Verify each of the Laws of Matrix Algebra using examples.

3. Let $A = \begin{pmatrix} 1 & 2 \\ 0 & -1 \end{pmatrix}$, $B = \begin{pmatrix} 3 & 7 & 6 \\ 2 & -1 & 5 \end{pmatrix}$, and $C = \begin{pmatrix} 0 & -2 & 4 \\ 7 & 1 & 1 \end{pmatrix}$.
   Compute the following as efficiently as possible by using any of the Laws of Matrix Algebra:

   (a) $AB + AC$

   (b) A^{-1}

   (c) $A(B + C)$

   (d) $\left(A^2\right)^{-1}$

   (e) $(C + B)^{-1}A^{-1}$

4. Let $A = \begin{pmatrix} 7 & 4 \\ 2 & 1 \end{pmatrix}$ and $B = \begin{pmatrix} 3 & 5 \\ 2 & 4 \end{pmatrix}$. Compute the following as efficiently as possible by using any of the Laws of Matrix Algebra:

    (a) AB

    (b) $A + B$

    (c) $A^2 + AB + BA + B^2$

    (d) $B^{-1}A^{-1}$

    (e) $A^2 + AB$

5. Let A and B be $n \times n$ matrices of real numbers. Is $A^2 - B^2 = (A-B)(A+B)$? Explain.

5.4 Matrix Oddities

5.4.1 Dissimilarities with elementary algebra

We have seen that matrix algebra is similar in many ways to elementary algebra. Indeed, if we want to solve the matrix equation $AX = B$ for the unknown X, we imitate the procedure used in elementary algebra for solving the equation $ax = b$. Notice how exactly the same properties are used in the following detailed solutions of both equations.

Equation in the real algebra		Equation in matrix algebra
$ax = b$		$AX = B$
$a^{-1}(ax) = a^{-1}b$ if $a \neq 0$		$A^{-1}(AX) = A^{-1}B$ if A^{-1} exists
$(a^{-1}a)x = a^{-1}b$	Associative Property	$(A^{-1}A)X = A^{-1}B$
$1x = a^{-1}b$	Inverse Property	$IX = A^{-1}B$
$x = a^{-1}b$	Identity Property	$X = A^{-1}B$

Table 5.4.1

Certainly the solution process for $AX = B$ is the same as that of $ax = b$.

The solution of $xa = b$ is $x = ba^{-1} = a^{-1}b$. In fact, we usually write the solution of both equations as $x = \frac{b}{a}$. In matrix algebra, the solution of $XA = B$ is $X = BA^{-1}$, which is not necessarily equal to $A^{-1}B$. So in matrix algebra, since the commutative law (under multiplication) is not true, we have to be more careful in the methods we use to solve equations.

It is clear from the above that if we wrote the solution of $AX = B$ as $X = \frac{B}{A}$, we would not know how to interpret $\frac{B}{A}$. Does it mean $A^{-1}B$ or BA^{-1}? Because of this, A^{-1} is never written as $\frac{1}{A}$.

Observation 5.4.2 Matrix Oddities. Some of the main dissimilarities between matrix algebra and elementary algebra are that in matrix algebra:

(1) AB may be different from BA.

(2) There exist matrices A and B such that $AB = \mathbf{0}$, and yet $A \neq \mathbf{0}$ and $B \neq \mathbf{0}$.

(3) There exist matrices A where $A \neq \mathbf{0}$, and yet $A^2 = \mathbf{0}$.

(4) There exist matrices A where $A^2 = A$ with $A \neq I$ and $A \neq \mathbf{0}$

(5) There exist matrices A where $A^2 = I$, where $A \neq I$ and $A \neq -I$

5.4.2 Exercises for Section 5.4

1. Discuss each of the "Matrix Oddities" with respect to elementary algebra.

2. Determine 2×2 matrices which show that each of the "Matrix Oddities" are true.

3. Prove or disprove the following implications.

 (a) $A^2 = A$ and $\det A \neq 0 \Rightarrow A = I$

 (b) $A^2 = I$ and $\det A \neq 0 \Rightarrow A = I$ or $A = -I$.

4. Let $M_{n \times n}(\mathbb{R})$ be the set of real $n \times n$ matrices. Let $P \subseteq M_{n \times n}(\mathbb{R})$ be the subset of matrices defined by $A \in P$ if and only if $A^2 = A$. Let $Q \subseteq P$ be defined by $A \in Q$ if and only if $\det A \neq 0$.

 (a) Determine the cardinality of Q.

 (b) Consider the special case $n = 2$ and prove that a sufficient condition for $A \in P \subseteq M_{2 \times 2}(\mathbb{R})$ is that A has a zero determinant (i.e., A is singular) and $tr(A) = 1$ where $tr(A) = a_{11} + a_{22}$ is the sum of the main diagonal elements of A.

 (c) Is the condition of part b a necessary condition?

5. Write each of the following systems in the form $AX = B$, and then solve the systems using matrices.

 (a) $\begin{aligned} 2x_1 + x_2 &= 3 \\ x_1 - x_2 &= 1 \end{aligned}$ (d) $\begin{aligned} 2x_1 + x_2 &= 1 \\ x_1 - x_2 &= -1 \end{aligned}$

 (b) $\begin{aligned} 2x_1 - x_2 &= 4 \\ x_1 - x_2 &= 0 \end{aligned}$ (e) $\begin{aligned} 3x_1 + 2x_2 &= 1 \\ 6x_1 + 4x_2 &= -1 \end{aligned}$

 (c) $\begin{aligned} 2x_1 + x_2 &= 1 \\ x_1 - x_2 &= 1 \end{aligned}$

6. Recall that $p(x) = x^2 - 5x + 6$ is called a polynomial, or more specifically, a polynomial over $\mathbb{R}$, where the coefficients are elements of $\mathbb{R}$ and $x \in \mathbb{R}$. Also, think of the method of solving, and solutions of, $x^2 - 5x + 6 = 0$. We would like to define the analogous situation for 2×2 matrices. First define where A is a 2×2 matrix $p(A) = A^2 - 5A + 6I$. Discuss the method of solving and the solutions of $A^2 - 5A + 6I = \mathbf{0}$.

7. For those who know calculus:

 (a) Write the series expansion for e^a centered around $a = 0$.

 (b) Use the idea of exercise 6 to write what would be a plausible definion of e^A where A is an $n \times n$ matrix.

(c) If $A = \begin{pmatrix} 1 & 1 \\ 0 & 0 \end{pmatrix}$ and $B = \begin{pmatrix} 0 & -1 \\ 0 & 0 \end{pmatrix}$, use the series in part (b)

to show that $e^A = \begin{pmatrix} e & e-1 \\ 0 & 1 \end{pmatrix}$ and $e^B = \begin{pmatrix} 1 & -1 \\ 0 & 1 \end{pmatrix}$.

(d) Show that $e^A e^B \neq e^B e^A$.

(e) Show that $e^{A+B} = \begin{pmatrix} e & 0 \\ 0 & 1 \end{pmatrix}$.

(f) Is $e^A e^B = e^{A+B}$?

Chapter 6

Relations

One understands a set of objects completely only if the structure of that set is made clear by the interrelationships between its elements. For example, the individuals in a crowd can be compared by height, by age, or through any number of other criteria. In mathematics, such comparisons are called relations. The goal of this chapter is to develop the language, tools, and concepts of relations.

6.1 Basic Definitions

In Chapter 1 we introduced the concept of the Cartesian product of sets. Let's assume that a person owns three shirts and two pairs of slacks. More precisely, let $A = \{$blue shirt, tan shirt, mint green shirt$\}$ and $B = \{$grey slacks, tan slacks$\}$. Then $A \times B$ is the set of all six possible combinations of shirts and slacks that the individual could wear. However, an individual may wish to restrict himself or herself to combinations which are color coordinated, or "related." This may not be all possible pairs in $A \times B$ but will certainly be a subset of $A \times B$. For example, one such subset may be $\{$(blue shirt, grey slacks), (blue shirt, tan slacks), (mint green shirt, tan slack

6.1.1 Relations between two sets

Definition 6.1.1 Relation. Let A and B be sets. A relation from A into B is any subset of $A \times B$. $\qquad\qquad\Diamond$

Example 6.1.2 A simple example. Let $A = \{1, 2, 3\}$ and $B = \{4, 5\}$. Then $\{(1,4), (2,4), (3,5)\}$ is a relation from A into B. Of course, there are many others we could describe; 64, to be exact. $\qquad\qquad\square$

Example 6.1.3 Divisibility Example. Let $A = \{2, 3, 5, 6\}$ and define a relation r from A into A by $(a, b) \in r$ if and only if a divides evenly into b. The set of pairs that qualify for membership is $r = \{(2,2), (3,3), (5,5), (6,6), (2,6), (3,6)\}$. $\qquad\square$

6.1.2 Relations on a Set

Definition 6.1.4 Relation on a Set. A relation from a set A into itself is called a relation on A. $\qquad\qquad\Diamond$

The relation "divides" in Example 6.1.3 will appear throughout the book. Here is a general definition on the whole set of integers.

Definition 6.1.5 Divides. Let $a, b \in \mathbb{Z}$, $a \neq 0$. We say that a divides b, denoted $a \mid b$, if and only if there exists an integer k such that $ak = b$. $\Diamond$

Be very careful in writing about the relation "divides." The vertical line symbol use for this relation, if written carelessly, can look like division. While $a \mid b$ is either true or false, a/b is a number.

Based on the equation $ak = b$, we can say that $a|b$ is equivalent to $k = \frac{b}{a}$, or a divides evenly into b. In fact the "divides" is short for "divides evenly into." You might find the equation $k = \frac{b}{a}$ initially easier to understand, but in the long run we will find the equation $ak = b$ more convenient.

Sometimes it is helpful to illustrate a relation with a graph. Consider Example 6.1.2. A graph of r can be drawn as in Figure 6.1.6. The arrows indicate that 1 is related to 4 under r. Also, 2 is related to 4 under r, and 3 is related to 5, while the upper arrow denotes that r is a relation from the whole set A into the set B.

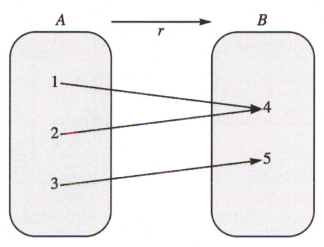

Figure 6.1.6: The graph of a relation

A typical element in a relation r is an ordered pair (x, y). In some cases, r can be described by actually listing the pairs which are in r, as in the previous examples. This may not be convenient if r is relatively large. Other notations are used with certain well-known relations. Consider the "less than or equal" relation on the real numbers. We could define it as a set of ordered pairs this way:

$$\leq = \{(x, y)|x \leq y\}$$

However, the notation $x \leq y$ is clear and self-explanatory; it is a more natural, and hence preferred, notation to use than $(x, y) \in \leq$.

Many of the relations we will work with "resemble" the relation $\leq$, so xsy is a common way to express the fact that x is related to y through the relation s.

Relation Notation Let s be a relation from a set A into a set B. Then the fact that $(x, y) \in s$ is frequently written xsy.

6.1.3 Composition of Relations

With $A = \{2, 3, 5, 8\}$, $B = \{4, 6, 16\}$, and $C = \{1, 4, 5, 7\}$, let r be the relation "divides," from A into B, and let s be the relation $\leq$ from B into C. So $r = \{(2, 4), (2, 6), (2, 16), (3, 6), (8, 16)\}$ and $s = \{(4, 4), (4, 5), (4, 7), (6, 7)\}$.

Notice that in Figure 6.1.8 that we can, for certain elements of A, go through elements in B to results in C. That is:

$$2|4 \text{ and } 4 \leq 4$$
$$2|4 \text{ and } 4 \leq 5$$
$$2|4 \text{ and } 4 \leq 7$$
$$2|6 \text{ and } 6 \leq 7$$
$$3|6 \text{ and } 6 \leq 7$$

Table 6.1.7

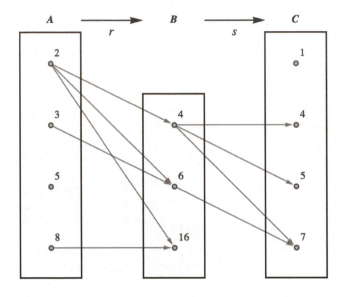

Figure 6.1.8: Relation Composition - a graphical view

Based on this observation, we can define a new relation, call it rs, from A into C. In order for (a, c) to be in rs, it must be possible to travel along a path in Figure 6.1.8 from a to c. In other words, $(a, c) \in rs$ if and only if $(\exists b)_B (arb \text{ and } bsc)$. The name rs was chosen because it reminds us that this new relation was formed by the two previous relations r and s. The complete listing of all elements in rs is $\{(2, 4), (2, 5), (2, 7), (3, 7)\}$. We summarize in a definition.

Definition 6.1.9 Composition of Relations. Let r be a relation from a set A into a set B, and let s be a relation from B into a set C. The composition of r with s, written rs, is the set of pairs of the form $(a, c) \in A \times C$, where $(a, c) \in rs$ if and only if there exists $b \in B$ such that $(a, b) \in r$ and $(b, c) \in s$.
◇

Remark: A word of warning to those readers familiar with composition of functions. (For those who are not, disregard this remark. It will be repeated at an appropriate place in the next chapter.) As indicated above, the traditional way of describing a composition of two relations is rs where r is the first relation and s the second. However, function composition is traditionally expressed in the opposite order: $s \circ r$, where r is the first function and s is the second.

6.1.4 Exercises for Section 6.1

1. For each of the following relations r defined on $\mathbb{P}$, determine which of the given ordered pairs belong to r

(a) xry iff $x|y$; (2, 3), (2, 4), (2, 8), (2, 17)

(b) xry iff $x \leq y$; (2, 3), (3, 2), (2, 4), (5, 8)

(c) xry iff $y = x^2$; (1,1), (2, 3), (2, 4), (2, 6)

2. The following relations are on $\{1, 3, 5\}$. Let r be the relation xry iff $y = x + 2$ and s the relation xsy iff $x \leq y$.

(a) List all elements in rs.

(b) List all elements in sr.

(c) Illustrate rs and sr via a diagram.

(d) Is the relation rs equal to the relation sr?

3. Let $A = \{1, 2, 3, 4, 5\}$ and define r on A by xry iff $x + 1 = y$. We define $r^2 = rr$ and $r^3 = r^2 r$. Find:

(a) r

(b) r^2

(c) r^3

4. Given s and t, relations on $\mathbb{Z}$, $s = \{(1, n) : n \in \mathbb{Z}\}$ and $t = \{(n, 1) : n \in \mathbb{Z}\}$, what are st and ts? Hint: Even when a relation involves infinite sets, you can often get insights into them by drawing partial graphs.

5. Let ρ be the relation on the power set, $\mathcal{P}(S)$, of a finite set S of cardinality n defined ρ by $(A, B) \in \rho$ iff $A \cap B = \emptyset$.

(a) Consider the specific case $n = 3$, and determine the cardinality of the set ρ.

(b) What is the cardinality of ρ for an arbitrary n? Express your answer in terms of n. (Hint: There are three places that each element of S can go in building an element of ρ.)

6. Let r_1, r_2, and r_3 be relations on any set A. Prove that if $r_1 \subseteq r_2$ then $r_1 r_3 \subseteq r_2 r_3$.

6.2 Graphs of Relations on a Set

In this section we introduce directed graphs as a way to visualize relations on a set.

6.2.1 Digraphs

Let $A = \{0, 1, 2, 3\}$, and let

$$r = \{(0, 0), (0, 3), (1, 2), (2, 1), (3, 2), (2, 0)\}$$

In representing this relation as a graph, elements of A are called the vertices of the graph. They are typically represented by labeled points or small circles. We connect vertex a to vertex b with an arrow, called an edge, going from vertex a to vertex b if and only if arb. This type of graph of a relation r is called a **directed graph** or **digraph**. Figure 6.2.1 is a digraph for r. Notice that since 0 is related to itself, we draw a "self-loop" at 0.

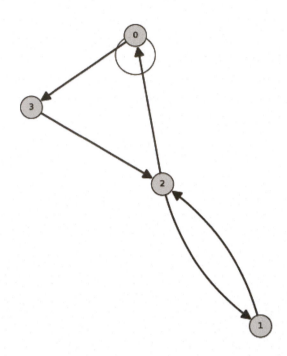

Figure 6.2.1: Digraph of a relation

The actual location of the vertices in a digraph is immaterial. The actual location of vertices we choose is called an **embedding of a graph**. The main idea is to place the vertices in such a way that the graph is easy to read. After drawing a rough-draft graph of a relation, we may decide to relocate the vertices so that the final result will be neater. Figure 6.2.1 could also be presented as in Figure 6.2.2.

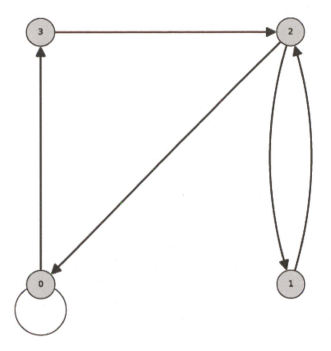

Figure 6.2.2: Alternate embedding of the previous directed graph

A vertex of a graph is also called a node, point, or a junction. An edge of a graph is also referred to as an arc, a line, or a branch. Do not be concerned if two graphs of a given relation look different as long as the connections between vertices are the same in two graphs.

Example 6.2.3 Another directed graph. Consider the relation s whose digraph is Figure 6.2.4. What information does this give us? The graph tells us that s is a relation on $A = \{1, 2, 3\}$ and that $s = \{(1, 2), (2, 1), (1, 3), (3, 1), (2, 3), (3, 3)\}$.

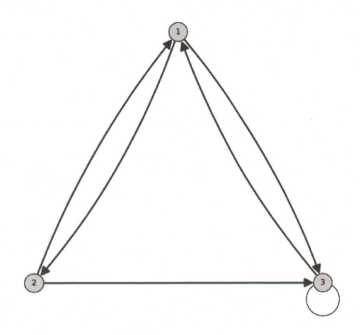

Figure 6.2.4: Digraph of the relation s

☐

We will be building on the next example in the following section.

Example 6.2.5 Ordering subsets of a two element universe. Let $B = \{1, 2\}$, and let $A = \mathcal{P}(B) = \{\emptyset, \{1\}, \{2\}, \{1, 2\}\}$. Then $\subseteq$ is a relation on A whose digraph is Figure 6.2.6.

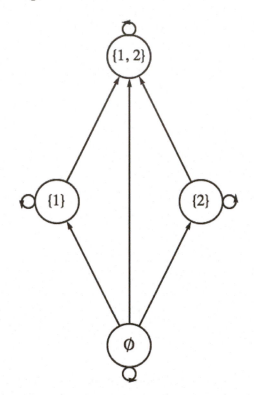

Figure 6.2.6: Graph for set containment on subsets of $\{1, 2\}$

We will see in the next section that since $\subseteq$ has certain structural properties that describe "partial orderings." We will be able to draw a much simpler type graph than this one, but for now the graph above serves our purposes. ☐

6.2.2 Exercises for Section 6.2

1. Let $A = \{1, 2, 3, 4\}$, and let r be the relation $\leq$ on A. Draw a digraph for r.

2. Let $B = \{1, 2, 3, 4, 6, 8, 12, 24\}$, and let s be the relation "divides" on B. Draw a digraph for s.

3. Let $A = \{1, 2, 3, 4, 5\}$. Define t on A by atb if and only if $b - a$ is even. Draw a digraph for t.

4.

    (a) Let A be the set of strings of 0's and 1's of length 3 or less. Define the relation of d on A by xdy if x is contained within y. For example, $01d101$. Draw a digraph for this relation.

    (b) Do the same for the relation p defined by xpy if x is a prefix of y. For example, $10p101$, but $01p101$ is false.

5. Recall the relation in Exercise 5 of Section 6.1, ρ defined on the power set, $\mathcal{P}(S)$, of a set S. The definition was $(A, B) \in \rho$ iff $A \cap B = \emptyset$. Draw the digraph for ρ where $S = \{a, b\}$.

6. Let $C = \{1, 2, 3, 4, 6, 8, 12, 24\}$ and define t on C by atb if and only if a and b share a common divisor greater than 1. Draw a digraph for t.

6.3 Properties of Relations

6.3.1 Individual Properties

Consider the set $B = \{1, 2, 3, 4, 6, 12, 36, 48\}$ and the relations "divides" and $\leq$ on B. We notice that these two relations on B have three properties in common:

- Every element in B divides itself and is less than or equal to itself. This is called the reflexive property.

- If we search for two elements from B where the first divides the second and the second divides the first, then we are forced to choose the two numbers to be the same. In other words, no two *different* numbers are related in both directions. The reader can verify that a similar fact is true for the relation $\leq$ on B. This is called the antisymmetric property.

- Next if we choose three values (not necessarily distinct) from B such that the first divides the second and the second divides the third, then we always find that the first number divides the third. Again, the same is true if we replace "divides" with "is less than or equal to." This is called the transitive property.

Relations that satisfy these properties are of special interest to us. Formal definitions of the properties follow.

Definition 6.3.1 Reflexive Relation. Let A be a set and let r be a relation on A. Then r is **reflexive** if and only if ara for all $a \in A$. $\diamond$

Definition 6.3.2 Antisymmetric Relation. Let A be a set and let r be a relation on A. Then r is **antisymmetric** if and only if whenever arb and $a \neq b$ then bra is false. $\diamond$

An equivalent condition for antisymmetry is that if arb and bra then $a = b$. You are encouraged to convince yourself that this is true. This condition is often more convenient to prove than the definition, even though the definition is probably easier to understand.

A word of warning about antisymmetry: Students frequently find it difficult to understand this definition. Keep in mind that this term is defined through an "If...then..." statement. The question that you must ask is: Is it true that whenever there are elements a and b from A where arb and $a \neq b$, it follows that b is not related to a? If so, then the relation is antisymmetric.

Another way to determine whether a relation is antisymmetric is to examine (or imagine) its digraph. The relation is not antisymmetric if there exists a pair of vertices that are connected by edges in both directions.

Definition 6.3.3 Transitive Relation. Let A be a set and let r be a relation on A. r is **transitive** if and only if whenever arb and brc then arc. $\diamond$

6.3.2 Partial Orderings

Not all relations have all three of the properties discussed above, but those that do are a special type of relation.

Definition 6.3.4 Partial Ordering. A relation on a set A that is reflexive, antisymmetric, and transitive is called a **partial ordering** on A. A set on which there is a partial ordering relation defined is called a **partially ordered set** or **poset**. $\Diamond$

Example 6.3.5 Set Containment as a Partial Ordering. Let A be a set. Then $\mathcal{P}(A)$ together with the relation $\subseteq$ (set containment) is a poset. To prove this we observe that the three properties hold, as discussed in Chapter 4.

- Let $B \in \mathcal{P}(A)$. The fact that $B \subseteq B$ follows from the definition of subset. Hence, set containment is reflexive.

- Let $B_1, B_2 \in \mathcal{P}(A)$ and assume that $B_1 \subseteq B_2$ and $B_1 \neq B_2$. Could it be that $B_2 \subseteq B_1$? No. There must be some element $a \in A$ such that $a \notin B_1$, but $a \in B_2$. This is exactly what we need to conclude that B_2 is not contained in B_1. Hence, set containment is antisymmetric.

- Let $B_1, B_2, B_3 \in \mathcal{P}(A)$ and assume that $B_1 \subseteq B_2$ and $B_2 \subseteq B_3$. Does it follow that $B_1 \subseteq B_3$? Yes, if $a \in B_1$, then $a \in B_2$ because $B_1 \subseteq B_2$. Now that we have $a \in B_2$ and we have assumed $B_2 \subseteq B_3$, we conclude that $a \in B_3$. Therefore, $B_1 \subseteq B_3$ and so set containment is transitive.

Figure 6.2.6 is the graph for the "set containment" relation on the power set of $\{1, 2\}$. $\square$

Figure 6.2.6 is helpful insofar as it reminds us that each set is a subset of itself and shows us at a glance the relationship between the various subsets in $\mathcal{P}(\{1, 2\})$. However, when a relation is a partial ordering, we can streamline a graph like this one. The streamlined form of a graph is called a **Hasse diagram** or **ordering diagram**. A Hasse diagram takes into account the following facts.

- By the reflexive property, each vertex must be related to itself, so the arrows from a vertex to itself (called "self-loops") are not drawn in a Hasse diagram. They are simply assumed.

- By the antisymmetry property, connections between two distinct elements in a directed graph can only go one way, if at all. When there is a connection, we agree to always place the second element above the first (as we do above with the connection from $\{1\}$ to $\{1, 2\}$). For this reason, we can just draw a connection without an arrow, just a line.

- By the transitive property, if there are edges connecting one element up to a second element and the second element up to a third element, then there will be a direct connection from the first to the third. We see this in Figure 6.2.6 with $\emptyset$ connected to $\{1\}$ and then $\{1\}$ connected to $\{1, 2\}$. Notice the edge connecting $\emptyset$ to $\{1, 2\}$. Whenever we identify this situation, remove the connection from the first to the third in a Hasse diagram and simply observe that an upward path of any length implies that the lower element is related to the upper one.

Using these observations as a guide, we can draw a Hasse diagram for $\subseteq$ on $\{1,2\}$ as in Figure 6.3.6.

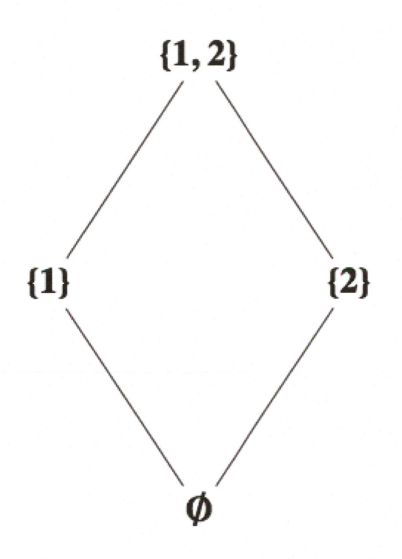

Figure 6.3.6: Hasse diagram for set containment on subsets of $\{1,2\}$

Example 6.3.7 Definition of a relation using a Hasse diagram. Consider the partial ordering relation s whose Hasse diagram is Figure 6.3.8.

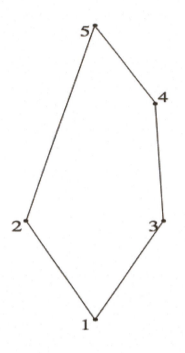

Figure 6.3.8: Hasse diagram for the pentagonal poset

How do we read this diagram? What is A? What is s? What does the digraph of s look like? Certainly $A = \{1, 2, 3, 4, 5\}$ and $1s2$, $3s4$, $1s4$, $1s5$, etc., Notice that $1s5$ is implied by the fact that there is a path of length three upward from 1 to 5. This follows from the edges that are shown and the transitive property that is presumed in a poset. Since $1s3$ and $3s4$, we know that $1s4$. We then combine $1s4$ with $4s5$ to infer $1s5$. Without going into details why, here is a complete list of pairs defined by s.

$$s = \{(1, 1), (2, 2), (3, 3), (4, 4), (5, 5), (1, 3), (1, 4), (1, 5), (1, 2), (3, 4), (3, 5), (4, 5), (2, 5)\}$$

A digraph for s is Figure 6.3.9. It is certainly more complicated to read and difficult to draw than the Hasse diagram.

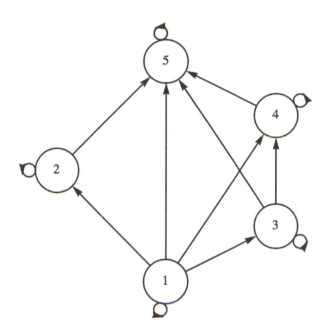

Figure 6.3.9: Digraph for the pentagonal poset

□

A classic example of a partial ordering relation is $\leq$ on the real numbers, $\mathbb{R}$. Indeed, when graphing partial ordering relations, it is natural to "plot" the elements from the given poset starting with the "least" element to the "greatest" and to use terms like "least," "greatest," etc. Because of this the reader should be forewarned that some texts use the symbol $\leq$ for arbitrary partial orderings. This can be quite confusing for the novice, so we continue to use generic letters r, s, etc.

6.3.3 Equivalence Relations

Another common property of relations is symmetry.

Definition 6.3.10 Symmetric Relation. Let r be a relation on a set A. r is **symmetric** if and only if whenever arb, it follows that bra.   ◇

Consider the relation of equality defined on any set A. Certainly $a = b$ implies that $b = a$ so equality is a symmetric relation on A.

Surprisingly, equality is also an antisymmetric relation on A. This is due to the fact that the condition that defines the antisymmetry property, $a = b$ and $a \neq b$, is a contradiction. Remember, a conditional proposition is always true when the condition is false. So a relation can be both symmetric and antisymmetric on a set! Again recall that these terms are *not* negatives of one other. That said, there are very few important relations other than equality that are both symmetric and antisymmetric.

Definition 6.3.11 Equivalence Relation. A relation r on a set A is called an equivalence relation if and only if it is reflexive, symmetric, and transitive.
                                                                                    ◇

The classic example of an equivalence relation is equality on a set A. In fact, the term equivalence relation is used because those relations which satisfy the definition behave quite like the equality relation. Here is another important equivalence relation.

Example 6.3.12   Equivalent Fractions.   Let $\mathbb{Z}*$ be the set of nonzero integers. One of the most basic equivalence relations in mathematics is the relation q on $\mathbb{Z} \times \mathbb{Z}^*$ defined by $(a, b)q(c, d)$ if and only if $ad = bc$. We will leave it to the reader to, verify that q is indeed an equivalence relation. Be aware that since the elements of $\mathbb{Z} \times \mathbb{Z}^*$ are ordered pairs, proving symmetry involves four numbers and transitivity involves six numbers. Two ordered pairs, (a, b) and (c, d), are related if the fractions $\frac{a}{b}$ and $\frac{c}{d}$ are numerically equal.   □

Our next example involves the following fundamental relations on the set of integers.

Definition 6.3.13   Congruence Modulo m.   Let m be a positive integer, $m \geq 2$. We define **congruence modulo m** to be the relation $\equiv_m$ defined on the integers by

$$a \equiv_m b \Leftrightarrow m \mid (a - b)$$

◊

We observe the following about congruence modulo m:

- This relation is reflexive, for if $a \in \mathbb{Z}$, $m \mid (a - a) \Rightarrow a \equiv_m a$.

- This relation is symmetric. We can prove this through the following chain of implications.

$$a \equiv_m b \Rightarrow m \mid (a - b)$$
$$\Rightarrow \text{For some } k \in \mathbb{Z}, a - b = mk$$
$$\Rightarrow b - a = m(-k)$$
$$\Rightarrow m \mid (b - a)$$
$$\Rightarrow b \equiv_m a$$

- Finally, this relation is transitive. We leave it to the reader to prove that if $a \equiv_m b$ and $b \equiv_m c$, then $a \equiv_m c$.

Frequently, you will see the equivalent notation $a \equiv b (\text{mod } m)$ for congruence modulo m.

Example 6.3.14 Random Relations usually have no properties.   Consider the relation s described by the digraph in Figure 6.3.15. This was created by randomly selecting whether or not two elements from $\{a, b, c\}$ were related or not. Convince yourself that the following are true:

- This relation is not reflexive.

- It is not antisymmetric.

- Also, it is not symmetric.

- It is not transitive.

- Is s an equivalence relation or a partial ordering?

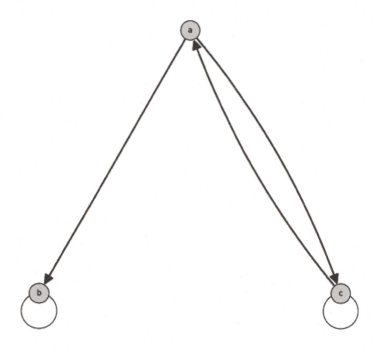

Figure 6.3.15: Digraph of a random relation r

Not every random choice of a relation will be so totally negative, but as the underlying set increases, the likelihood any of the properties are true begins to vanish. □

6.3.4 Exercises for Section 6.3

1.

    (a) Let $B = \{a, b\}$ and $U = \mathcal{P}(B)$. Draw a Hasse diagram for $\subseteq$ on U.

    (b) Let $A = \{1, 2, 3, 6\}$. Show that divides, |, is a partial ordering on A.

    (c) Draw a Hasse diagram for divides on A.

    (d) Compare the graphs of parts a and c.

2.   Repeat Exercise 1 with $B = \{a, b, c\}$ and $A = \{1, 2, 3, 5, 6, 10, 15, 30\}$.

    Hint.   Here is a Hasse diagram for the part (a).

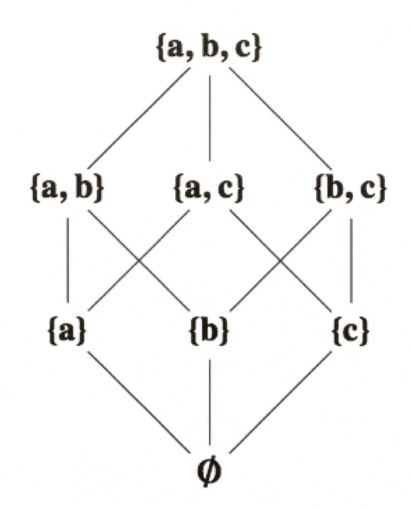

Figure 6.3.16

3.   Consider the relations defined by the digraphs in Figure 6.3.17.

    (a) Determine whether the given relations are reflexive, symmetric, anti-
        symmetric, or transitive. Try to develop procedures for determining
        the validity of these properties from the graphs,

    (b) Which of the graphs are of equivalence relations or of partial order-
        ings?

(i)

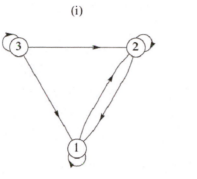

(ii)

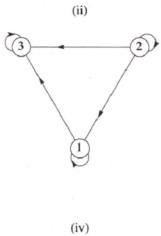

(iii)

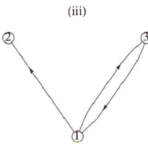

(iv)

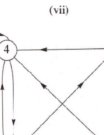

(v)

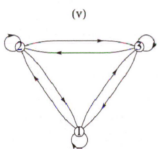

(vi)

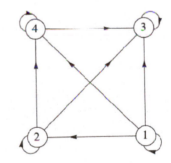

(vii)

Figure 6.3.17: Some digraphs of relations

4.   Determine which of the following are equivalence relations and/or partial

ordering relations for the given sets:

(a) $A = \{$ lines in the plane$\}$, and r defined by xry if and only if x is parallel to y. Assume every line is parallel to itself.

(b) $A = \mathbb{R}$ and r defined by xry if and only if $|x - y| \leq 7$.

5. Consider the relation on $\{1, 2, 3, 4, 5, 6\}$ defined by $r = \{(i, j) : \ |i - j| = 2\}$.

(a) Is r reflexive?

(b) Is r symmetric?

(c) Is r transitive?

(d) Draw a graph of r.

6. For the set of cities on a map, consider the relation xry if and only if city x is connected by a road to city y. A city is considered to be connected to itself, and two cities are connected even though there are cities on the road between them. Is this an equivalence relation or a partial ordering? Explain.

7. **Equivalence Classes.** Let $A = \{0, 1, 2, 3\}$ and let

$$r = \{(0,0), (1,1), (2,2), (3,3), (1,2), (2,1), (3,2), (2,3), (3,1), (1,3)\}$$

(a) Verify that r is an equivalence relation on A.

(b) Let $a \in A$ and define $c(a) = \{b \in A \mid arb\}$. $c(a)$ is called the **equivalence class of a under** r. Find $c(a)$ for each element $a \in A$.

(c) Show that $\{c(a) \mid a \in A\}$ forms a partition of A for this set A.

(d) Let r be an equivalence relation on an arbitrary set A. Prove that the set of all equivalence classes under r constitutes a partition of A.

8. Define r on the power set of $\{1, 2, 3\}$ by $ArB \Leftrightarrow |A| = |B|$. Prove that r is an equivalence relation. What are the equivalence classes under r?

9. Consider the following relations on $\mathbb{Z}_8 = \{0, 1, ..., 7\}$. Which are equivalence relations? For the equivalence relations, list the equivalence classes.

(a) arb iff the English spellings of a and b begin with the same letter.

(b) asb iff $a - b$ is a positive integer.

(c) atb iff $a - b$ is an even integer.

10. Building on Exercise 6.3.4.7:

(a) Prove that congruence modulo m is transitive.

(b) What are the equivalence classes under congruence modulo 2?

(c) What are the equivalence classes under congruence modulo 10?

11. In this exercise, we prove that implication is a partial ordering. Let A be any set of propositions.

   (a) Verify that $q \to q$ is a tautology, thereby showing that $\Rightarrow$ is a reflexive relation on A.

   (b) Prove that $\Rightarrow$ is antisymmetric on A. Note: we do not use $=$ when speaking of propositions, but rather equivalence, $\Leftrightarrow$.

   (c) Prove that $\Rightarrow$ is transitive on A.

   (d) Given that q_i is the proposition $n < i$ on $\mathbb{N}$, draw the Hasse diagram for the relation $\Rightarrow$ on $\{q_1, q_2, q_3, \ldots\}$.

12. Let $S = \{1, 2, 3, 4, 5, 6, 7\}$ be a poset $(S, \leq)$ with the Hasse diagram shown below. Another relation $r \subseteq S \times S$ is defined as follows: $(x, y) \in r$ if and only if there exists $z \in S$ such that $z < x$ and $z < y$ in the poset $(S, \leq)$.

   (a) Prove that r is reflexive.

   (b) Prove that r is symmetric.

   (c) A compatible with respect to relation r is any subset Q of set S such that $x \in Q$ and $y \in Q \Rightarrow (x, y) \in r$. A compatible g is a maximal compatible if Q is not a proper subset of another compatible. Give all maximal compatibles with respect to relation r defined above.

   (d) Discuss a characterization of the set of maximal compatibles for relation r when $(S, \leq)$ is a general finite poset. What conditions, if any, on a general finite poset $(S, \leq)$ will make r an equivalence relation?

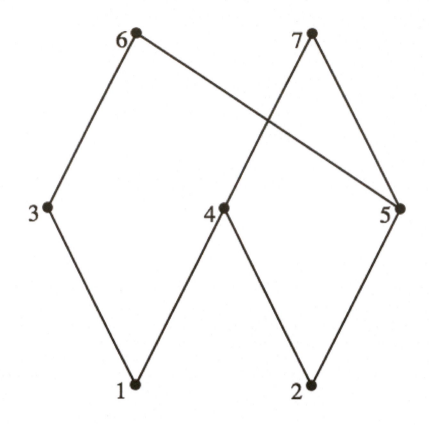

Figure 6.3.18: Hasse diagram for r in exercise 12.

6.4 Matrices of Relations

We have discussed two of the many possible ways of representing a relation, namely as a digraph or as a set of ordered pairs. In this section we will discuss the representation of relations by matrices.

6.4.1 Representing a Relation with a Matrix

Definition 6.4.1 Adjacency Matrix. Let $A = \{a_1, a_2, \ldots, a_m\}$ and $B = \{b_1, b_2, \ldots, b_n\}$ be finite sets of cardinality m and n, respectively. Let r be a relation from A into B. Then r can be represented by the $m \times n$ matrix R defined by

$$R_{ij} = \begin{cases} 1 & \text{if } a_i r b_j \\ 0 & \text{otherwise} \end{cases}$$

R is called the **adjacency matrix** (or the relation matrix) of r.   ◊

Example 6.4.2 A simple example. Let $A = \{2, 5, 6\}$ and let r be the relation $\{(2,2), (2,5), (5,6), (6,6)\}$ on A. Since r is a relation from A into the same set A (the B of the definition), we have $a_1 = 2$, $a_2 = 5$, and $a_3 = 6$, while $b_1 = 2$, $b_2 = 5$, and $b_3 = 6$. Next, since

- $2r2$, we have $R_{11} = 1$

- $2r5$, we have $R_{12} = 1$

- $5r6$, we have $R_{23} = 1$

- $6r6$, we have $R_{33} = 1$

All other entries of R are zero, so

$$R = \begin{pmatrix} 1 & 1 & 0 \\ 0 & 0 & 1 \\ 0 & 0 & 1 \end{pmatrix}$$

□

6.4.2 Composition as Matrix Multiplication

From the definition of r and of composition, we note that

$$r^2 = \{(2,2),(2,5),(2,6),(5,6),(6,6)\}$$

The adjacency matrix of r^2 is

$$R^2 = \begin{pmatrix} 1 & 1 & 1 \\ 0 & 0 & 1 \\ 0 & 0 & 1 \end{pmatrix}.$$

We do not write R^2 only for notational purposes. In fact, R^2 can be obtained from the matrix product RR; however, we must use a slightly different form of arithmetic.

Definition 6.4.3   Boolean Arithmetic. Boolean arithmetic is the arithmetic defined on $\{0,1\}$ using Boolean addition and Boolean multiplication, defined by

$$\begin{array}{lll} 0+0=0 & 0+1=1+0=1 & 1+1=1 \\ 0\cdot0=0 & 0\cdot1=1\cdot0=0 & 1\cdot1=1 \end{array}$$

Table 6.4.4

◊

Notice that from Chapter 3, this is the "arithmetic of logic," where $+$ replaces "or" and $\cdot$ replaces "and."

Example 6.4.5   Composition by Multiplication. Suppose that $R = \begin{pmatrix} 0 & 1 & 0 & 0 \\ 1 & 0 & 1 & 0 \\ 0 & 1 & 0 & 1 \\ 0 & 0 & 1 & 0 \end{pmatrix}$ and $S = \begin{pmatrix} 0 & 1 & 1 & 1 \\ 0 & 0 & 1 & 1 \\ 0 & 0 & 0 & 1 \\ 0 & 0 & 0 & 0 \end{pmatrix}$. Then using Boolean arithmetic, $RS = \begin{pmatrix} 0 & 0 & 1 & 1 \\ 0 & 1 & 1 & 1 \\ 0 & 0 & 1 & 1 \\ 0 & 0 & 0 & 1 \end{pmatrix}$ and $SR = \begin{pmatrix} 1 & 1 & 1 & 1 \\ 0 & 1 & 1 & 1 \\ 0 & 0 & 1 & 0 \\ 0 & 0 & 0 & 0 \end{pmatrix}$. □

Theorem 6.4.6   Composition is Matrix Multiplication. *Let A_1, A_2, and A_3 be finite sets where r_1 is a relation from A_1 into A_2 and r_2 is a relation from A_2 into A_3. If R_1 and R_2 are the adjacency matrices of r_1 and r_2, respectively, then the product R_1R_2 using Boolean arithmetic is the adjacency matrix of the composition r_1r_2.*

Remark: A convenient help in constructing the adjacency matrix of a relation from a set A into a set B is to write the elements from A in a column preceding the first column of the adjacency matrix, and the elements of B in a row above the first row. Initially, R in Example 2 would be

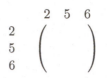

$$\begin{array}{c} \\ 2 \\ 5 \\ 6 \end{array} \begin{array}{ccc} 2 & 5 & 6 \\ \left(\right) \end{array}$$

To fill in the matrix, R_{ij} is 1 if and only if $(a_i, b_j) \in r$. So that, since the pair $(2,5) \in r$, the entry of R corresponding to the row labeled 2 and the column labeled 5 in the matrix is a 1.

Example 6.4.7 Relations and Information. This final example gives an insight into how relational data base programs can systematically answer questions pertaining to large masses of information. Matrices R (on the left) and S (on the right) define the relations r and s where arb if software a can be run with operating system b, and bsc if operating system b can run on computer c.

$$
\begin{array}{c}
\\ P1 \\ P2 \\ P3 \\ P4
\end{array}
\begin{array}{cccc}
OS1 & OS2 & OS3 & OS4 \\
1 & 0 & 1 & 0 \\
1 & 1 & 0 & 0 \\
0 & 0 & 0 & 1 \\
0 & 0 & 1 & 1
\end{array}
\qquad
\begin{array}{c}
\\ OS1 \\ OS2 \\ OS3 \\ OS4
\end{array}
\begin{array}{ccc}
C1 & C2 & C3 \\
1 & 1 & 0 \\
0 & 1 & 0 \\
0 & 0 & 1 \\
0 & 1 & 1
\end{array}
$$

Although the relation between the software and computers is not implicit from the data given, we can easily compute this information. The matrix of rs is RS, which is

$$
\begin{array}{c}
\\ P1 \\ P2 \\ P3 \\ P4
\end{array}
\begin{array}{ccc}
C1 & C2 & C3 \\
1 & 1 & 1 \\
1 & 1 & 0 \\
0 & 1 & 1 \\
0 & 1 & 1
\end{array}
$$

This matrix tells us at a glance which software will run on the computers listed. In this case, all software will run on all computers with the exception of program P2, which will not run on the computer C3, and program P4, which will not run on the computer C1. □

6.4.3 Exercises for Section 6.4

1. Let $A_1 = \{1, 2, 3, 4\}$, $A_2 = \{4, 5, 6\}$, and $A_3 = \{6, 7, 8\}$. Let r_1 be the relation from A_1 into A_2 defined by $r_1 = \{(x, y) \mid y - x = 2\}$, and let r_2 be the relation from A_2 into A_3 defined by $r_2 = \{(x, y) \mid y - x = 1\}$.

    (a) Determine the adjacency matrices of r_1 and r_2.

    (b) Use the definition of composition to find $r_1 r_2$.

    (c) Verify the result in part b by finding the product of the adjacency matrices of r_1 and r_2.

2.

    (a) Determine the adjacency matrix of each relation given via the digraphs in Exercise 3 of Section 6.3.

(b) Using the matrices found in part (a) above, find r^2 of each relation in Exercise 3 of Section 6.3.

(c) Find the digraph of r^2 directly from the given digraph and compare your results with those of part (b).

3.   Suppose that the matrices in Example 6.4.5 are relations on $\{1, 2, 3, 4\}$. What relations do R and S describe?

4.   Let D be the set of weekdays, Monday through Friday, let W be a set of employees $\{1, 2, 3\}$ of a tutoring center, and let V be a set of computer languages for which tutoring is offered, $\{A(PL), B(asic), C(++), J(ava), L(isp), P(ython)\}$. We define s (schedule) from D into W by dsw if w is scheduled to work on day d. We also define r from W into V by wrl if w can tutor students in language l. If s and r are defined by matrices

$$S = \begin{array}{c} M \\ T \\ W \\ R \\ F \end{array} \begin{pmatrix} 1 & 2 & 3 \\ 1 & 0 & 1 \\ 0 & 1 & 1 \\ 1 & 0 & 1 \\ 0 & 1 & 0 \\ 1 & 1 & 0 \end{pmatrix} \quad \text{and } R = \begin{array}{c} 1 \\ 2 \\ 3 \end{array} \begin{pmatrix} A & B & C & J & L & P \\ 0 & 1 & 1 & 0 & 0 & 1 \\ 1 & 1 & 0 & 1 & 0 & 1 \\ 0 & 1 & 0 & 0 & 1 & 1 \end{pmatrix}$$

(a) compute SR using Boolean arithmetic and give an interpretation of the relation it defines, and

(b) compute SR using regular arithmetic and give an interpretation of what the result describes.

5.   How many different reflexive, symmetric relations are there on a set with three elements?

Hint.   Consider the possible matrices.

6.   Let $A = \{a, b, c, d\}$. Let r be the relation on A with adjacency matrix

$$\begin{array}{c} a \\ b \\ c \\ c \end{array} \begin{array}{cccc} a & b & c & d \\ \begin{pmatrix} 1 & 0 & 0 & 0 \\ 0 & 1 & 0 & 0 \\ 1 & 1 & 1 & 0 \\ 0 & 1 & 0 & 1 \end{pmatrix} \end{array}$$

(a) Explain why r is a partial ordering on A.

(b) Draw its Hasse diagram.

7.   Define relations p and q on $\{1, 2, 3, 4\}$ by $p = \{(a, b) \mid |a - b| = 1\}$ and $q = \{(a, b) \mid a - b \text{ is even}\}$.

(a) Represent p and q as both graphs and matrices.

(b) Determine pq, p^2, and q^2; and represent them clearly in any way.

8.

(a) Prove that if r is a transitive relation on a set A, then $r^2 \subseteq r$.

(b) Find an example of a transitive relation for which $r^2 \neq r$.

9. We define $\leq$ on the set of all $n \times n$ relation matrices by the rule that if R and S are any two $n \times n$ relation matrices, $R \leq S$ if and only if $R_{ij} \leq S_{ij}$ for all $1 \leq i, j \leq n$.

  (a) Prove that $\leq$ is a partial ordering on all $n \times n$ relation matrices.

  (b) Prove that $R \leq S \Rightarrow R^2 \leq S^2$, but the converse is not true.

  (c) If R and S are matrices of equivalence relations and $R \leq S$, how are the equivalence classes defined by R related to the equivalence classes defined by S?

6.5 Closure Operations on Relations

In Section 6.1, we studied relations and one important operation on relations, namely composition. This operation enables us to generate new relations from previously known relations. In Section 6.3, we discussed some key properties of relations. We now wish to consider the situation of constructing a new relation r^+ from an existing relation r where, first, r^+ contains r and, second, r^+ satisfies the transitive property.

6.5.1 Transitive Closure

Consider a telephone network in which the main office a is connected to, and can communicate to, individuals b and c. Both b and c can communicate to another person, d; however, the main office cannot communicate with d. Assume communication is only one way, as indicated. This situation can be described by the relation $r = \{(a,b), (a,c), (b,d), (c,d)\}$. We would like to change the system so that the main office a can communicate with person d and still maintain the previous system. We, of course, want the most economical system.

This can be rephrased as follows; Find the smallest relation r^+ which contains r as a subset and which is transitive; $r^+ = \{(a,b), (a,c), (b,d), (c,d), (a,d)\}$.

Definition 6.5.1 Transitive Closure. Let A be a set and r be a relation on A. The transitive closure of r, denoted by r^+, is the smallest transitive relation that contains r as a subset. $\diamond$

Let $A = \{1, 2, 3, 4\}$, and let $\mathcal{S} = \{(1,2), (2,3), (3,4)\}$ be a relation on A. This relation is called the successor relation on A since each element is related to its successor. How do we compute $\mathcal{S}^+$? By inspection we note that $(1,3)$ must be in $\mathcal{S}^+$. Let's analyze why. This is so because $(1,2) \in \mathcal{S}$ and $(2,3) \in \mathcal{S}$, and the transitive property forces $(1,3)$ to be in $\mathcal{S}^+$.

In general, it follows that if $(a,b) \in \mathcal{S}$ and $(b,c) \in S$, then $(a,c) \in \mathcal{S}^+$. This condition is exactly the membership requirement for the pair (a,c) to be in the composition $\mathcal{S}\mathcal{S} = \mathcal{S}^2$. So every element in $\mathcal{S}^2$ must be an element in $\mathcal{S}^+$. So we now know that, $\mathcal{S}^+$ contains at least $\mathcal{S} \cup \mathcal{S}^2$. In particular, for this example, since $\mathcal{S} = \{(1,2), (2,3), (3,4)\}$ and $\mathcal{S}^2 = \{(1,3), (2,4)\}$, we have

$$\mathcal{S} \cup \mathcal{S}^2 = \{(1,2), (2,3), (3,4), (1,3), (2,4)\}$$

Is the relation $\mathcal{S} \cup \mathcal{S}^2$ transitive? Again, by inspection, $(1,4)$ is not an element of $\mathcal{S} \cup \mathcal{S}^2$, but $(1,3) \in \mathcal{S}^2$ and $(3,4) \in \mathcal{S}$. Therefore, the composition $\mathcal{S}^2\mathcal{S} = \mathcal{S}^3$ produces $(1,4)$, and it must be an element of $\mathcal{S}^+$ since $(1,3)$ and

$(3, 4)$ are required to be in $\mathcal{S}^+$. This shows that $\mathcal{S}^3 \subseteq \mathcal{S}^+$. This process must be continued until the resulting relation is transitive. If A is finite, as is true in this example, the transitive closure will be obtained in a finite number of steps. For this example,

$$\mathcal{S}^+ = \mathcal{S} \cup \mathcal{S}^2 \cup \mathcal{S}^3 = \{(1,2),(2,3),(3,4),(1,3),(2,4),(1,4)\}$$

Theorem 6.5.2 Transitive Closure on a Finite Set. *If r is a relation on a set A and $|A| = n$, then the transitive closure of r is the union of the first n powers of r. That is,*

$$r^+ = r \cup r^2 \cup r^3 \cup \cdots \cup r^n.$$

Let's now consider the matrix analogue of the transitive closure.
Consider the relation

$$r = \{(1,4),(2,1),(2,2),(2,3),(3,2),(4,3),(4,5),(5,1)\}$$

on the set $A = \{1,2,3,4,5\}$. The matrix of r is

$$R = \begin{pmatrix} 0 & 0 & 0 & 1 & 0 \\ 1 & 1 & 1 & 0 & 0 \\ 0 & 1 & 0 & 0 & 0 \\ 0 & 0 & 1 & 0 & 1 \\ 1 & 0 & 0 & 0 & 0 \end{pmatrix}$$

Recall that $r^2, r^3, \ldots$ can be determined through computing the matrix powers $R^2, R^3, \ldots$. For our example,

$$R^2 = \begin{pmatrix} 0 & 0 & 1 & 0 & 1 \\ 1 & 1 & 1 & 1 & 0 \\ 1 & 1 & 1 & 0 & 0 \\ 1 & 1 & 0 & 0 & 0 \\ 0 & 0 & 0 & 1 & 0 \end{pmatrix} \quad R^3 = \begin{pmatrix} 1 & 1 & 0 & 0 & 0 \\ 1 & 1 & 1 & 1 & 1 \\ 1 & 1 & 1 & 1 & 0 \\ 1 & 1 & 1 & 1 & 0 \\ 0 & 0 & 1 & 0 & 1 \end{pmatrix}$$

$$R^4 = \begin{pmatrix} 1 & 1 & 1 & 1 & 0 \\ 1 & 1 & 1 & 1 & 1 \\ 1 & 1 & 1 & 1 & 1 \\ 1 & 1 & 1 & 1 & 1 \\ 1 & 1 & 0 & 0 & 0 \end{pmatrix} \quad R^5 = \begin{pmatrix} 1 & 1 & 1 & 1 & 1 \\ 1 & 1 & 1 & 1 & 1 \\ 1 & 1 & 1 & 1 & 1 \\ 1 & 1 & 1 & 1 & 1 \\ 1 & 1 & 1 & 1 & 0 \end{pmatrix}$$

Table 6.5.3

How do we relate $\bigcup\limits_{i=1}^{5} r^i$ to the powers of R?

Theorem 6.5.4 Matrix of a Transitive Closure. *Let r be a relation on a finite set and let R^+ be the matrix of r^+, the transitive closure of r. Then $R^+ = R + R^2 + \cdots + R^n$, using Boolean arithmetic.*

Using this theorem, we find R^+ is the 5×5 matrix consisting of all $1's$, thus, r^+ is all of $A \times A$.

6.5.2 Algorithms for computing transitive closure

Let r be a relation on the set $\{1,2,\ldots,n\}$ with relation matrix R. The matrix of the transitive closure R^+, can be computed by the equation $R^+ = R + R^2 + \cdots + R^n$. By using ordinary polynomial evaluation methods, you can compute R^+ with $n-1$ matrix multiplications:

$$R^+ = R(I + R(I + (\cdots R(I + R)\cdots)))$$

For example, if $n = 3$, $R^+ = R(I + R(I + R))$.

We can make use of the fact that if T is a relation matrix, $T + T = T$ due to the fact that $1 + 1 = 1$ in Boolean arithmetic. Let $S_k = R + R^2 + \cdots + R^k$. Then

$$R = S_1$$
$$S_1(I + S_1) = R(I + R) = R + R^2 = S_2$$
$$S_2(I + S_2) = (R + R^2)(I + R + R^2)$$
$$= (R + R^2) + (R^2 + R^3) + (R^3 + R^4)$$
$$= R + R^2 + R^3 + R^4 = S_4$$

Similarly,

$$S_4(I + S_4) = S_8$$

and by induction we can prove

$$S_{2^k}(I + S_{2^k}) = S_{2^{k+1}}$$

Notice how each matrix multiplication doubles the number of terms that have been added to the sum that you currently have computed. In algorithmic form, we can compute R^+ as follows.

Algorithm 6.5.5 Transitive Closure Algorithm. *Let R be a relation matrix and let R^+ be its transitive closure matrix, which is to be computed as matrix T*

```
1.0. S = R
2.0  T= S*(I+S)
3.0 While T != S
              3.1 S = T
              3.2 T= S*(I+S) // using Boolean arithmetic
4.0 Return T
```

Listing 6.5.6

Note 6.5.7

- Often the higher-powered terms in S_n do not contribute anything to R^+. When the condition $T = S$ becomes true in Step 3, this is an indication that no higher-powered terms are needed.

- To compute R^+ using this algorithm, you need to perform no more than $\lceil \log_2 n \rceil$ matrix multiplications, where $\lceil x \rceil$ is the least integer that is greater than or equal to x. For example, if r is a relation on 25 elements, no more than $\lceil \log_2 25 \rceil = 5$ matrix multiplications are needed.

A second algorithm, Warshall's Algorithm, reduces computation time to the time that it takes to multiply two square matrices with the same order as the relation matrix in question.

Algorithm 6.5.8 Warshall's Algorithm. *Let R be an $n \times n$ relation matrix and let R^+ be its transitive closure matrix, which is to be computed as matrix T using Boolean arithmetic*

```
1.0 T = R
2.0 for k = 1 to n:
      for i = 1 to n:
        for j = 1 to n:
          T[i,j]= T[i,j] + T[i,k] * T[k,j]
3.0 Return T
```

Listing 6.5.9

6.5.3 Exercises for Section 6.5

1. Let $A = \{0, 1, 2, 3, 4\}$ and $\mathcal{S} = \{(0, 1), (1, 3), (2, 3), (3, 4), (4, 1)\}$. Compute $\mathcal{S}^+$ using the matrix representation of $\mathcal{S}$. Verify your results by checking against the result obtained directly from the definition of transitive closure.

2. Let $A = \{1, 2, 3, 4, 6, 12\}$ and $t = \{(a, b) \mid b/a \text{ is a prime number}\}$. Determine t^+ by any means. Represent your answer as a matrix.

3.

   (a) Draw digraphs of the relations $\mathcal{S}$, $\mathcal{S}^2$, $\mathcal{S}^3$, and $\mathcal{S}^+$ where $\mathcal{S}$ is defined in the first exercise above.

   (b) Verify that in terms of the graph of $\mathcal{S}$, $a\mathcal{S}^+b$ if and only if b is reachable from a along a path of any finite nonzero length.

4. Let r be the relation represented by the following digraph.

   (a) Find r^+ using the definition based on order pairs.

   (b) Determine the digraph of r^+ directly from the digraph of r.

   (c) Verify your result in part (b) by computing the digraph from your result in part (a).

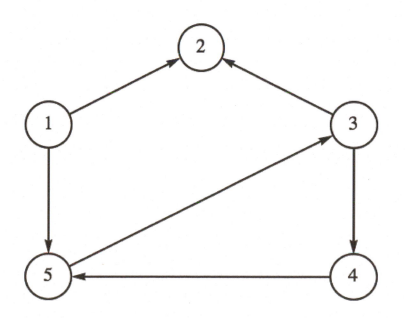

Figure 6.5.10: Digraph of r in exercise 4.

5.

(a) Define reflexive closure and symmetric closure by imitating the definition of transitive closure.

(b) Use your definitions to compute the reflexive and symmetric closures of examples in the text.

(c) What are the transitive reflexive closures of these examples?

(d) Convince yourself that the reflexive closure of the relation $<$ on the set of positive integers $\mathbb{P}$ is $\leq$.

6. What common relations on $\mathbb{Z}$ are the transitive closures of the following relations?

(a) aSb if and only if $a + 1 = b$.

(b) aRb if and only if $|a - b| = 2$.

7.

(a) Let A be any set and r a relation on A, prove that $(r^+)^+ = r^+$.

(b) Is the transitive closure of a symmetric relation always both symmetric and reflexive? Explain.

8. The definition of the Transitive Closure of r refers to the "smallest transitive relation that contains r as a subset." Show that the intersection of all transitive relations on A containing r is a transitive relation containing r and is precisely r^+.

Chapter 7

Functions

countably infinite

A countably infinite set
Is as simple as things like this get.
Just start counting at 1,
Then continue—it's fun!
I'll check back when you're done, so don't sweat.

Chris Doyle, The Omnificent English Dictionary In Limerick Form

In this chapter we will consider some basic concepts of the relations that are called functions. A large variety of mathematical ideas and applications can be more completely understood when expressed through the function concept.

7.1 Definition and Notation

7.1.1 Fundamentals

Definition 7.1.1 Function. A function from a set A into a set B is a relation from A into B such that each element of A is related to exactly one element of the set B. The set A is called the **domain** of the function and the set B is called the **codomain**. ◇

The reader should note that a function f is a relation from A into B with two important restrictions:

- Each element in the set A, the domain of f, must be related to some element of B, the codomain.

- The phrase "is related to exactly one element of the set B" means that if $(a, b) \in f$ and $(a, c) \in f$, then $b = c$.

Example 7.1.2 A function as a list of ordered pairs. Let $A = \{-2, -1, 0, 1, 2\}$ and $B = \{0, 1, 2, 3, 4\}$, and if $s = \{(-2, 4), (-1, 1), (0, 0), (1, 1), (2, 4)\}$, then s is a function from A into B. □

Example 7.1.3 A function as a set of ordered pairs in set-builder notation. Let $\mathbb{R}$ be the real numbers. Then $L = \{(x, 3x) \mid x \in \mathbb{R}\}$ is a function from $\mathbb{R}$ into $\mathbb{R}$, or, more simply, L is a function on $\mathbb{R}$. □

It is customary to use a different system of notation for functions than the one we used for relations. If f is a function from the set A into the set B, we will write $f : A \to B$.

The reader is probably more familiar with the notation for describing functions that is used in basic algebra or calculus courses. For example, $y = \frac{1}{x}$ or $f(x) = \frac{1}{x}$ both define the function $\left\{ \left(x, \frac{1}{x}\right) \mid x \in \mathbb{R}, x \neq 0 \right\}$. Here the domain was assumed to be those elements of $\mathbb{R}$ whose substitutions for x make sense, the nonzero real numbers, and the codomain was assumed to be $\mathbb{R}$. In most cases, we will make a point of listing the domain and codomain in addition to describing what the function does in order to define a function.

The terms **mapping**, **map**, and **transformation** are also used for functions.

Definition 7.1.4 The Set of Functions Between Two Sets. Given two sets, A and B, the set of all function from A into B is denoted B^A. $\Diamond$

The notation used for sets of functions makes sense in light of Exercise 7.

One way to imagine a function and what it does is to think of it as a machine. The machine could be mechanical, electronic, hydraulic, or abstract. Imagine that the machine only accepts certain objects as raw materials or input. The possible raw materials make up the domain. Given some input, the machine produces a finished product that depends on the input. The possible finished products that we imagine could come out of this process make up the codomain.

Example 7.1.5 A definition based on images. We can define a function based on specifying the codomain element to which each domain element is related. For example, $f : \mathbb{R} \to \mathbb{R}$ defined by $f(x) = x^2$ is an alternate description of $f = \left\{ \left(x, x^2\right) \mid x \in \mathbb{R} \right\}$. $\square$

Definition 7.1.6 Image of an element under a function. Let $f : A \to B$, read "Let f be a function from the set A into the set B." If $a \in A$, then $f(a)$ is used to denote that element of B to which a is related. $f(a)$ is called the **image** of a, or, more precisely, the image of a under f. We write $f(a) = b$ to indicate that the image of a is b. $\Diamond$

In Example 7.1.5, the image of 2 under f is 4; that is, $f(2) = 4$. In Example 7.1.2, the image of -1 under s is 1; that is, $s(-1) = 1$.

Definition 7.1.7 Range of a Function. The range of a function is the set of images of its domain. If $f : X \to Y$, then the range of f is denoted $f(X)$, and

$$f(X) = \{f(a) \mid a \in X\} = \{b \in Y \mid \exists a \in X \text{ such that } f(a) = b\}.$$

$\Diamond$

Note that the range of a function is a subset of its codomain. $f(X)$ is also read as "the image of the set X under the function f" or simply "the image of f."

In Example 7.1.2, $s(A) = \{0, 1, 4\}$. Notice that 2 and 3 are not images of any element of A. In addition, note that both 1 and 4 are related to more than one element of the domain: $s(1) = s(-1) = 1$ and $s(2) = s(-2) = 4$. This does not violate the definition of a function. Go back and read the definition if this isn't clear to you.

In Example 7.1.3, the range of L is equal to its codomain, $\mathbb{R}$. If b is any real number, we can demonstrate that it belongs to $L(\mathbb{R})$ by finding a real number

x for which $L(x) = b$. By the definition of L, $L(x) = 3x$, which leads us to the equation $3x = b$. This equation always has a solution, $\frac{b}{3}$; thus $L(\mathbb{R}) = \mathbb{R}$.

The formula that we used to describe the image of a real number under L, $L(x) = 3x$, is preferred over the set notation for L due to its brevity. Any time a function can be described with a rule or formula, we will use this form of description. In Example 7.1.2, the image of each element of A is its square. To describe that fact, we write $s(a) = a^2$ ($a \in A$), or $S : A \to B$ defined by $S(a) = a^2$.

There are many ways that a function can be described. Many factors, such as the complexity of the function, dictate its representation.

Example 7.1.8 Data as a function. Suppose a survey of 1,000 persons is done asking how many hours of television each watches per day. Consider the function $W : \{0, 1, \ldots, 24\} \to \{0, 1, 2, \ldots, 1000\}$ defined by

$$W(t) = \text{the number of persons who gave a response of } t \text{ hours}$$

This function will probably have no formula such as the ones for s and L above.

$\square$

Example 7.1.9 Conditional definition of a function. Consider the function $m : \mathbb{P} \to \mathbb{Q}$ defined by the set

$$m = \{(1, 1), (2, 1/2), (3, 9), (4, 1/4), (5, 25), \ldots\}$$

No simple single formula could describe m, but if we assume that the pattern given continues, we can write

$$m(x) = \begin{cases} x^2 & \text{if } x \text{ is odd} \\ 1/x & \text{if } x \text{ is even} \end{cases}$$

$\square$

7.1.2 Functions of Two Variables

If the domain of a function is the Cartesian product of two sets, then our notation and terminology changes slightly. For example, consider the function $C : \mathbb{N} \times \mathbb{N} \to \mathbb{N}$ defined by $C((n_1, n_2)) = n_1^2 + n_2^2 - n_1 n_2 + 10$. For this function, we would drop one set of parentheses and write $C(4, 2) = 22$, not $C((4, 2)) = 22$. We call C a function of two variables. From one point of view, this function is no different from any others that we have seen. The elements of the domain happen to be slightly more complicated. On the other hand, we can look at the individual components of the ordered pairs as being separate. If we interpret C as giving us the cost of producing quantities of two products, we can imagine varying n_1 while n_2 is fixed, or vice versa.

7.1.3 SageMath Note

There are several ways to define a function in Sage. The simplest way to implement f is as follows.

```
f(x)=x^2
f
```

```
x |--> x^2
```

```
[f(4),f(1.2)]
```

```
[16,1.44000000000000]
```

Sage is built upon the programming language Python, which is a *strongly typed language* and so you can't evaluate expressions such as f('Hello'). However a function such as f, as defined above, will accept any type of number, so a bit more work is needed to restrict the inputs of f to the integers.

A second way to define a function in Sage is based on Python syntax.

```
def fa(x):
        return x^2
[fa(2),fa(1.2)]
```

```
[16,1.44000000000000]
```

7.1.4 Non-Functions

We close this section with two examples of relations that are not functions.

Example 7.1.10 A non-function. Let $A = B = \{1, 2, 3\}$ and let $f = \{(1, 2), (2, 3)\}$. Here f is not a function from A into B since f does not act on, or "use," all elements of A. □

Example 7.1.11 Another non-function. Let $A = B = \{1, 2, 3\}$ and let $g = \{(1, 2), (2, 3), (2, 1), (3, 2)\}$. We note that g acts on all of A. However, g is still not a function since $(2, 3) \in g$ and $(2, 1) \in g$ and the condition on each domain being related to exactly one element of the codomain is violated. □

7.1.5 Exercises for Section 7.1

1. Let $A = \{1, 2, 3, 4\}$ and $B = \{a, b, c, d\}$. Determine which of the following are functions. Explain.

    (a) $f \subseteq A \times B$, where $f = \{(1, a), (2, b), (3, c), (4, d)\}$.

    (b) $g \subseteq A \times B$, where $g = \{(1, a), (2, a), (3, b), (4, d)\}$.

    (c) $h \subseteq A \times B$, where $h = \{(1, a), (2, b), (3, c)\}$.

    (d) $k \subseteq A \times B$, where $k = \{(1, a), (2, b), (2, c), (3, a), (4, a)\}$.

    (e) $L \subseteq A \times A$, where $L = \{(1, 1), (2, 1), (3, 1), (4, 1)\}$.

2. Let A be a set and let S be any subset of A. Let $\chi_S : A \to \{0, 1\}$ be defined by

$$\chi_S(x) = \begin{cases} 1 & \text{if } x \in S \\ 0 & \text{if } x \notin S \end{cases}$$

    The function χ_S is called the **characteristic function** of S.

    (a) If $A = \{a, b, c\}$ and $S = \{a, b\}$, list the elements of χ_S .

    (b) If $A = \{a, b, c, d, e\}$ and $S = \{a, c, e\}$, list the elements of χ_S.

    (c) If $A = \{a, b, c\}$, what are $\chi_\emptyset$ and χ_A?

3. Find the ranges of each of the relations that are functions in Exercise 1.

4. Find the ranges of the following functions on $\mathbb{Z}$:

    (a) $g = \{(x, 4x+1)|x \in \mathbb{Z}\}$.

    (b) $h(x) =$ the least integer that is greater than or equal to $\sqrt{|x|}$.

    (c) $P(x) = x + 10$.

5. Let $f : \mathbb{P} \to \mathbb{P}$, where $f(a)$ is the largest power of two that evenly divides a; for example, $f(12) = 4, f(9) = 1$, and $f(8) = 8$. Describe the equivalence classes of the kernel of f.

6. Let U be a set with subsets A and B.

    (a) Show that $g : U \to \{0,1\}$ defined by $g(a) = \min\left(C_A(a), C_B(a)\right)$ is the characteristic function of $A \cap B$.

    (b) What characteristic function is $h : U \to \{0,1\}$ defined by $h(a) = \max\left(C_A(a), C_B(a)\right)$?

    (c) How are the characteristic functions of A and A^c related?

7. If A and B are finite sets, how many different functions are there from A into B?

8. Let f be a function with domain A and codomain B. Consider the relation $K \subseteq A \times A$ defined on the domain of f by $(x, y) \in K$ if and only if $f(x) = f(y)$. The relation K is called \textit{ the kernel of f}.

    (a) Prove that K is an equivalence relation.

    (b) For the specific case of $A = \mathbb{Z}$, where $\mathbb{Z}$ is the set of integers, let $f : \mathbb{Z} \to \mathbb{Z}$ be defined by $f(x) = x^2$. Describe the equivalence classes of the kernel for this specific function.

7.2 Properties of Functions

7.2.1 Properties

Consider the following functions:

    Let $A = \{1,2,3,4\}$ and $B = \{a,b,c,d\}$, and define $f : A \to B$ by

$$f(1) = a, f(2) = b, f(3) = c \text{ and } f(4) = d$$

    Let $A = \{1,2,3,4\}$ and $B = \{a,b,c,d\}$, and define $g : A \to B$ by

$$g(1) = a, g(2) = b, g(3) = a \text{ and } g(4) = b.$$

    The first function, f, gives us more information about the set B than the second function, g. Since A clearly has four elements, f tells us that B contains at least four elements since each element of A is mapped onto a different element of B. The properties that f has, and g does not have, are the most basic properties that we look for in a function. The following definitions summarize the basic vocabulary for function properties.

Definition 7.2.1 Injective Function, Injection. A function $f : A \to B$ is injective if

$$\forall a, b \in A, a \neq b \Rightarrow f(a) \neq f(b)$$

An injective function is called an injection, or a one-to-one function. $\Diamond$

Notice that the condition for an injective function is logically equivalent to

$$f(a) = f(b) \Rightarrow a = b.$$

for all $a, b \in A$. This is often a more convenient condition to prove than what is given in the definition.

Definition 7.2.2 Surjective Function, Surjection. A function $f : A \to B$ is surjective if its range, $f(A)$, is equal to its codomain, B. A surjective function is called a surjection, or an onto function. $\Diamond$

Notice that the condition for a surjective function is equivalent to

For all $b \in B$, there exists $a \in A$ such that $f(a) = b$.

Definition 7.2.3 Bijective Function, Bijection. A function $f : A \to B$ is bijective if it is both injective and surjective. Bijective functions are also called one-to-one, onto functions. $\Diamond$

The function f that we opened this section with is bijective. The function g is neither injective nor surjective.

Example 7.2.4 Injective but not surjective function. Let $A = \{1, 2, 3\}$ and $B = \{a, b, c, d\}$, and define $f : A \to B$ by $f(1) = b$, $f(2) = c$, and $f(3) = a$. Then f is injective but not surjective. $\square$

Example 7.2.5 Characteristic Functions. The characteristic function, χ_S, in Exercise 7.1.5.2 is surjective if S is a proper subset of A, but never injective if $|A| > 2$. $\square$

7.2.2 Counting

Example 7.2.6 Seating Students. Let A be the set of students who are sitting in a classroom, let B be the set of seats in the classroom, and let s be the function which maps each student into the chair he or she is sitting in. When is s one to one? When is it onto? Under normal circumstances, s would always be injective since no two different students would be in the same seat. In order for s to be surjective, we need all seats to be used, so s is a surjection if the classroom is filled to capacity. $\square$

Functions can also be used for counting the elements in large finite sets or in infinite sets. Let's say we wished to count the occupants in an auditorium containing 1,500 seats. If each seat is occupied, the answer is obvious, 1,500 people. What we have done is to set up a one-to-one correspondence, or bijection, from seats to people. We formalize in a definition.

Definition 7.2.7 Cardinality. Two sets are said to have the same cardinality if there exists a bijection between them. If a set has the same cardinality as the set $\{1, 2, 3, \ldots, n\}$, then we say its cardinality is n. $\Diamond$

The function f that opened this section serves to show that the two sets $A = \{1, 2, 3, 4\}$ and $B = \{a, b, c, d\}$ have the same cardinality. Notice in applying the definition of cardinality, we don't actually appear to count either set, we just match up the elements. However, matching the letters in B with the numbers 1, 2, 3, and 4 is precisely how we count the letters.

Definition 7.2.8 Countable Set. If a set is finite or has the same cardinality as the set of positive integers, it is called a countable set. $\Diamond$

Example 7.2.9 Counting the Alphabet. The alphabet $\{A, B, C, ..., Z\}$ has cardinality 26 through the following bijection into the set $\{1, 2, 3, \ldots, 26\}$.

$$
\begin{array}{ccccc}
A & B & C & \cdots & Z \\
\downarrow & \downarrow & \downarrow & \cdots & \downarrow \\
1 & 2 & 3 & \cdots & 26
\end{array} .
$$

$\square$

Example 7.2.10 As many evens as all positive integers. Recall that $2\mathbb{P} = \{b \in \mathbb{P} \mid b = 2k \text{ for some } k \in \mathbb{P}\}$. Paradoxically, $2\mathbb{P}$ has the same cardinality as the set $\mathbb{P}$ of positive integers. To prove this, we must find a bijection from $\mathbb{P}$ to $2\mathbb{P}$. Such a function isn't unique, but this one is the simplest: $f : \mathbb{P} \to 2\mathbb{P}$ where $f(m) = 2m$. Two statements must be proven to justify our claim that f is a bijection:

- f is one-to-one.

  Proof: Let $a, b \in \mathbb{P}$ and assume that $f(a) = f(b)$. We must prove that $a = b$.
  $$f(a) = f(b) \implies 2a = 2b \implies a = b.$$

- f is onto.

  Proof: Let $b \in 2\mathbb{P}$. We want to show that there exists an element $a \in \mathbb{P}$ such that $f(a) = b$. If $b \in 2\mathbb{P}$, $b = 2k$ for some $k \in \mathbb{P}$ by the definition of $2\mathbb{P}$. So we have $f(k) = 2k = b$. Hence, each element of $2\mathbb{P}$ is the image of some element of $\mathbb{P}$.

$\square$

Another way to look at any function with $\mathbb{P}$ as its domain is creating a list of the form $f(1), f(2), f(3), \ldots$. In the previous example, the list is $2, 4, 6, \ldots$. This infinite list clearly has no duplicate entries and every even positive integer appears in the list eventually.

A function $f : \mathbb{P} \to A$ is a bijection if the infinite list $f(1), f(2), f(3), \ldots$ contains no duplicates, and every element of A appears on in the list. In this case, we say the A is **countably infinite**, or simply **countable**

Readers who have studied real analysis should recall that the set of rational numbers is a countable set, while the set of real numbers is not a countable set. See the exercises at the end of this section for an another example of such a set.

We close this section with a theorem called the Pigeonhole Principle, which has numerous applications even though it is an obvious, common-sense statement. Never underestimate the importance of simple ideas. The Pigeonhole Principle states that if there are more pigeons than pigeonholes, then two or more pigeons must share the same pigeonhole. A more rigorous mathematical statement of the principle follows.

Theorem 7.2.11 The Pigeonhole Principle. *Let f be a function from a finite set X into a finite set Y. If $n \geq 1$ and $|X| > n|Y|$, then there exists an element of Y that is the image under f of at least $n + 1$ elements of X.*

Proof. Assume no such element exists. For each $y \in Y$, let $A_y = \{x \in X \mid f(x) = y\}$. Then it must be that $|A_y| \leq n$. Furthermore, the set of nonempty A_y form a partition of X. Therefore,

$$|X| = \sum_{y \in Y} |A_y| \leq n|Y|$$

which is a contradiction. ∎

Example 7.2.12 A duplicate name is assured. Assume that a room contains four students with the first names John, James, and Mary. Prove that two students have the same first name. We can visualize a mapping from the set of students to the set of first names; each student has a first name. The pigeonhole principle applies with $n = 1$, and we can conclude that at least two of the students have the same first name. □

7.2.3 Exercises for Section 7.2

1. Determine which of the functions in Exercise 7.1.5.1 of Section 7.1 are one-to-one and which are onto.

2.

    (a) Determine all bijections from the $\{1, 2, 3\}$ into $\{a, b, c\}$.

    (b) Determine all bijections from $\{1, 2, 3\}$ into $\{a, b, c, d\}$.

3. Which of the following are one-to-one, onto, or both?

    (a) $f_1 : \mathbb{R} \to \mathbb{R}$ defined by $f_1(x) = x^3 - x$.

    (b) $f_2 : \mathbb{Z} \to \mathbb{Z}$ defined by $f_2(x) = -x + 2$.

    (c) $f_3 : \mathbb{N} \times \mathbb{N} \to \mathbb{N}$ defined by $f_3(j, k) = 2^j 3^k$.

    (d) $f_4 : \mathbb{P} \to \mathbb{P}$ defined by $f_4(n) = \lceil n/2 \rceil$, where $\lceil x \rceil$ is the ceiling of x, the smallest integer greater than or equal to x.

    (e) $f_5 : \mathbb{N} \to \mathbb{N}$ defined by $f_5(n) = n^2 + n$.

    (f) $f_6 : \mathbb{N} \to \mathbb{N} \times \mathbb{N}$ defined by $f_6(n) = (2n, 2n + 1)$.

4. Which of the following are injections, surjections, or bijections on $\mathbb{R}$, the set of real numbers?

    (a) $f(x) = -2x$.

    (b) $g(x) = x^2 - 1$.

    (c) $h(x) = \begin{cases} x & x < 0 \\ x^2 & x \geq 0 \end{cases}$

    (d) $q(x) = 2^x$

    (e) $r(x) = x^3$

    (f) $s(x) = x^3 - x$

5. Suppose that m pairs of socks are mixed up in your sock drawer. Use the Pigeonhole Principle to explain why, if you pick $m + 1$ socks at random, at least two will make up a matching pair.

6. In your own words explain the statement "The sets of integers and even integers have the same cardinality."

7. Let $A = \{1, 2, 3, 4, 5\}$. Find functions, if they exist that have the properties specified below.

    (a) A function that is one-to-one and onto.

(b) A function that is neither one-to-one nor onto.

(c) A function that is one-to-one but not onto.

(d) A function that is onto but not one-to-one.

8.

(a) Define functions, if they exist, on the positive integers, $\mathbb{P}$, with the same properties as in Exercise 7 (if possible).

(b) Let A and B be finite sets where $|A| = |B|$. Is it possible to define a function $f : A \to B$ that is one-to-one but not onto? Is it possible to find a function $g : A \to B$ that is onto but not one-to-one?

9.

(a) Prove that the set of natural numbers is countable.

(b) Prove that the set of integers is countable.

(c) Prove that the set of rational numbers is countable.

10.

(a) Prove that the set of finite strings of 0's and 1's is countable.

(b) Prove that the set of odd integers is countable.

(c) Prove that the set $\mathbb{N} \times \mathbb{N}$ is countable.

11. Use the Pigeonhole Principle to prove that an injection cannot exist between a finite set A and a finite set B if the cardinality of A is greater than the cardinality of B.

12. The important properties of relations are not generally of interest for functions. Most functions are not reflexive, symmetric, antisymmetric, or transitive. Can you give examples of functions that do have these properties?

13. Prove that the set of all infinite sequences of 0's and 1's is not a countable set.

13. Prove that the set of all functions on the integers is an uncountable set.

14. Given five points on the unit square, $\{(x,y) \mid 0 \le x, y \le 1\}$, prove that there are two of the points a distance of no more than $\frac{\sqrt{2}}{2}$ from one another.

7.3 Function Composition

Now that we have a good understanding of what a function is, our next step is to consider an important operation on functions. Our purpose is not to develop the algebra of functions as completely as we did for the algebras of logic, matrices, and sets, but the reader should be aware of the similarities between the algebra of functions and that of matrices. We first define equality of functions.

7.3.1 Function Equality

Definition 7.3.1 Equality of Functions. Let $f, g : A \to B$; that is, let f and g both be functions from A into B. Then f is equal to g (denoted $f = g$) if and only if $f(x) = g(x)$ for all $x \in A$. $\Diamond$

Two functions that have different domains cannot be equal. For example, $f : \mathbb{Z} \to \mathbb{Z}$ defined by $f(x) = x^2$ and $g : \mathbb{R} \to \mathbb{R}$ defined by $g(x) = x^2$ are not equal even though the formula that defines them is the same.

On the other hand, it is not uncommon for two functions to be equal even though they are defined differently. For example consider the functions h and k, where $h : \{-1, 0, 1, 2\} \to \{0, 1, 2\}$ is defined by $h(x) = |x|$ and $k : \{-1, 0, 1, 2\} \to \{0, 1, 2\}$ is defined by $k(x) = -\frac{x^3}{3} + x^2 + \frac{x}{3}$ appear to be very different functions. However, they are equal because $h(x) = k(x)$ for $x = -1, 0, 1,$ and 2.

7.3.2 Function Composition

One of the most important operations on functions is that of composition.

Definition 7.3.2 Composition of Functions. Let $f : A \to B$ and $g : B \to C$. Then the composition of f followed by g, written $g \circ f$, is a function from A into C defined by $(g \circ f)(x) = g(f(x))$, which is read "g of f of x." ◊

The reader should note that it is traditional to write the composition of functions from right to left. Thus, in the above definition, the first function performed in computing $g \circ f$ is f. On the other hand, for relations, the composition rs is read from left to right, so that the first relation is r.

Example 7.3.3 A basic example. Let $f : \{1, 2, 3\} \to \{a, b\}$ be defined by $f(1) = a$, $f(2) = a$, and $f(3) = b$. Let $g : \{a, b\} \to \{5, 6, 7\}$ be defined by $g(a) = 5$ and $g(b) = 7$. Then $g \circ f : \{1, 2, 3\} \to \{5, 6, 7\}$ is defined by $(g \circ f)(1) = 5$, $(g \circ f)(2) = 5$, and $(g \circ f)(3) = 7$. For example, $(g \circ f)(1) = g(f(l)) = g(a) = 5$. Note that $f \circ g$ is not defined. Why?

Let $f : \mathbb{R} \to \mathbb{R}$ be defined by $f(x) = x^3$ and let $g : \mathbb{R} \to \mathbb{R}$ be defined by $g(x) = 3x + 1$. Then, since

$$(g \circ f)(x) = g(f(x)) = g\left(x^3\right) = 3x^3 + 1$$

we have $g \circ f : \mathbb{R} \to \mathbb{R}$ is defined by $(g \circ f)(x) = 3x^3 + 1$. Here $f \circ g$ is also defined and $f \circ g : \mathbb{R} \to \mathbb{R}$ is defined by $(f \circ g)(x) = (3x + 1)^3$. Moreover, since $3x^3 + 1 \neq (3x + 1)^3$ for at least one real number, $g \circ f \neq f \circ g$. Therefore, the commutative law is not true for functions under the operation of composition. However, the associative law is true for functions under the operation of composition. □

Theorem 7.3.4 Function composition is associative. *If $f : A \to B$, $g : B \to C$, and $h : C \to D$, then $h \circ (g \circ f) = (h \circ g) \circ f$.*

Proof. Note: In order to prove that two functions are equal, we must use the definition of equality of functions. Assuming that the functions have the same domain, they are equal if, for each domain element, the images of that element under the two functions are equal.

We wish to prove that $(h \circ (g \circ f))(x) = ((h \circ g) \circ f)(x)$ for all $x \in A$, which is the domain of both functions.

$$(h \circ (g \circ f))(x) = h((g \circ f)(x)) \text{ by the definition of composition}$$
$$= h(g(f(x))) \text{ by the definition of composition}$$

Similarly,

$$((h \circ g) \circ f)(x) = (h \circ g)(f(x)) \text{ by the definition of composition}$$
$$= h(g(f(x))) \text{ by the definition of composition}$$

Notice that no matter how the functions the expression $h \circ g \circ f$ is grouped, the final image of any element of $x \in A$ is $h(g(f(x)))$ and so $h \circ (g \circ f) = (h \circ g) \circ f$. ∎

If f is a function on a set A, then the compositions $f \circ f$, $f \circ f \circ f, \ldots$ are valid, and we denote them as f^2, $f^3, \ldots$. Repeated compositions of f with itself can be defined recursively. We will discuss this form of definition in detail in Section 8.1.

Definition 7.3.5 Powers of Functions. Let $f : A \to A$.

- $f^1 = f$; that is, $f^1(a) = f(a)$, for $a \in A$.

- For $n \geq 1$, $f^{n+1} = f \circ f^n$; that is, $f^{n+1}(a) = f(f^n(a))$ for $a \in A$.

$\Diamond$

Two useful theorems concerning composition are given below. The proofs are left for the exercises.

Theorem 7.3.6 The composition of injections is an injection. *If* $f : A \to B$ *and* $g : B \to C$ *are injections, then* $g \circ f : A \to C$ *is an injection.*

Theorem 7.3.7 The composition of surjections is a surjection. *If* $f : A \to B$ *and* $g : B \to C$ *are surjections, then* $g \circ f : A \to C$ *is a surjection.*

We would now like to define the concepts of identity and inverse for functions under composition. The motivation and descriptions of the definitions of these terms come from the definitions of the terms in the set of real numbers and for matrices. For real numbers, the numbers 0 and 1 play the unique role that $x + 0 = 0 + x = x$ and $x \cdot 1 = 1 \cdot x = x$ for any real number x. 0 and 1 are the identity elements for the reals under the operations of addition and multiplication, respectively. Similarly, the $n \times n$ zero matrix 0 and the $n \times n$ identity matrix I are such that for any $n \times n$ matrix A, $A + 0 = 0 + A = A$ and $AI = IA = I$. Hence, an elegant way of defining the identity function under the operation of composition would be to imitate the above well-known facts.

Definition 7.3.8 Identity Function. For any set A, the identity function on A is a function from A onto A, denoted by i (or, more specifically, i_A) such that $i(a) = a$ for all $a \in A$. $\Diamond$

Based on the definition of i, we can show that for all functions $f : A \to A$, $f \circ i = i \circ f = f$.

Example 7.3.9 The identity function on $\{1, 2, 3\}$. If $A = \{1, 2, 3\}$, then the identity function $i : A \to A$ is defined by $i(1) = 1$, $i(2) = 2$, and $i(3) = 3$. □

Example 7.3.10 The identity function on $\mathbb{R}$. The identity function on $\mathbb{R}$ is $i : \mathbb{R} \to \mathbb{R}$ defined by $i(x) = x$. □

7.3.3 Inverse Functions

We will introduce the inverse of a function with a special case: the inverse of a function on a set. After you've taken the time to understand this concept, you can read about the inverse of a function from one set into another. The reader is encouraged to reread the definition of the inverse of a matrix in Section 5.2 (Definition 5.2.5) to see that the following definition of the inverse function is a direct analogue of that definition.

Definition 7.3.11 Inverse of a Function on a Set. Let $f : A \to A$. If there exists a function $g : A \to A$ such that $g \circ f = f \circ g = i$, then g is called the inverse of f and is denoted by f^{-1} , read "f inverse." ◇

Notice that in the definition we refer to "the inverse" as opposed to "an inverse." It can be proven that a function can never have more than one inverse (see exercises).

An alternate description of the inverse of a function, which can be proven from the definition, is as follows: Let $f : A \to A$ be such that $f(a) = b$. Then when it exists, f^{-1} is a function from A to A such that $f^{-1}(b) = a$. Note that f^{-1} "undoes" what f does.

Example 7.3.12 The inverse of a function on $\{1,2,3\}$. Let $A = \{1,2,3\}$ and let f be the function defined on A such that $f(1) = 2$, $f(2) = 3$, and $f(3) = 1$. Then $f^{-1} : A \to A$ is defined by $f^{-1}(1) = 3$, $f^{-1}(2) = 1$, and $f^{-1}(3) = 2$. □

Example 7.3.13 Inverse of a real function. If $g : \mathbb{R} \to \mathbb{R}$ is defined by $g(x) = x^3$, then g^{-1} is the function that undoes what g does. Since g cubes real numbers, g^{-1} must be the "reverse" process, namely, takes cube roots. Therefore, $g^{-1} : \mathbb{R} \to \mathbb{R}$ is defined by $g^{-1}(x) = \sqrt[3]{x}$. We should show that $g^{-1} \circ g = i$ and $g \circ g^{-1} = i$. We will do the first, and the reader is encouraged to do the second.

$$
\begin{aligned}
\left(g^{-1} \circ g\right)(x) &= g^{-1}(g(x)) \quad \text{Definition of composition} \\
&= g^{-1}\left(x^3\right) \quad \text{Definition of } g \\
&= \sqrt[3]{x^3} \quad \text{Definition of } g^{-1} \\
&= x \quad \text{Definition of cube root} \\
&= i(x) \quad \text{Definition of the identity function}
\end{aligned}
$$

Therefore, $g^{-1} \circ g = i$. Why? □

The definition of the inverse of a function alludes to the fact that not all functions have inverses. How do we determine when the inverse of a function exists?

Theorem 7.3.14 Bijections have inverses. *Let $f : A \to A$. f^{-1} exists if and only if f is a bijection; i. e. f is one-to-one and onto.*

Proof. ($\Rightarrow$) In this half of the proof, assume that f^{-1} exists and we must prove that f is one-to-one and onto. To do so, it is convenient for us to use the relation notation, where $f(s) = t$ is equivalent to $(s,t) \in f$. To prove that f is one-to-one, assume that $f(a) = f(b) = c$. Alternatively, that means (a, c) and (b, c) are elements of f . We must show that $a = b$. Since $(a, b), (c, b) \in f$, (c, a) and (c, b) are in f^{-1}. By the fact that f^{-1} is a function and c cannot have two images, a and b must be equal, so f is one-to-one.

Next, to prove that f is onto, observe that for f^{-1} to be a function, it must use all of its domain, namely A. Let b be any element of A. Then b has an image under f^{-1} , $f^{-1}(b)$. Another way of writing this is $\left(b, f^{-1}(b)\right) \in f^{-1}$, By the definition of the inverse, this is equivalent to $\left(f^{-1}(b), b\right) \in f$. Hence, b is in the range of f. Since b was chosen arbitrarily, this shows that the range of f must be all of A.

($\Leftarrow$) Assume f is one-to-one and onto and we are to prove f^{-1} exists. We leave this half of the proof to the reader. □ ∎

Definition 7.3.15 Permutation. A bijection of a set A into itself is called a permutation of A. ◇

Next, we will consider the functions for which the domain and codomain are not necessarily equal. How do we define the inverse in this case?

Definition 7.3.16 Inverse of a Function (General Case). Let $f : A \to B$, If there exists a function $g : B \to A$ such that $g \circ f = i_A$ and $f \circ g = i_B$, then g is called the inverse of f and is denoted by f^{-1} , read "f inverse." ◇

Note the slightly more complicated condition for the inverse in this case because the domains of $f \circ g$ and $g \circ f$ are different if A and B are different. The proof of the following theorem isn't really very different from the special case where $A = B$.

Theorem 7.3.17 When does a function have an inverse? *Let* $f : A \to B$. f^{-1} *exists if and only if* f *is a bijection.*

Example 7.3.18 Another inverse. Let $A = \{1, 2, 3\}$ and $B = \{a, b, c\}$. Define $f : A \to B$ by $f(1) = a$, $f(2) = b$, and $f(3) = c$. Then $g : B \to A$ defined by $g(a) = 1$, $g(b) = 2$, and $g(c) = 3$ is the inverse of f.

$$\left.\begin{array}{l}(g \circ f)(1) = 1 \\ (g \circ f)(2) = 2 \\ (g \circ f)(3) = 3\end{array}\right\} \Rightarrow g \circ f = i_A \text{ and } \left.\begin{array}{l}(f \circ g)(a) = a \\ (f \circ g)(b) = b \\ (f \circ g)(c) = c\end{array}\right\} \Rightarrow f \circ g = i_B$$

□

7.3.4 Exercises for Section 7.3

1. Let $A = \{1, 2, 3, 4, 5\}$, $B = \{a, b, c, d, e, f\}$, and $C = \{+, -\}$. Define $f : A \to B$ by $f(k)$ equal to the kth letter in the alphabet, and define $g : B \to C$ by $g(\alpha) = +$ if α is a vowel and $g(\alpha) = -$ if α is a consonant.

   (a) Find $g \circ f$.

   (b) Does it make sense to discuss $f \circ g$? If not, why not?

   (c) Does f^{-1} exist? Why?

   (d) Does g^{-1} exist? Why?

2. Let $A = \{1, 2, 3\}$. Define $f : A \to A$ by $f(1) = 2$, $f(2) = 1$, and $f(3) = 3$. Find f^2, f^3, f^4 and f^{-1}.

3. Let $A = \{1, 2, 3\}$.

   (a) List all permutations of A.

   (b) Find the inverse and square of each of the permutations of part a, where the square of a permutation, f, is the composition $f \circ f$.

   (c) Show that the composition of any two permutations of A is a permutation of A.

   (d) Prove that if A is any set where $|A| = n$, then the number of permutations of A is $n!$.

4. Define s, u, and d, all functions on the integers, by $s(n) = n^2$, $u(n) = n+1$, and $d(n) = n - 1$. Determine:

   (a) $u \circ s \circ d$

   (b) $s \circ u \circ d$

(c) $d \circ s \circ u$

5. Based on the definition of the identity function, show that for all functions $f : A \to A$, $f \circ i = i \circ f = f$.

6. **Inverse images.** If f is any function from A into B, we can describe the inverse image as a function from B into $\mathcal{P}(A)$, which is also commonly denoted f^{-1}. If $b \in B$, $f^{-1}(b) = \{a \in A \mid f(a) = b\}$. If f does have an inverse, the inverse image of b is $\{f^{-1}(b)\}$.

    (a) Let $g : \mathbb{R} \to \mathbb{R}$ be defined by $g(x) = x^2$. What are $g^{-1}(4)$, $g^{-1}(0)$ and $g^{-1}(-1)$?

    (b) If $r : \mathbb{R} \to \mathbb{Z}$, where $r(x) = \lceil x \rceil$, what is $r^{-1}(1)$?

7. Let f, g, and h all be functions from $\mathbb{Z}$ into $\mathbb{Z}$ defined by $f(n) = n + 5$, $g(n) = n - 2$, and $h(n) = n^2$. Define:

    (a) $f \circ g$

    (b) f^3

    (c) $f \circ h$

8. Define the following functions on the integers by $f(k) = k + 1$, $g(k) = 2k$, and $h(k) = \lceil k/2 \rceil$

    (a) Which of these functions are one-to-one?

    (b) Which of these functions are onto?

    (c) Express in simplest terms the compositions $f \circ g$, $g \circ f$, $g \circ h$, $h \circ g$, and h^2 ,

9. Let A be a nonempty set. Prove that if f is a bijection on A and $f \circ f = f$, then f is the identity function, i

    Hint. You have seen a similar proof in matrix algebra.

10. For the real matrix $A = \begin{pmatrix} a & b \\ c & d \end{pmatrix}$, $\det(A) = ad - bc$.

    Recall that a **bijection** from a set to itself is also referred to as a **permutation** of the set. Let π be a permutation of $\{a, b, c, d\}$ such that a becomes $\pi(a)$, b becomes $\pi(b)$, etc.

    Let $B = \begin{pmatrix} \pi(a) & \pi(b) \\ \pi(c) & \pi(d) \end{pmatrix}$. How many permutations of π leave the determinant of A invariant, that is, $\det A = \det B$?

11. State and prove a theorem on inverse functions analogous to the one that says that if a matrix has an inverse, that inverse is unique.

12. Let f and g be functions whose inverses exist. Prove that $(f \circ g)^{-1} = g^{-1} \circ f^{-1}$.

    Hint. See Exercise 3 of Section 5.4.

13. Prove Theorem 7.3.6 and Theorem 7.3.7.

14. Prove the second half of Theorem 7.3.14.

15. Prove by induction that if $n \geq 2$ and f_1, f_2 , \ldots , f_n are invertible functions on some nonempty set A, then $\{ \}(f_1 \circ f_2 \circ \cdots \circ f_n)^{-1} = f_n^{-1} \circ \cdots \circ f_2^{-1} \circ f_1^{-1}$. The basis has been taken care of in Exercise 10.

16.

    (a) Our definition of cardinality states that two sets, A and B, have the same cardinality if there exists a bijection between the two sets. Why does it not matter whether the bijection is from A into B or B into A?

    (b) Prove that "has the same cardinality as" is an equivalence relation on sets.

17. Construct a table listing as many "Laws of Function Composition" as you can identify. Use previous lists of laws as a guide.

Chapter 8

Recursion and Recurrence Relations

An essential tool that anyone interested in computer science must master is how to think recursively. The ability to understand definitions, concepts, algorithms, etc., that are presented recursively and the ability to put thoughts into a recursive framework are essential in computer science. One of our goals in this chapter is to help the reader become more comfortable with recursion in its commonly encountered forms.

A second goal is to discuss recurrence relations. We will concentrate on methods of solving recurrence relations, including an introduction to generating functions.

8.1 The Many Faces of Recursion

Consider the following definitions, all of which should be somewhat familiar to you. When reading them, concentrate on how they are similar.

8.1.1 Binomial Coefficients

Here is a recursive definition of binomial coefficients, which we introduced in Chapter 2.

Definition 8.1.1 Binomial Coefficient - Recursion Definition. Assume $n \geq 0$ and $n \geq k \geq 0$. We define $\binom{n}{k}$ by

- $\binom{n}{0} = 1$

- $\binom{n}{n} = 1$ and

- $\binom{n}{k} = \binom{n-1}{k} + \binom{n-1}{k-1}$ if $n > k > 0$

$\Diamond$

Observation 8.1.2 A word about definitions: Strictly speaking, when mathematical objects such as binomial coefficents are defined, they should be defined just once. Since we defined binomial coefficients earlier, in Definition 2.4.3, other statements describing them should be theorems. The theorem, in this case, would be that the "definition" above is consistent with the original definition. Our point in this chapter in discussing recursion is to observe alternative definitions that have a recursive nature. In the exercises, you will have the opportunity to prove that the two definitions are indeed equivalent.

Here is how we can apply the recursive definition to compute $\binom{5}{2}$.

$$
\begin{aligned}
\binom{5}{2} &= \binom{4}{2} + \binom{4}{1} \\
&= (\binom{3}{2} + \binom{3}{1}) + (\binom{3}{1} + \binom{3}{0}) \\
&= \binom{3}{2} + 2\binom{3}{1} + 1 \\
&= (\binom{2}{2} + \binom{2}{1}) + 2(\binom{2}{1} + \binom{2}{0}) + 1 \\
&= (1 + \binom{2}{1}) + 2(\binom{2}{1} + 1) + 1 \\
&= 3\binom{2}{1} + 4 \\
&= 3(\binom{1}{1} + \binom{1}{0}) + 4 \\
&= 3(1 + 1) + 4 = 10
\end{aligned}
$$

8.1.2 Polynomials and Their Evaluation

Definition 8.1.3 Polynomial Expression in x over S (Non-Recursive). Let n be an integer, $n \geq 0$. An nth degree polynomial in x is an expression of the form $a_n x^n + a_{n-1} x^{n-1} + \cdots + a_1 x + a_0$, where $a_n, a_{n-1}, \ldots, a_1, a_0$ are elements of some designated set of numbers, S, called the set of coefficients and $a_n \neq 0$. ◇

We refer to x as a variable here, although the more precise term for x is an *indeterminate*. There is a distinction between the terms indeterminate and variable, but that distinction will not come into play in our discussions.

Zeroth degree polynomials are called constant polynomials and are simply elements of the set of coefficients.

This definition is often introduced in algebra courses to describe expressions such as $f(n) = 4n^3 + 2n^2 - 8n + 9$, a third-degree, or cubic, polynomial in n. This definition has a drawback when the variable is given a value and the expression must be evaluated. For example, suppose that $n = 7$. Your first impulse is likely to do this:

$$
\begin{aligned}
f(7) &= 4 \cdot 7^3 + 2 \cdot 7^2 - 8 \cdot 7 + 9 \\
&= 4 \cdot 343 + 2 \cdot 49 - 8 \cdot 7 + 9 \\
&= 1423
\end{aligned}
$$

A count of the number of operations performed shows that five multiplications and three additions/subtractions were performed. The first two multiplications compute 7^2 and 7^3, and the last three multiply the powers of 7 times the coefficients. This gives you the four terms; and adding/subtracting a list of k numbers requires $k - 1$ addition/subtractions. The following definition of a polynomial expression suggests another more efficient method of evaluation.

Definition 8.1.4 Polynomial Expression in x over S (Recursive). Let S be a set of coefficients and x a variable.

(a) A zeroth degree polynomial expression in x over S is a nonzero element of S.

(b) For $n \geq 1$, an nth degree polynomial expression in x over S is an expression of the form $p(x)x + a$ where $p(x)$ is an $(n-1)^{st}$ degree polynomial expression in x and $a \in S$.

◊

We can easily verify that $f(n) = 4n^3 + 2n^2 - 8n + 9$ is a third-degree polynomial expression in n over $\mathbb{Z}$ based on this definition:

$$f(n) = 4n^3 + 2n^2 - 8n + 9 = ((4n + 2)n - 8)n + 9$$

Notice that 4 is a zeroth degree polynomial since it is an integer. Therefore $4n + 2$ is a first-degree polynomial; therefore, $(4n + 2)n - 8$ is a second-degree polynomial in n over $\mathbb{Z}$; therefore, $f(n)$ is a third-degree polynomial in n over $\mathbb{Z}$. The final expression for $f(n)$ is called its **telescoping form**. If we use it to calculate $f(7)$, we need only three multiplications and three additions/subtractions. This is called **Horner's method** for evaluating a polynomial expression.

Example 8.1.5 More Telescoping Polynomials.

(a) The telescoping form of $p(x) = 5x^4 + 12x^3 - 6x^2 + x + 6$ is $(((5x + 12)x - 6)x + 1)x + 6$. Using Horner's method, computing the value of $p(c)$ requires four multiplications and four additions/subtractions for any real number c.

(b) $g(x) = -x^5 + 3x^4 + 2x^2 + x$ has the telescoping form $((((-x + 3)x)x + 2)x + 1)x$.

□

Many computer languages represent polynomials as lists of coefficients, usually starting with the constant term. For example, $g(x) = -x^5 + 3x^4 + 2x^2 + x$ would be represented with the list $\{0, 1, 2, 0, 3, -1\}$. In both Mathematica and Sage, polynomial expressions can be entered and manipulated, so the list representation is only internal. Some programming languages require users to program polynomial operations with lists. We will leave these programming issues to another source.

8.1.3 Recursive Searching - The Binary Search

Next, we consider a recursive algorithm for a binary search within a sorted list of items. Suppose $r = \{r(1), r(2), \ldots, r(n)\}$ represent a list of n items sorted by a numeric key in descending order. The jth item is denoted $r(j)$ and its key value by $r(j)$.key. For example, each item might contain data on the buildings in a city and the key value might be the height of the building. Then $r(1)$ would be the item for the tallest building and $r(1)$.key would be its height. The algorithm $\mathsf{BinarySearch}(j, k)$ can be applied to search for an item in r with key value C. This would be accomplished by the execution of $\mathsf{BinarySearch}(1, n)$. When the algorithm is completed, the variable Found will have a value of true if an item with the desired key value was found, and the value of location will be the index of an item whose key is C. If Found keeps the value `false`, no such item exists in the list. The general idea behind the algorithm is illustrated in Figure 8.1.6

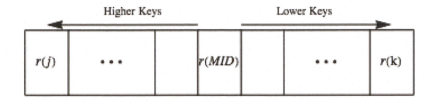

Figure 8.1.6: General Scheme for a Binary Search

In the following implementation of the Binary Search in SageMath, we search within a sorted list of integers. Therefore, the items themselves are the keys.

```
def BinarySearch(r,j,k,C):
    found = False
    if j <= k:
        mid = floor((j + k)/2)
        print 'probing_at_position_'+str(mid)
        if r[mid] == C:
            location = mid
            found = True
            print 'found_in_position_'+str(location)
            return location
        else:
            if r[mid] > C:
                BinarySearch(r,j, mid - 1,C)
            else:
                BinarySearch(r,mid + 1,k,C)
    else:
        print 'not_found'
        return False
s=[1,9,13,16,30,31,32,33,36,37,38,45,49,50,52,61,63,64,69,77,79,80,81,83,86,90,93,96]
BinarySearch(s,0,len(s)-1,30)
```

```
probing at position 13
probing at position 6
probing at position 2
probing at position 4
found in position 4
```

8.1.4 Recursively Defined Sequences

For the next two examples, consider a sequence of numbers to be a list of numbers consisting of a zeroth number, first number, second number, ... If a sequence is given the name S, the kth number of S is usually written S_k or $S(k)$.

Example 8.1.7 Geometric Growth Sequence. Define the sequence of numbers B by

$$B_0 = 100 \text{ and}$$

$$B_k = 1.08B_{k-1} \text{ for } k \geq 1.$$

These rules stipulate that each number in the list is 1.08 times the previous number, with the starting number equal to 100. For example

$$B_3 = 1.08 B_2$$
$$= 1.08 \left(1.08 B_1\right)$$
$$= 1.08 \left(1.08 \left(1.08 B_0\right)\right)$$
$$= 1.08(1.08(1.08 \cdot 100))$$
$$= 1.08^3 \cdot 100 = 125.971$$

$\square$

Example 8.1.8 The Fibonacci Sequence. The Fibonacci sequence is the sequence F defined by

$$F_0 = 1, \ F_1 = 1 \text{ and}$$
$$F_k = F_{k-2} + F_{k-1} \text{ for } k \geq 2$$

$\square$

8.1.5 Recursion

All of the previous examples were presented recursively. That is, every "object" is described in one of two forms. One form is by a simple definition, which is usually called the basis for the recursion. The second form is by a recursive description in which objects are described in terms of themselves, with the following qualification. What is essential for a proper use of recursion is that the objects can be expressed in terms of simpler objects, where "simpler" means closer to the basis of the recursion. To avoid what might be considered a circular definition, the basis must be reached after a finite number of applications of the recursion.

To determine, for example, the fourth item in the Fibonacci sequence we repeatedly apply the recursive rule for F until we are left with an expression involving F_0 and F_1:

$$F_4 = F_2 + F_3$$
$$= (F_0 + F_1) + (F_1 + F_2)$$
$$= (F_0 + F_1) + (F_1 + (F_0 + F_1))$$
$$= (1 + 1) + (1 + (1 + 1))$$
$$= 5$$

8.1.6 Iteration

On the other hand, we could compute a term in the Fibonacci sequence such as F_5 by starting with the basis terms and working forward as follows:

$$F_2 = F_0 + F_1 = 1 + 1 = 2$$
$$F_3 = F_1 + F_2 = 1 + 2 = 3$$
$$F_4 = F_2 + F_3 = 2 + 3 = 5$$
$$F_5 = F_3 + F_4 = 3 + 5 = 8$$

Table 8.1.9

This is called an iterative computation of the Fibonacci sequence. Here we start with the basis and work our way forward to a less simple number, such as 5. Try to compute F_5 using the recursive definition for F as we did for F_4. It will take much more time than it would have taken to do the computations above. Iterative computations usually tend to be faster than computations that apply recursion. Therefore, one useful skill is being able to convert a recursive formula into a nonrecursive formula, such as one that requires only iteration or a faster method, if possible.

An iterative formula for $\binom{n}{k}$ is also much more efficient than an application of the recursive definition. The recursive definition is not without its merits, however. First, the recursive equation is often useful in manipulating algebraic expressions involving binomial coefficients. Second, it gives us an insight into the combinatoric interpretation of $\binom{n}{k}$. In choosing k elements from $\{1, 2, ..., n\}$, there are $\binom{n-1}{k}$ ways of choosing all k from $\{1, 2, ..., n-1\}$, and there are $\binom{n-1}{k-1}$ ways of choosing the k elements if n is to be selected and the remaining $k-1$ elements come from $\{1, 2, ..., n-1\}$. Note how we used the Law of Addition from Chapter 2 in our reasoning.

BinarySearch Revisited. In the binary search algorithm, the place where recursion is used is easy to pick out. When an item is examined and the key is not the one you want, the search is cut down to a sublist of no more than half the number of items that you were searching in before. Obviously, this is a simpler search. The basis is hidden in the algorithm. The two cases that complete the search can be thought of as the basis. Either you find an item that you want, or the sublist that you have been left to search in is empty, when $j > k$.

BinarySearch can be translated without much difficulty into any language that allows recursive calls to its subprograms. The advantage to such a program is that its coding would be much shorter than a nonrecursive program that does a binary search. However, in most cases the recursive version will be slower and require more memory at execution time.

8.1.7 Induction and Recursion

The definition of the positive integers in terms of Peano's Postulates is a recursive definition. The basis element is the number 1 and the recursion is that if n is a positive integer, then so is its successor. In this case, n is the simple object and the recursion is of a forward type. Of course, the validity of an induction proof is based on our acceptance of this definition. Therefore, the appearance of induction proofs when recursion is used is no coincidence.

Example 8.1.10 Proof of a formula for B. A formula for the sequence B in Example 8.1.7 is $B = 100(1.08)^k$ for $k \geq 0$. A proof by induction follow.

If $k = 0$, then $B = 100(1.08)^0 = 100$, as defined. Now assume that for some $k \geq 1$, the formula for B_k is true.

$$B_{k+1} = 1.08 B_k \text{ by the recursive definition}$$
$$= 1.08 \left(100(1.08)^k\right) \text{ by the induction hypothesis}$$
$$= 100(1.08)^{k+1}$$

hence the formula is true for $k + 1$

The formula that we have just proven for B is called a closed form expression. It involves no recursion or summation signs.   □

Definition 8.1.11 Closed Form Expression. Let $E = E(x_1, x_2, \ldots, x_n)$ be an algebraic expression involving variables $x_1, x_2, \ldots, x_n$ which are allowed to take on values from some predetermined set. E is a **closed form expression**

if there exists a number T such that the evaluation of E with any allowed values of the variables will take no more than T operations (alternatively, T time units). ◇

Example 8.1.12 Reducing a summation to closed form. The sum $E(n) = \sum_{k=1}^{n} k$ is not a closed form expression because the number of additions needed to evaluate $E(n)$ grows indefinitely with n. A closed form expression that computes the value of $E(n)$ is $\frac{n(n+1)}{2}$, which only requires $T = 3$ operations. □

8.1.8 Exercises for Section 8.1

1. By the recursive definition of binomial coefficients, $\binom{7}{2} = \binom{6}{2} + \binom{6}{1}$. Continue expanding $\binom{7}{2}$ to express it in terms of quantities defined by the basis. Check your result by applying the factorial definition of $\binom{n}{k}$.

2. Define the sequence L by $L_0 = 5$ and for $k \geq 1$, $L_k = 2L_{k-1} - 7$. Determine L_4 and prove by induction that $L_k = 7 - 2^{k+1}$.

3. Let $p(x) = x^5 + 3x^4 - 15x^3 + x - 10$.

   (a) Write $p(x)$ in telescoping form.

   (b) Use a calculator to compute $p(3)$ using the original form of $p(x)$.

   (c) Use a calculator to compute $p(3)$ using the telescoping form of $p(x)$.

   (d) Compare your speed in parts b and c.

4. Suppose that a list of nine items, $(r(l), r(2), ..., r(9))$, is sorted by key in decending order so that $r(3).\mathsf{key} = 12$ and $r(4).\mathsf{key} = 10$. List the executions of the BinarySearch algorithms that would be needed to complete BinarySearch(1,9) when:
       (a) The search key is C $= 12$ (b) The search key is C $= 11$

   Assume that distinct items have distinct keys.

5. What is wrong with the following definition of $f : \mathbb{R} \to \mathbb{R}$? $f(0) = 1$ and $f(x) = f(x/2)/2$ if $x \neq 0$.

6. Prove the two definitions of binomials coefficients, Definition 2.4.3 and Definition 8.1.1, are equivalent.

7. Prove by induction that if $n \geq 0$, $\sum_{k=0}^{n} \binom{n}{k} = 2^n$

8.2 Sequences

8.2.1 Sequences and Ways They Are Defined

Definition 8.2.1 Sequence. A sequence is a function from the natural numbers into some predetermined set. The image of any natural number k can be written as $S(k)$ or S_k and is called the kth *term* of S. The variable k is called the *index* or *argument* of the sequence. ◇

For example, a sequence of integers would be a function $S : \mathbb{N} \to \mathbb{Z}$.

Example 8.2.2 Three sequences defined in different ways.

(a) The sequence A defined by $A(k) = k^2 - k$, $k \geq 0$, is a sequence of integers.

(b) The sequence B defined recursively by $B(0) = 2$ and $B(k) = B(k-1) + 3$ for $k \geq 1$ is a sequence of integers. The terms of B can be computed either by applying the recursion formula or by iteration. For example,

$$
\begin{aligned}
B(3) &= B(2) + 3 \\
&= (B(1) + 3) + 3 \\
&= ((B(0) + 3) + 3) + 3 \\
&= ((2 + 3) + 3) + 3 = 11
\end{aligned}
$$

or

$$
\begin{aligned}
B(1) &= B(0) + 3 = 2 + 3 = 5 \\
B(2) &= B(1) + 3 = 5 + 3 = 8 \\
B(3) &= B(2) + 3 = 8 + 3 = 11
\end{aligned}
$$

(c) Let C_r be the number of strings of 0's and 1's of length r having no consecutive zeros. These terms define a sequence C of integers.

$\square$

Remarks:

(1) A sequence is often called a *discrete function*.

(2) Although it is important to keep in mind that a sequence is a function, another useful way of visualizing a sequence is as a list. For example, the sequence A in the previous example could be written as $(0, 0, 2, 6, 12, 20, \ldots)$. Finite sequences can appear much the same way when they are the input to or output from a computer. The index of a sequence can be thought of as a time variable. Imagine the terms of a sequence flashing on a screen every second. Then s_k would be what you see in the kth second. It is convenient to use terminology like this in describing sequences. For example, the terms that precede the kth term of A would be $A(0), A(1), ..., A(k-1)$. They might be called the earlier terms.

8.2.2 A Fundamental Problem

Given the definition of any sequence, a fundamental problem that we will concern ourselves with is to devise a method for determining any specific term in a minimum amount of time. Generally, time can be equated with the number of operations needed. In counting operations, the application of a recursive formula would be considered an operation.

(a) The terms of A in Example 8.2.2 are very easy to compute because of the closed form expression. No matter what term you decide to compute, only three operations need to be performed.

(b) How to compute the terms of B is not so clear. Suppose that you wanted to know $B(100)$. One approach would be to apply the definition recursively:

$$B(100) = B(99) + 3 = (B(98) + 3) + 3 = \cdots$$

The recursion equation for B would be applied 100 times and 100 additions would then follow. To compute $B(k)$ by this method, $2k$ operations are needed. An iterative computation of $B(k)$ is an improvement:
$B(1) = B(0) + 3 = 2 + 3 = 5$

$B(2) = B(1) + 3 = 5 + 3 = 8$

etc. Only k additions are needed. This still isn't a good situation. As k gets large, we take more and more time to compute $B(k)$. The formula $B(k) = B(k-1) + 3$ is called a recurrence relation on B. The process of finding a closed form expression for $B(k)$, one that requires no more than some fixed number of operations, is called solving the recurrence relation.

(c) The determination of C_k is a standard kind of problem in combinatorics. One solution is by way of a recurrence relation. In fact, many problems in combinatorics are most easily solved by first searching for a recurrence relation and then solving it. The following observation will suggest the recurrence relation that we need to determine C_k. If $k \geq 2$, then every string of 0's and 1's with length k and no two consecutive 0's is either $1s_{k-1}$ or $01s_{k-2}$, where s_{k-1} and s_{k-2} are strings with no two consecutive 0's of length $k-1$ and $k-2$ respectively. From this observation we can see that $C_k = C_{k-2} + C_{k-1}$ for $k \geq 2$. The terms $C_0 = 1$ and $C_1 = 2$ are easy to determine by enumeration. Now, by iteration, any C_k can be easily determined. For example, $C_5 = 21$ can be computed with five additions. A closed form expression for C_k would be an improvement. Note that the recurrence relation for C_k is identical to the one for The Fibonacci Sequence. Only the basis is different.

8.2.3 Exercises for Section 8.2

1. Prove by induction that $B(k) = 3k + 2$, $k \geq 0$, is a closed form expression for the sequence B in Example 8.2.2

2.

(a) Consider sequence Q defined by $Q(k) = 2k + 9$, $k \geq 1$. Complete the table below and determine a recurrence relation that describes

k	$Q(k)$	$Q(k) - Q(k-1)$
2		
3		
4		
5		
6		
7		

$Q.$

(b) Let $A(k) = k^2 - k$, $k \geq 0$. Complete the table below and determine a recurrence relation for A.

k	$A(k)$	$A(k) - A(k-1)$	$A(k) - 2A(k-1) + A(k-2)$
2			
3			
4			
5			

3. Given k lines ($k \geq 0$) on a plane such that no two lines are parallel and no three lines meet at the same point, let $P(k)$ be the number of regions into which the lines divide the plane (including the infinite ones) (see Figure 8.2.3). Describe how the recurrence relation $P(k) = P(k-1) + k$ can be derived. Given that $P(0) = 1$, determine $P(5)$.

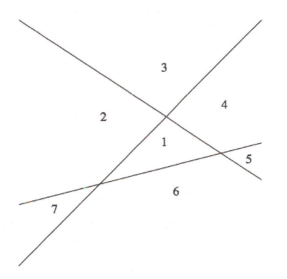

Figure 8.2.3: A general configuration of three lines

4. A sample of a radioactive substance is expected to decay by 0.15 percent each hour. If w_t, $t \geq 0$, is the weight of the sample t hours into an experiment, write a recurrence relation for w.

5. Let $M(n)$ be the number of multiplications needed to evaluate an nth degree polynomial. Use the recursive definition of a polynomial expression to define M recursively.

8.3 Recurrence Relations

In this section we will begin our study of recurrence relations and their solutions. Our primary focus will be on the class of finite order linear recurrence relations with constant coefficients (shortened to finite order linear relations). First, we will examine closed form expressions from which these relations arise. Second, we will present an algorithm for solving them. In later sections we will consider some other common relations (8.4) and introduce two additional tools for studying recurrence relations: generating functions (8.5) and matrix methods (Chapter 12).

8.3.1 Definition and Terminology

Definition 8.3.1 Recurrence Relation. Let S be a sequence of numbers. A recurrence relation on S is a formula that relates all but a finite number of terms of S to previous terms of S. That is, there is a k_0 in the domain of S such that if $k \geq k_0$, then $S(k)$ is expressed in terms of some (and possibly all) of the terms that precede $S(k)$. If the domain of S is $\{0, 1, 2, \ldots\}$, the terms $S(0), S(1), \ldots, S(k_0 - 1)$ are not defined by the recurrence formula. Their values are the initial conditions (or boundary conditions, or basis) that complete the definition of S. ◇

Example 8.3.2 Some Examples of Recurrence Relations.

(a) The Fibonacci sequence is defined by the recurrence relation $F_k = F_{k-2} + F_{k-1}$, $k \geq 2$, with the initial conditions $F_0 = 1$ and $F_1 = 1$. The recurrence relation is called a second-order relation because F_k depends on the two previous terms of F. Recall that the sequence C in Section 8.2, Example 8.2.2, can be defined with the same recurrence relation, but with different initial conditions.

(b) The relation $T(k) = 2T(k-1)^2 - kT(k-3)$ is a third-order recurrence relation. If values of $T(0)$, $T(1)$, and $T(2)$ are specified, then T is completely defined.

(c) The recurrence relation $S(n) = S(\lfloor n/2 \rfloor) + 5$, $n > 0$, with $S(0) = 0$ has infinite order. To determine $S(n)$ when n is even, you must go back $n/2$ terms. Since $n/2$ grows unbounded with n, no finite order can be given to S.

□

8.3.2 Solving Recurrence Relations

Sequences are often most easily defined with a recurrence relation; however, the calculation of terms by directly applying a recurrence relation can be time-consuming. The process of determining a closed form expression for the terms of a sequence from its recurrence relation is called solving the relation. There is no single technique or algorithm that can be used to solve all recurrence relations. In fact, some recurrence relations cannot be solved. The relation that defines T above is one such example. Most of the recurrence relations that you are likely to encounter in the future are classified as finite order linear recurrence relations with constant coefficients. This class is the one that we will spend most of our time with in this chapter.

Definition 8.3.3 n^{th} **Order Linear Recurrence Relation.** Let S be a sequence of numbers with domain $k \geq 0$. An nth order linear recurrence relation on S with constant coefficients is a recurrence relation that can be written in the form

$$S(k) + C_1 S(k-1) + ... + C_n S(k-n) = f(k) \text{ for } k \geq n$$

where $C_1, C_2, \ldots, C_n$ are constants and f is a numeric function that is defined for $k \geq n$. ◊

Note: We will shorten the name of this class of relations to n^{th} order linear relations. Therefore, in further discussions, $S(k) + 2kS(k-1) = 0$ would not be considered a first-order linear relation.

Example 8.3.4 Some Finite Order Linear Relations.

(a) The Fibonacci sequence is defined by the second-order linear relation because $F_k - F_{k-1} - F_{k-2} = 0$

(b) The relation $P(j) + 2P(j-3) = j^2$ is a third-order linear relation. In this case, $C_1 = C_2 = 0$.

(c) The relation $A(k) = 2(A(k-1)+k)$ can be written as $A(k) - 2A(k-1) = 2k$. Therefore, it is a first-order linear relation.

□

8.3.3 Recurrence relations obtained from "solutions"

Before giving an algorithm for solving finite order linear relations, we will examine recurrence relations that arise from certain closed form expressions. The closed form expressions are selected so that we will obtain finite order linear relations from them. This approach may seem a bit contrived, but if you were to write down a few simple algebraic expressions, chances are that most of them would be similar to the ones we are about to examine.

For our first example, consider D, defined by $D(k) = 5 \cdot 2^k$, $k \geq 0$. If $k \geq 1$, $D(k) = 5 \cdot 2^k = 2 \cdot 5 \cdot 2^{k-1} = 2D(k-1)$. Therefore, D satisfies the first order linear relation $D(k) - 2D(k-1) = 0$ and the initial condition $D(0) = 5$ serves as an initial condition for D.

As a second example, consider $C(k) = 3^{k-1} + 2^{k+1} + k$, $k \geq 0$. Quite a bit more algebraic manipulation is required to get our result:

$C(k) = 3^{k-1} + 2^{k+1} + k$	Original equation
$3C(k-1) = 3^{k-1} + 3 \cdot 2^k + 3(k-1)$	Substitute $k-1$ for k and multiply by 3
	Subtract the second equation from the first.
$C(k) - 3C(k-1) = -2^k - 2k + 3$	3^{k-1} term is eliminated. This is a first order relation.
$2C(k-1) - 6C(k-2) = -2^k - 2(2(k-1)) + 6$	Substitute $k-1$ for k in the third equation, multiply by 2. Subtract the 4th equation from the 3rd.
$C(k) - 5C(k-1) + 6C(k-2) = 2k - 7$	2^{k+1} term is eliminated. This is 2nd order relation.

Table 8.3.5

The recurrence relation that we have just obtained, defined for $k \geq 2$, together with the initial conditions $C(0) = 7/3$ and $C(1) = 6$, define C.

Table 8.3.6 summarizes our results together with a few other examples that we will let the reader derive. Based on these results, we might conjecture that any closed form expression for a sequence that combines exponential expressions and polynomial expressions will be solutions of finite order linear relations. Not only is this true, but the converse is true: a finite order linear relation defines a closed form expression that is similar to the ones that were just examined. The only additional information that is needed is a set of initial conditions.

Closed Form Expression	Recurrence Relation
$D(k) = 5 \cdot 2^k$	$D(k) - 2D(k-1) = 0$
$C(k) = 3^{k-1} + 2^{k+1} + k$	$C(k) - 2C(k-1) - 6C(k-2) = 2k - 7$
$Q(k) = 2k + 9$	$Q(k) - Q(k-1) = 2$
$A(k) = k^2 - k$	$A(k) - 2A(k-1) + A(k-2) = 2$
$B(k) = 2k^2 + 1$	$B(k) - 2B(k-1) + B(k-2) = 4$
$G(k) = 2 \cdot 4^k - 5(-3)^k$	$G(k) - G(k-1) + 12G(k-2) = 0$
$J(k) = (3 + k)2^k$	$J(k) - 4J(k-1) + 4J(k-2) = 0$

Table 8.3.6: Recurrence relations obtained from given sequences

Definition 8.3.7 Homogeneous Recurrence Relation. An nth order linear relation is homogeneous if $f(k) = 0$ for all k. For each recurrence relation $S(k) + C_1 S(k-1) + \ldots + C_n S(k-n) = f(k)$, the associated homogeneous relation is $S(k) + C_1 S(k-1) + \ldots + C_n S(k-n) = 0$ $\diamond$

Example 8.3.8 First Order Homogeneous Recurrence Relations. $D(k) - 2D(k-1) = 0$ is a first-order homogeneous relation. Since it can also be written as $D(k) = 2D(k-1)$, it should be no surprise that it arose from an expression that involves powers of 2. More generally, you would expect that the solution of $L(k) - aL(k-1)$ would involve a^k. Actually, the solution is $L(k) = L(0)a^k$, where the value of $L(0)$ is given by the initial condition. $\square$

Example 8.3.9 A Second Order Example. Consider the second-order homogeneous relation $S(k) - 7S(k-1) + 12S(k-2) = 0$ together with the initial conditions $S(0) = 4$ and $S(1) = 4$. From our discussion above, we can predict that the solution to this relation involves terms of the form ba^k, where b and a are nonzero constants that must be determined. If the solution were to equal this quantity exactly, then

$$S(k) = ba^k$$
$$S(k-1) = ba^{k-1}$$
$$S(k-2) = ba^{k-2}$$

Substitute these expressions into the recurrence relation to get

$$ba^k - 7ba^{k-1} + 12ba^{k-2} = 0$$

Each term on the left-hand side of this equation has a factor of ba^{k-2}, which is nonzero. Dividing through by this common factor yields

$$a^2 - 7a + 12 = (a-3)(a-4) = 0 \tag{8.3.1}$$

Therefore, the only possible values of a are 3 and 4. Equation (8.3.1) is called the characteristic equation of the recurrence relation. The fact is that our original recurrence relation is true for any sequence of the form $S(k) = b_1 3^k + b_2 4^k$, where b_1 and b_2 are real numbers. This set of sequences is called the general solution of the recurrence relation. If we didn't have initial conditions for S, we would stop here. The initial conditions make it possible for us to find definite values for b_1 and b_2.

$$\left\{ \begin{array}{l} S(0) = 4 \\ S(1) = 4 \end{array} \right\} \Rightarrow \left\{ \begin{array}{l} b_1 3^0 + b_2 4^0 = 4 \\ b_1 3^1 + b_2 4^1 = 4 \end{array} \right\} \Rightarrow \left\{ \begin{array}{l} b_1 + b_2 = 4 \\ 3b_1 + 4b_2 = 4 \end{array} \right\}$$

The solution of this set of simultaneous equations is $b_1 = 12$ and $b_2 = -8$ and so the solution is $S(k) = 12 \cdot 3^k - 8 \cdot 4^k$. $\square$

Definition 8.3.10 Characteristic Equation. The characteristic equation of the homogeneous n^{th} order linear relation $S(k) + C_1 S(k-1) + \ldots + C_n S(k-n) = 0$ is the nth degree polynomial equation

$$a^n + \sum_{j=1}^{n} C_j a^{n-j} = a^n + C_1 a^{n-1} + \cdots + C_{n-1} a + C_n = 0$$

The left-hand side of this equation is called the characteristic polynomial. The roots of the characteristic polynomial are called the characteristic roots of the equation. $\diamond$

Example 8.3.11 Some characteristic equations.

(a) The characteristic equation of $F(k) - F(k-1) - F(k-2) = 0$ is $a^2 - a - 1 = 0$.

(b) The characteristic equation of $Q(k) + 2Q(k-1) - 3Q(k-2) - 6Q(k-4) = 0$ is $a^4 + 2a^3 - 3a^2 - 6 = 0$. Note that the absence of a $Q(k-3)$ term means that there is not an $x^{4-3} = x$ term appearing in the characteristic equation.

$\square$

Algorithm 8.3.12 Algorithm for Solving Homogeneous Finite-order Linear Relations.

(a) *Write out the characteristic equation of the relation $S(k) + C_1 S(k-1) + \ldots + C_n S(k-n) = 0$, which is $a^n + C_1 a^{n-1} + \cdots + C_{n-1}a + C_n = 0$.*

(b) *Find all roots of the characteristic equation, the characteristic roots.*

(c) *If there are n distinct characteristic roots, $a_1, a_2, \ldots a_n$, then the general solution of the recurrence relation is $S(k) = b_1 a_1{}^k + b_2 a_2{}^k + \cdots + b_n a_n{}^k$. If there are fewer than n characteristic roots, then at least one root is a multiple root. If a_j is a double root, then the $b_j a_j{}^k$ term is replaced with $(b_{j0} + b_{j1}k) a_j^k$. In general, if a_j is a root of multiplicity p, then the $b_j a_j{}^k$ term is replaced with $\left(b_{j0} + b_{j1}k + \cdots + b_{j(p-1)}k^{p-1} \right) a_j^k$.*

(d) *If n initial conditions are given, we get n linear equations in n unknowns (the b_j's from Step 3) by substitution. If possible, solve these equations to determine a final form for $S(k)$.*

Although this algorithm is valid for all values of n, there are limits to the size of n for which the algorithm is feasible. Using just a pencil and paper, we can always solve second-order equations. The quadratic formula for the roots of $ax^2 + bx + c = 0$ is

$$x = \frac{-b \pm \sqrt{b^2 - 4ac}}{2a}$$

The solutions of $a^2 + C_1 a + C_2 = 0$ are then

$$\frac{1}{2} \left(-C_1 + \sqrt{C_1{}^2 - 4C_2} \right) \text{ and } \frac{1}{2} \left(-C_1 - \sqrt{C_1{}^2 - 4C_2} \right)$$

Although cubic and quartic formulas exist, they are too lengthy to introduce here. For this reason, the only higher-order relations ($n \geq 3$) that you could be expected to solve by hand are ones for which there is an easy factorization of the characteristic polynomial.

Example 8.3.13 A solution using the algorithm. Suppose that T is defined by $T(k) = 7T(k-1) - 10T(k-2)$, with $T(0) = 4$ and $T(1) = 17$. We can solve this recurrence relation with Algorithm 8.3.12:

(a) Note that we have written the recurrence relation in "nonstandard" form. To avoid errors in this easy step, you might consider a rearrangement of the equation to, in this case, $T(k) - 7T(k-1) + 10T(k-2) = 0$. Therefore, the characteristic equation is $a^2 - 7a + 10 = 0$.

(b) The characteristic roots are $\frac{1}{2} \left(7 + \sqrt{49 - 40} \right) = 5$ and $\frac{1}{2} \left(7 - \sqrt{49 - 40} \right) = 2$. These roots can be just as easily obtained by factoring the characteristic polynomial into $(a - 5)(a - 2)$.

(c) The general solution of the recurrence relation is $T(k) = b_1 2^k + b_2 5^k$.

(d) $\left\{ \begin{array}{l} T(0) = 4 \\ T(1) = 17 \end{array} \right\} \Rightarrow \left\{ \begin{array}{l} b_1 2^0 + b_2 5^0 = 4 \\ b_1 2^1 + b_2 5^1 = 17 \end{array} \right\} \Rightarrow \left\{ \begin{array}{l} b_1 + b_2 = 4 \\ 2b_1 + 5b_2 = 17 \end{array} \right\}$

The simultaneous equations have the solution $b_1 = 1$ and $b_2 = 3$. Therefore, $T(k) = 2^k + 3 \cdot 5^k$.

□

Here is one rule that might come in handy: If the coefficients of the characteristic polynomial are all integers, with the constant term equal to m, then the only possible rational characteristic roots are divisors of m (both positive and negative).

With the aid of a computer (or possibly only a calculator), we can increase n. Approximations of the characteristic roots can be obtained by any of several well-known methods, some of which are part of standard software packages. There is no general rule that specifies the values of n for which numerical approximations will be feasible. The accuracy that you get will depend on the relation that you try to solve. (See Exercise 17 of this section.)

Example 8.3.14 Solution of a Third Order Recurrence Relation. Solve $S(k) - 7S(k - 2) + 6S(k - 3) = 0$, where $S(0) = 8$, $S(1) = 6$, and $S(2) = 22$.

(a) The characteristic equation is $a^3 - 7a + 6 = 0$.

(b) The only rational roots that we can attempt are $\pm 1, \pm 2, \pm 3,$ and ± 6. By checking these, we obtain the three roots 1, 2, and -3.

(c) The general solution is $S(k) = b_1 1^k + b_2 2^k + b_3 (-3)^k$. The first term can simply be written b_1 .

(d) $\left\{ \begin{array}{l} S(0) = 8 \\ S(1) = 6 \\ S(2) = 22 \end{array} \right\} \Rightarrow \left\{ \begin{array}{l} b_1 + b_2 + b_3 = 8 \\ b_1 + 2b_2 - 3b_3 = 6 \\ b_1 + 4b_2 + 9b_3 = 22 \end{array} \right\}$ You can solve this system by elimination to obtain $b_1 = 5$, $b_2 = 2$, and $b_3 = 1$. Therefore, $S(k) = 5 + 2 \cdot 2^k + (-3)^k = 5 + 2^{k+1} + (-3)^k$

□

Example 8.3.15 Solution with a Double Characteristic Root. Solve $D(k) - 8D(k - 1) + 16D(k - 2) = 0$, where $D(2) = 16$ and $D(3) = 80$.

(a) Characteristic equation: $a^2 - 8a + 16 = 0$.

(b) $a^2 - 8a + 16 = (a - 4)^2$. Therefore, there is a double characteristic root, 4.

(c) General solution: $D(k) = (b_{10} + b_{11}k) 4^k$.

(d) $\left\{ \begin{array}{l} D(2) = 16 \\ D(3) = 80 \end{array} \right\} \Rightarrow \left\{ \begin{array}{l} (b_{10} + b_{11}2) 4^2 = 16 \\ (b_{10} + b_{11}3) 4^3 = 80 \end{array} \right\}$

$\Rightarrow \left\{ \begin{array}{l} 16b_{10} + 32b_{11} = 16 \\ 64b_{10} + 192b_{11} = 80 \end{array} \right\} \Rightarrow \left\{ \begin{array}{l} b_{10} = \frac{1}{2} \\ b_{11} = \frac{1}{4} \end{array} \right\}$

Therefore $D(k) = (1/2 + (1/4)k)4^k = (2 + k)4^{k-1}$.

□

8.3.4 Solution of Nonhomogeneous Finite Order Linear Relations

Our algorithm for nonhomogeneous relations will not be as complete as for the homogeneous case. This is due to the fact that different right-hand sides ($f(k)$'s) call for different rules in obtaining a particular solution.

Algorithm 8.3.16 Algorithm for Solving Nonhomogeneous Finite-order Linear Relations. *To solve the recurrence relation $S(k) + C_1 S(k - 1) + \ldots + C_n S(k - n) = f(k)$*

(1) *Write the associated homogeneous relation and find its general solution (Steps (a) through (c) of Algorithm 8.3.12). Call this the homogeneous solution, $S^{(h)}(k)$.*

(2) *Start to obtain what is called a particular solution, $S^{(p)}(k)$ of the recurrence relation by taking an educated guess at the form of a particular solution. For a large class of right-hand sides, this is not really a guess, since the particular solution is often the same type of function as $f(k)$ (see Table 8.3.17).*

(3) *Substitute your guess from Step 2 into the recurrence relation. If you made a good guess, you should be able to determine the unknown coefficients of your guess. If you made a wrong guess, it should be apparent from the result of this substitution, so go back to Step 2.*

(4) *The general solution of the recurrence relation is the sum of the homogeneous and particular solutions. If no conditions are given, then you are finished. If n initial conditions are given, they will translate to n linear equations in n unknowns and solve the system to get a complete solution.*

Right Hand Side, $f(k)$	Form of Particular Solution, $S^{(p)}(k)$
Constant, q	Constant, d
Linear Function, $q_0 + q_1 k$	Linear Function, $d_0 + d_1 k$
m^{th} degree polynomial, $q_0 + q_1 k + \cdots + q_m k^m$	m^{th} degree polynomial, $d_0 + d_1 k + \cdots + d_m k^m$
exponential function, $q a^k$	exponential function, $d a^k$

Table 8.3.17: Particular solutions for given right-hand sides

Example 8.3.18 Solution of a Nonhomogeneous First Order Recurrence Relation. Solve $S(k) + 5S(k - 1) = 9$, with $S(0) = 6$.

(a) The associated homogeneous relation, $S(k) + 5S(k-1) = 0$ has the characteristic equation $a + 5 = 0$; therefore, $a = -5$. The homogeneous solution is $S^{(h)}(k) = b(-5)^k$.

(b) Since the right-hand side is a constant, we guess that the particular solution will be a constant, d.

(c) If we substitute $S^{(p)}(k) = d$ into the recurrence relation, we get $d + 5d = 9$, or $6d = 9$. Therefore, $S^{(p)}(k) = 1.5$.

(d) The general solution of the recurrence relation is $S(k) = S^{(h)}(k) + S^{(p)}(k) = b(-5)^k + 1.5$ The initial condition will give us one equation to solve in order to determine b. $S(0) = 6 \Rightarrow b(-5)^0 + 1.5 = 6 \Rightarrow b + 1.5 = 6$ Therefore, $b = 4.5$ and $S(k) = 4.5(-5)^k + 1.5$.

Example 8.3.19 Solution of a Nonhomogeneous Second Order Recurrence Relation. Consider $T(k) - 7T(k-1) + 10T(k-2) = 6 + 8k$ with $T(0) = 1$ and $T(1) = 2$.

(a) From Example 8.3.13, we know that $T^{(h)}(k) = b_1 2^k + b_2 5^k$. Caution:Don't apply the initial conditions to $T^{(h)}$ until you add $T^{(p)}$!

(b) Since the right-hand side is a linear polynomial, $T^{(p)}$ is linear; that is, $T^{(p)}(k) = d_0 + d_1 k$.

(c) Substitution into the recurrence relation yields: $(d_0 + d_1 k) - 7(d_0 + d_1(k-1)) + 10(d_0 + d_1(k-2)) = 6 + 8k \Rightarrow (4d_0 - 13d_1) + (4d_1)k = 6 + 8k$ Two polynomials are equal only if their coefficients are equal. Therefore,
$$\left\{ \begin{array}{c} 4d_0 - 13d_1 = 6 \\ 4d_1 = 8 \end{array} \right\} \Rightarrow \left\{ \begin{array}{c} d_0 = 8 \\ d_1 = 2 \end{array} \right\}$$

(d) Use the general solution $T(k) = b_1 2^k + b_2 5^k + 8 + 2k$ and the initial conditions to get a final solution:
$$\left\{ \begin{array}{c} T(0) = 1 \\ T(1) = 2 \end{array} \right\} \Rightarrow \left\{ \begin{array}{c} b_1 + b_2 + 8 = 1 \\ 2b_1 + 5b_2 + 10 = 2 \end{array} \right\}$$
$$\Rightarrow \left\{ \begin{array}{c} b_1 + b_2 = -7 \\ 2b_1 + 5b_2 = -8 \end{array} \right\}$$
$$\Rightarrow \left\{ \begin{array}{c} b_1 = -9 \\ b_2 = 2 \end{array} \right\}$$
Therefore, $T(k) = -9 \cdot 2^k + 2 \cdot 5^k + 8 + 2k$.

Note 8.3.20 A quick note on interest rates. When a quantity, such as a savings account balance, is increased by some fixed percent, it is most easily computed with a multiplier. In the case of an 8% increase, the multiplier is 1.08 because any original amount A, has $0.08A$ added to it, so that the new balance is $A + 0.08A = (1 + 0.08)A = 1.08A$.

Another example is that if the interest rate is 3.5%, the multiplier would be 1.035. This presumes that the interest is applied at the end of year for 3.5% annual interest, often called **simple interest**. If the interest is applied monthly, and we assume a simplifed case where each month has the same length, the multiplier after every month would be $\left(1 + \frac{0.035}{12}\right) \approx 1.00292$. After a year passes, this multiplier would be applied 12 times, which is the same as multiplying by $1.00292^{12} \approx 1.03557$. That increase from 1.035 to 1.03557 is the effect of **compound interest**.

Example 8.3.21 A Sort of Annuity. Suppose you open a savings account that pays an annual interest rate of 8%. In addition, suppose you decide to deposit one dollar when you open the account, and you intend to double your deposit each year. Let $B(k)$ be your balance after k years. B can be described by the relation $B(k) = 1.08B(k-1) + 2^k$, with $S(0) = 1$. If, instead of doubling the deposit each year, you deposited a constant amount, q, the 2^k term would be replaced with q. A sequence of regular deposits such as this is called a simple annuity.

Returning to the original situation,

(a) $B^{(h)}(k) = b_1(1.08)^k$

(b) $B^{(p)}(k)$ should be of the form $d2^k$.

(c)

$$d2^k = 1.08d2^{k-1} + 2^k \Rightarrow (2d)2^{k-1} = 1.08d2^{k-1} + 2 \cdot 2^{k-1}$$
$$\Rightarrow 2d = 1.08d + 2$$
$$\Rightarrow .92d = 2$$
$$\Rightarrow d = 2.174 \text{ to the nearest thousandth)}$$

Therefore $B^{(p)}(k) = 2.174 \cdot 2^k$.

(d) $B(0) = 1 \Rightarrow b_1 + 2.174 = 1$

$\Rightarrow b_1 = -1.174$

Therefore, $B(k) = -1.174 \cdot 1.08^k + 2.174 \cdot 2^k$.

□

Example 8.3.22 Matching Roots. Find the general solution to $S(k) - 3S(k-1) - 4S(k-2) = 4^k$.

(a) The characteristic roots of the associated homogeneous relation are -1 and 4. Therefore, $S^{(h)}(k) = b_1(-1)^k + b_2 4^k$.

(b) A function of the form $d4^k$ will not be a particular solution of the nonhomogeneous relation since it solves the associated homogeneous relation. When the right-hand side involves an exponential function with a base that equals a characteristic root, you should multiply your guess at a particular solution by k. Our guess at $S^{(p)}(k)$ would then be $dk4^k$. See Observation 8.3.23 for a more complete description of this rule.

(c) Substitute $dk4^k$ into the recurrence relation for $S(k)$:

$$dk4^k - 3d(k-1)4^{k-1} - 4d(k-2)4^{k-2} = 4^k$$
$$16dk4^{k-2} - 12d(k-1)4^{k-2} - 4d(k-2)4^{k-2} = 4^k$$

Each term on the left-hand side has a factor of 4^{k-2}

$$16dk - 12d(k-1) - 4d(k-2) = 4^2 20d = 16 \Rightarrow d = 0.8$$

Therefore, $S^{(p)}(k) = 0.8k4^k$.

(d) The general solution to the recurrence relation is

$$S(k) = b_1(-1)^k + b_2 4^k + 0.8k4^k$$

□

Observation 8.3.23 When the base of right-hand side is equal to a characteristic root. If the right-hand side of a nonhomogeneous relation involves an exponential with base a, and a is also a characteristic root of multiplicity p, then multiply your guess at a particular solution as prescribed in Table 8.3.17 by k^p, where k is the index of the sequence.

Example 8.3.24 Examples of matching bases.

(a) If $S(k) - 9S(k-1) + 20S(k-2) = 2 \cdot 5^k$, the characteristic roots are 4 and 5. Since 5 matches the base of the right side, $S^{(p)}(k)$ will take the form $dk5^k$.

(b) If $S(n) - 6S(n-1) + 9S(n-2) = 3^{n+1}$ the only characteristic root is 3, but it is a double root (multiplicity 2). Therefore, the form of the particular solution is $dn^2 3^n$.

(c) If $Q(j) - Q(j-1) - 12Q(j-2) = (-3)^j + 6 \cdot 4^j$, the characteristic roots are -3 and 4. The form of the particular solution will be $d_1 j(-3)^j + d_2 j \cdot 4^j$.

(d) If $S(k) - 9S(k-1) + 8S(k-2) = 9k + 1 = (9k+1)1^k$, the characteristic roots are 1 and 8. If the right-hand side is a polynomial, as it is in this case, then the exponential factor 1^k can be introduced. The particular solution will take the form $k(d_0 + d_1 k)$.

$\square$

We conclude this section with a comment on the situation in which the characteristic equation gives rise to complex roots. If we restrict the coefficients of our finite order linear relations to real numbers, or even to integers, we can still encounter characteristic equations whose roots are complex. Here, we will simply take the time to point out that our algorithms are still valid with complex characteristic roots, but the customary method for expressing the solutions of these relations is different. Since an understanding of these representations requires some background in complex numbers, we will simply suggest that an interested reader can refer to a more advanced treatment of recurrence relations (see also difference equations).

8.3.5 Exercises for Section 8.3

Solve the following sets of recurrence relations and initial conditions:

1. $S(k) - 10S(k-1) + 9S(k-2) = 0$, $S(0) = 3$, $S(1) = 11$
2. $S(k) - 9S(k-1) + 18S(k-2) = 0$, $S(0) = 0$, $S(1) = 3$
3. $S(k) - 0.25S(k-1) = 0$, $S(0) = 6$
4. $S(k) - 20S(k-1) + 100S(k-2) = 0$, $S(0) = 2$, $S(1) = 50$
5. $S(k) - 2S(k-1) + S(k-2) = 2$, $S(0) = 25$, $S(1) = 16$
6. $S(k) - S(k-1) - 6S(k-2) = -30$, $S(0) = 7$, $S(1) = 10$
7. $S(k) - 5S(k-1) = 5^k$, $S(0) = 3$
8. $S(k) - 5S(k-1) + 6S(k-2) = 2$, $S(0) = -1$, $S(1) = 0$
9. $S(k) - 4S(k-1) + 4S(k-2) = 3k + 2^k$, $S(0) = 1, S(1) = 1$
10. $S(k) = rS(k-1) + a$, $S(0) = 0, r, a \geq 0, r \neq 1$
11. $S(k) - 4S(k-1) - 11S(k-2) + 30S(k-3) = 0$, $S(0) = 0, S(1) = -35$, $S(2) = -85$

12. Find a closed form expression for $P(k)$ in Exercise 3 of Section 8.2.

13.

(a) Find a closed form expression for the terms of the Fibonacci sequence (see Example 8.1.8).

(b) The sequence C was defined by C_r = the number of strings of zeros and ones with length r having no consecutive zeros (Example 8.2.2(c)). Its recurrence relation is the same as that of the Fibonacci sequence. Determine a closed form expression for C_r, $r \geq 1$.

14. If $S(n) = \sum_{j=1}^{n} g(j), n \geq 1$, then S can be described with the recurrence relation $S(n) = S(n-1) + g(n)$. For each of the following sequences that are defined using a summation, find a closed form expression:

   (a) $S(n) = \sum_{j=1}^{n} j$, $n \geq 1$

   (b) $Q(n) = \sum_{j=1}^{n} j^2$, $n \geq 1$

   (c) $P(n) = \sum_{j=1}^{n} \left(\frac{1}{2}\right)^j$, $n \geq 0$

   (d) $T(n) = \sum_{j=1}^{n} j^3$, $n \geq 1$

15. Let $D(n)$ be the number of ways that the set $\{1, 2, ..., n\}$, $n \geq 1$, can be partitioned into two nonempty subsets.

   (a) Find a recurrence relation for D. (Hint: It will be a first-order linear relation.)

   (b) Solve the recurrence relation.

16. If you were to deposit a certain amount of money at the end of each year for a number of years, this sequence of payments would be called an annuity (see Example 8.3.21).

   (a) Find a closed form expression for the balance or value of an annuity that consists of payments of q dollars at a rate of interest of i. Note that for a normal annuity, the first payment is made after one year.

   (b) With an interest rate of 5.5 percent, how much would you need to deposit into an annuity to have a value of one million dollars after 18 years?

   (c) The payment of a loan is a form of annuity in which the initial value is some negative amount (the amount of the loan) and the annuity ends when the value is raised to zero. How much could you borrow if you can afford to pay $5,000 per year for 25 years at 11 percent interest?

17. Suppose that C is a small positive number. Consider the recurrence relation $B(k) - 2B(k-1) + \left(1 - C^2\right) B(k-2) = C^2$, with initial conditions $B(0) = 1$ and $B(1) = 1$. If C is small enough, we might consider approximating the relation by replacing $1 - C^2$ with 1 and C^2 with 0. Solve the original relation and its approximation. Let B_a a be the solution of the approximation. Compare closed form expressions for $B(k)$ and $B_a(k)$. Their forms are very different because the characteristic roots of the original relation were close together and the approximation resulted in one double characteristic root.If characteristic roots of a relation are relatively far apart, this problem will not occur. For example, compare the general solutions of $S(k) + 1.001S(k-1) - 2.004002S(k-2) = 0.0001$ and $S_a(k) + S_a(k-1) - 2S_a(k-2) = 0$.

8.4 Some Common Recurrence Relations

In this section we intend to examine a variety of recurrence relations that are not finite-order linear with constant coefficients. For each part of this section, we will consider a concrete example, present a solution, and, if possible, examine a more general form of the original relation.

8.4.1 A First Basic Example

Consider the homogeneous first-order linear relation without constant coefficients, $S(n) - nS(n-1) = 0$, $n \geq 1$, with initial condition $S(0) = 1$. Upon close examination of this relation, we see that the nth term is n times the $(n-1)^{st}$ term, which is a property of n factorial. $S(n) = n!$ is a solution of this relation, for if $n \geq 1$,

$$S(n) = n! = n \cdot (n-1)! = n \cdot S(n-1)$$

In addition, since $0! = 1$, the initial condition is satisfied. It should be pointed out that from a computational point of view, our "solution" really isn't much of an improvement since the exact calculation of $n!$ takes $n-1$ multiplications.

If we examine a similar relation, $G(k) - 2^k G(k-1)$, $k \geq 1$ with $G(0) = 1$, a table of values for G suggests a possible solution:

k	0	1	2	3	4	5
$G(k)$	1	2	2^3	2^6	2^{10}	2^{15}

The exponent of 2 in $G(k)$ is growing according to the relation $E(k) = E(k-1) + k$, with $E(0) = 0$. Thus $E(k) = \frac{k(k+1)}{2}$ and $G(k) = 2^{k(k+1)/2}$. Note that $G(k)$ could also be written as $2^0 2^1 2^2 \cdots 2^k$, for $k \geq 0$, but this is not a closed form expression.

In general, the relation $P(n) = f(n)P(n-1)$ for $n \geq 1$ with $P(0) = f(0)$, where f is a function that is defined for all $n \geq 0$, has the "solution"

$$P(n) = \prod_{k=0}^{n} f(k)$$

This product form of $P(n)$ is not a closed form expression because as n grows, the number of multiplications grow. Thus, it is really not a true solution. Often, as for $G(k)$ above, a closed form expression can be derived from the product form.

8.4.2 An Analysis of the Binary Search Algorithm

8.4.2.1

Suppose you intend to use a binary search algorithm (see Subsection 8.1.3) on lists of zero or more sorted items, and that the items are stored in an array, so that you have easy access to each item. A natural question to ask is "How much time will it take to complete the search?" When a question like this is asked, the time we refer to is often the so-called worst-case time. That is, if we were to search through n items, what is the longest amount of time that we will need to complete the search? In order to make an analysis such as this independent of the computer to be used, time is measured by counting the number of steps that are executed. Each step (or sequence of steps) is assigned an absolute time, or weight; therefore, our answer will not be in seconds, but in absolute time units. If the steps in two different algorithms are assigned weights that are consistent, then analyses of the algorithms can be used to compare their relative efficiencies. There are two major steps that must be executed in a call of the binary search algorithm:

(1) If the lower index is less than or equal to the upper index, then the middle of the list is located and its key is compared to the value that you are searching for.

(2) In the worst case, the algorithm must be executed with a list that is roughly half as large as in the previous execution. If we assume that Step 1 takes one time unit and $T(n)$ is the worst-case time for a list of n items, then

$$T(n) = 1 + T(\lfloor n/2 \rfloor), \quad n > 0 \tag{8.4.1}$$

For simplicity, we will assume that

$$T(0) = 0 \tag{8.4.2}$$

even though the conditions of Step 1 must be evaluated as false if $n = 0$. You might wonder why $n/2$ is truncated in (8.4.1). If n is odd, then $n = 2k + 1$ for some $k \geq 0$, the middle of the list will be the $(k+1)^{st}$ item, and no matter what half of the list the search is directed to, the reduced list will have $k = \lfloor n/2 \rfloor$ items. On the other hand, if n is even, then $n = 2k$ for $k > 0$. The middle of the list will be the kth item, and the worst case will occur if we are directed to the k items that come after the middle (the $(k+1)^{st}$ through $(2k)^{th}$ items). Again the reduced list has $\lfloor n/2 \rfloor$ items.

Solution to (8.4.1) and (8.4.2). To determine $T(n)$, the easiest case is when n is a power of two. If we compute $T(2^m)$, $m \geq 0$, by iteration, our results are

$$T(1) = 1 + T(0) = 1$$
$$T(2) = 1 + T(1) = 2$$
$$T(4) = 1 + T(2) = 3$$
$$T(8) = 1 + T(4) = 4$$

The pattern that is established makes it clear that $T(2^m) = m+1$. This result would seem to indicate that every time you double the size of your list, the search time increases by only one unit.

A more complete solution can be obtained if we represent n in binary form. For each $n \geq 1$, there exists a non-negative integer r such that

$$2^{r-1} \leq n < 2^r \tag{8.4.3}$$

For example, if $n = 21$, $2^4 \leq 21 < 2^5$; therefore, $r = 5$. If n satisfies (8.4c), its binary representation requires r digits. For example, $21_{ten} = 10101_{two}$.

In general, $n = (a_1 a_2 \ldots a_r)_{two}$. where $a_1 = 1$. Note that in this form, $\lfloor n/2 \rfloor$ is easy to describe: it is the $r-1$ digit binary number $(a_1 a_2 \ldots a_{r-1})_{two}$. Therefore,

$$\begin{aligned}
T(n) &= T(a_1 a_2 \ldots a_r) \\
&= 1 + T(a_1 a_2 \ldots a_{r-1}) \\
&= 1 + (1 + T(a_1 a_2 \ldots a_{r-2})) \\
&= 2 + T(a_1 a_2 \ldots a_{r-2}) \\
&\;\;\vdots \\
&= (r-1) + T(a_1) \\
&= (r-1) + 1 \quad \text{since } T(1) = 1 \\
&= r
\end{aligned}$$

From the pattern that we've just established, $T(n)$ reduces to r. A formal inductive proof of this statement is possible. However, we expect that most readers would be satisfied with the argument above. Any skeptics are invited to provide the inductive proof.

For those who prefer to see a numeric example, suppose $n = 21$.

$$
\begin{aligned}
T(21) &= T(10101) \\
&= 1 + T(1010) \\
&= 1 + (1 + T(101)) \\
&= 1 + (1 + (1 + T(10))) \\
&= 1 + (1 + (1 + (1 + T(1)))) \\
&= 1 + (1 + (1 + (1 + (1 + T(0))))) \\
&= 5
\end{aligned}
$$

Our general conclusion is that the solution to (8.4.1) and (8.4.2) is that for $n \geq 1$, $T(n) = r$, where $2^{r-1} \leq n < 2^r$.

A less cumbersome statement of this fact is that $T(n) = \lfloor \log_2 n \rfloor + 1$. For example, $T(21) = \lfloor \log_2 21 \rfloor + 1 = 4 + 1 = 5$.

8.4.2.2   Review of Logarithms

Any discussion of logarithms must start by establishing a base, which can be any positive number other than 1. With the exception of Theorem 5, our base will be 2. We will see that the use of a different base (10 and $e \approx 2.171828$ are the other common ones) only has the effect of multiplying each logarithm by a constant. Therefore, the base that you use really isn't very important. Our choice of base 2 logarithms is convenient for the problems that we are considering.

Definition 8.4.1   Base 2 logarithm. The base 2 logarithm of a positive number represents an exponent and is defined by the following equivalence for any positive real numbers a.

$$
\log_2 a = x \quad \Leftrightarrow \quad 2^x = a.
$$

$\Diamond$

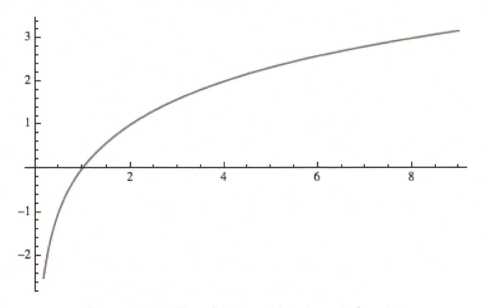

Figure 8.4.2: Plot of the logarithm, bases 2, function

For example, $\log_2 8 = 3$ because $2^3 = 8$ and $\log_2 1.414 \approx 0.5$ because $2^{0.5} \approx 1.414$. A graph of the function $f(x) = \log_2 x$ in Figure 8.4.2 shows that if $a < b$, the $\log_2 a < \log_2 b$; that is, when x increases, $\log_2 x$ also increases. However, if we move x from $2^{10} = 1024$ to $2^{11} = 2048$, $\log_2 x$ only increases from 10 to 11. This slow rate of increase of the logarithm function is an important point to remember. An algorithm acting on n pieces of data that can be executed in $\log_2 n$ time units can handle significantly larger sets of data than an algorithm that can be executed in $n/100$ or $\sqrt{n}$ time units. The graph of $T(n) = \lfloor \log_2 n \rfloor + 1$ would show the same behavior.

A few more properties that we will use in subsequent discussions involving logarithms are summarized in the following theorem.

Theorem 8.4.3 Fundamental Properties of Logarithms. *Let a and b be positive real numbers, and r a real number.*

$$\log_2 1 = 0 \tag{8.4.4}$$

$$\log_2 ab = \log_2 a + \log_2 b \tag{8.4.5}$$

$$\log_2 \frac{a}{b} = \log_2 a - \log_2 b \tag{8.4.6}$$

$$\log_2 a^r = r \log_2 a \tag{8.4.7}$$

$$2^{\log_2 a} = a \tag{8.4.8}$$

Definition 8.4.4 Logarithms base b. If $b > 0$, $b \neq 1$, then for $a > 0$,

$$\log_b a = x \Leftrightarrow b^x = a$$

$\Diamond$

Theorem 8.4.5 How logarithms with different bases are related. *Let $b > 0$, $b \neq 1$. Then for all $a > 0$, $\log_b a = \frac{\log_2 a}{\log_2 b}$. Therefore, if $b > 1$, base b logarithms can be computed from base 2 logarithms by dividing by the positive scaling factor $\log_2 b$. If $b < 1$, this scaling factor is negative.*

Proof. By an analogue of (8.4.8), $a = b^{\log_b a}$. Therefore, if we take the base 2 logarithm of both sides of this equality we get:

$$\log_2 a = \log_2 \left(b^{\log_b a} \right) \Rightarrow \log_2 a = \log_b a \cdot \log_2 b$$

Finally, divide both sides of the last equation by $\log_2 b$.   ∎

Note 8.4.6 $\log_2 10 \approx 3.32192$ and $\log_2 e \approx 1.4427$.

8.4.2.3

Returning to the binary search algorithm, we can derive the final expression for $T(n)$ using the properties of logarithms, including that the logarithm function is increasing so that inequalities are maintained when taking logarithms of numbers.

$$
\begin{aligned}
T(n) = r &\Leftrightarrow 2^{r-1} \leq n < 2^r \\
&\Leftrightarrow \log_2 2^{r-1} \leq \log_2 n < \log_2 2^r \\
&\Leftrightarrow r - 1 \leq \log_2 n < r \\
&\Leftrightarrow r - 1 = \lfloor \log_2 n \rfloor \\
&\Leftrightarrow T(n) = r = \lfloor \log_2 n \rfloor + 1
\end{aligned}
$$

We can apply several of these properties of logarithms to get an alternate expression for $T(n)$:

$$
\begin{aligned}
\lfloor \log_2 n \rfloor + 1 &= \lfloor \log_2 n + 1 \rfloor \\
&= \lfloor \log_2 n + \log_2 2 \rfloor \\
&= \lfloor \log_2 2n \rfloor
\end{aligned}
$$

If the time that was assigned to Step 1 of the binary search algorithm is changed, we wouldn't expect the form of the solution to be very different. If $T(n) = a + T(\lfloor n/2 \rfloor)$ with $T(0) = c$, then $T(n) = c + a \lfloor \log_2 2n \rfloor$.

A further generalization would be to add a coefficient to $T(\lfloor n/2 \rfloor)$: $T(n) = a + bT(\lfloor n/2 \rfloor)$ with $T(0) = c$, where $a, b, c \in \mathbb{R}$, and $b \neq 0$ is not quite as simple to derive. First, if we consider values of n that are powers of 2:

$$
\begin{aligned}
T(1) &= a + bT(0) = a + bc \\
T(2) &= a + b(a + bc) = a + ab + cb^2 \\
T(4) &= a + b\left(a + ab + cb^2\right) = a + ab + ab^2 + cb^3 \\
&\qquad\qquad \vdots \\
T(2^r) &= a + ab + ab^2 + \cdots + ab^r + cb^{r+1}
\end{aligned}
$$

If n is not a power of 2, by reasoning that is identical to what we used to (8.4.1) and (8.4.2),

$$T(n) = \sum_{k=0}^{r} ab^k + cb^{r+1}$$

where $r = \lfloor \log_2 n \rfloor$.

The first term of this expression is a geometric sum, which can be written in closed form. Let x be that sum:

$$
\begin{aligned}
x &= a + ab + ab^2 + \cdots + ab^r \\
bx &= ab + ab^2 + \cdots + ab^r + ab^{r+1}
\end{aligned}
$$

We've multiplied each term of x by b and aligned the identical terms in x and bx. Now if we subtract the two equations,

$$x - bx = a - ab^{r+1} \Rightarrow x(1-b) = a\left(1 - b^{r+1}\right)$$

Therefore, $x = a\frac{b^{r+1}-1}{b-1}$.

A closed form expression for $T(n)$ is

$$T(n) = a\frac{b^{r+1}-1}{b-1} + cb^{r+1} \text{ where } r = \lfloor \log_2 n \rfloor$$

8.4.3 Analysis of Bubble Sort and Merge Sort

The efficiency of any search algorithm such as the binary search relies on fact that the search list is sorted according to a key value and that the search is based on the key value. There are several methods for sorting a list. One example is the bubble sort. You might be familiar with this one since it is a popular "first sorting algorithm." A time analysis of the algorithm shows that if $B(n)$ is the worst-case time needed to complete the bubble sort on n items, then $B(n) = (n-1) + B(n-1)$ and $B(1) = 0$. The solution of this relation is a quadratic function $B(n) = \frac{1}{2}\left(n^2 - n\right)$. The growth rate of a quadratic function such as this one is controlled by its squared term. Any other terms are dwarfed by it as n gets large. For the bubble sort, this means that if we double the size of the list that we are to sort, n changes to $2n$ and so n^2 becomes $4n^2$. Therefore, the time needed to do a bubble sort is quadrupled. One alternative to bubble sort is the merge sort. Here is a simple version of this algorithm for sorting $F = \{r(1), r(2), \ldots, r(n)\}$, $n \geq 1$. If $n = 1$, the list is sorted trivially. If $n \geq 2$ then:

(1) Divide F into $F_1 = \{r(1), \ldots, r(\lfloor n/2 \rfloor)\}$ and $F_2 = \{r(\lfloor n/2 \rfloor + 1), \ldots, r(n)\}$.

(2) Sort F_1 and F_2 using a merge sort.

(3) Merge the sorted lists F_1 and F_2 into one sorted list. If the sort is to be done in descending order of key values, you continue to choose the higher key value from the fronts of F_1 and F_2 and place them in the back of F.

Note that F_1 will always have $\lfloor n/2 \rfloor$ items and F_2 will have $\lceil n/2 \rceil$ items; thus, if n is odd, F_2 gets one more item than F_1. We will assume that the time required to perform Step 1 of the algorithm is insignificant compared to the other steps; therefore, we will assign a time value of zero to this step. Step 3 requires roughly n comparisons and n movements of items from F_1 and F_2 to F; thus, its time is proportional to n. For this reason, we will assume that Step 3 takes n time units. Since Step 2 requires $T(\lfloor n/2 \rfloor) + T(\lceil n/2 \rceil)$ time units,

$$T(n) = n + T(\lfloor n/2 \rfloor) + T(\lceil n/2 \rceil) \tag{8.4.9}$$

with the initial condition

$$T(1) = 0 \tag{8.4.10}$$

Instead of an exact solution of these equations, we will be content with an estimate for $T(n)$. First, consider the case of $n = 2^r$, $r \geq 1$:

$$T\left(2^1\right) = T(2) = 2 + T(1) + T(1) = 2 = 1 \cdot 2$$
$$T\left(2^2\right) = T(4) = 4 + T(2) + T(2) = 8 = 2 \cdot 4$$
$$T\left(2^3\right) = T(8) = 8 + T(4) + T(4) = 24 = 3 \cdot 8$$
$$\vdots$$
$$T\left(2^r\right) = r2^r = 2^r \log_2 2^r$$

Thus, if n is a power of 2, $T(n) = n \log_2 n$. Now if, for some $r \geq 2$, $2^{r-1} \leq n \leq 2^r$, then $(r-1)2^{r-1} \leq T(n) < r2^r$. This can be proved by induction on r. As n increases from 2^{r-1} to 2^r, $T(n)$ increases from $(r-1)2^{r-1}$ to $r2^r$ and is slightly larger than $\lfloor n \log_2 n \rfloor$. The discrepancy is small enough so that $T_e(n) = \lfloor n \log_2 n \rfloor$ can be considered a solution of (8.4.9) and (8.4.10) for the purposes of comparing the merge sort with other algorithms. Table 8.4.7 compares $B(n)$ with $T_e(n)$ for selected values of n.

n	$B(n)$	$T_e(n)$
10	45	34
50	1225	283
100	4950	665
500	124750	4483
1000	499500	9966

Table 8.4.7: Comparison of Times for Bubble Sort and Merge Sort

8.4.4 Derangements

A derangement is a permutation on a set that has no "fixed points". Here is a formal definition:

Definition 8.4.8 Derangement. A derangement of a nonempty set A is a permutation of A (i.e., a bijection from A into A) such that $f(a) \neq a$ for all $a \in A$. $\diamond$

If $A = \{1, 2, ..., n\}$, an interesting question might be "How many derangements are there of A?" We know that our answer is bounded above by $n!$. We can also expect our answer to be quite a bit smaller than $n!$ since n is the image of itself for $(n-1)!$ of the permutations of A.

Let $D(n)$ be the number of derangements of $\{1, 2, ..., n\}$. Our answer will come from discovering a recurrence relation on D. Suppose that $n \geq 3$. If we are to construct a derangement of $\{1, 2, \ldots, n\}$, f, then $f(n) = k \neq n$. Thus, the image of n can be selected in $n-1$ different ways. No matter which of the $n-1$ choices we make, we can complete the definition of f in one of two ways. First, we can decide to make $f(k) = n$, leaving $D(n-2)$ ways of completing the definition of f, since f will be a derangement of $\{1, 2, \ldots, n\} - \{n, k\}$. Second, if we decide to select $f(k) \neq n$, each of the $D(n-1)$ derangements of $\{1, 2, \ldots, n-1\}$ can be used to define f. If g is a derangement of $\{1, 2, \ldots, n-1\}$ such that $g(p) = k$, then define f by

$$f(j) = \begin{cases} n & \text{if } j = p \\ k & \text{if } j = n \\ g(j) & \text{otherwise} \end{cases}$$

Note that with our second construction of f, $f(f(n)) = f(k) \neq n$, while in the first construction, $f(f(n)) = f(k) = n$. Therefore, no derangement of $\{1, 2, ..., n\}$ with $f(n) = k$ can be constructed by both methods.

To recap our result, we see that f is determined by first choosing one of $n - 1$ images of n and then constructing the remainder of f in one of $D(n - 2) + D(n - 1)$ ways. Therefore,

$$D(n) = (n - 1)(D(n - 2) + D(n - 1)) \qquad (8.4.11)$$

This homogeneous second-order linear relation with variable coefficients, together with the initial conditions $D(1) = 0$ and $D(2) = 1$, completely defines D. Instead of deriving a solution of this relation by analytical methods, we will give an empirical derivation of an approximation of $D(n)$. Since the derangements of $\{1, 2..., n\}$ are drawn from a pool of $n!$ permutations, we will see what percentage of these permutations are derangements by listing the values of $n!$, $D(n)$, and $\frac{D(n)}{n!}$. The results we observe will indicate that as n grows, $\frac{D(n)}{n!}$ hardly changes at all. If this quotient is computed to eight decimal places, for $n \geq 12$, $D(n)/n! = 0.36787944$. The reciprocal of this number, which $D(n)/n!$ seems to be tending toward, is, to eight places, 2.7182818. This number appears in so many places in mathematics that it has its own name, e. An approximate solution of our recurrence relation on D is then $D(n) \approx \frac{n!}{e}$.

```
def D(n):
        if n<=2:
                return n-1
        else:
                return (n-1)*(D(n-2)+D(n-1))

map(lambda
    k:[k,D(k),(D(k)/factorial(k)).n(digits=8)],range(1,16))
```

```
[[1,  0,  0.00000000],
 [2,  1,  0.50000000],
 [3,  2,  0.33333333],
 [4,  9,  0.37500000],
 [5,  44,  0.36666667],
 [6,  265,  0.36805556],
 [7,  1854,  0.36785714],
 [8,  14833,  0.36788194],
 [9,  133496,  0.36787919],
 [10,  1334961,  0.36787946],
 [11,  14684570,  0.36787944],
 [12,  176214841,  0.36787944],
 [13,  2290792932,  0.36787944],
 [14,  32071101049,  0.36787944],
 [15,  481066515734,  0.36787944]]
```

8.4.5 Exercises for Section 8.4

1. Solve the following recurrence relations. Indicate whether your solution is an improvement over iteration.

   (a) $nS(n) - S(n - 1) = 0$, $S(0) = 1$.

   (b) $T(k) + 3kT(k - 1) = 0$, $T(0) = 1$.

   (c) $U(k) - \frac{k-1}{k}U(k - 1) = 0$, $k \geq 2$, $U(1) = 1$.

2. Prove that if $n \geq 0$, $\lfloor n/2 \rfloor + \lceil n/2 \rceil = n$. (Hint: Consider the cases of n odd and n even separately.)

3. Solve as completely as possible:

   (a) $T(n) = 3 + T(\lfloor n/2 \rfloor)$, $T(0) = 0$.

   (b) $T(n) = 1 + \frac{1}{2}T(\lfloor n/2 \rfloor)$, $T(0) = 2$.

   (c) $V(n) = 1 + V\lfloor n/8 \rfloor)$, $V(0) = 0$. (Hint: Write n in octal form.)

4. Prove by induction that if $T(n) = 1 + T(\lfloor n/2 \rfloor)$, $T(0) = 0$, and $2^{r-1} \leq n < 2^r$, $r \geq 1$, then $T(n) = r$.

   Hint. Prove by induction on r.

5. Use the substitution $S(n) = T(n+1)/T(n)$ to solve $T(n)T(n-2) = T(n-1)^2$ for $n \geq 2$, with $T(0) = 1$, $T(1) = 6$, and $T(n) \geq 0$.

6. Use the substitution $G(n) = T(n)^2$ to solve $T(n)^2 - T(n-1)^2 = 1$ for $n \geq 1$, with $T(0) = 10$.

7. Solve as completely as possible:

   (a) $Q(n) = 1 + Q(\lfloor \sqrt{n} \rfloor)$, $n \geq 2$, $Q(1) = 0$.

   (b) $R(n) = n + R(\lfloor n/2 \rfloor)$, $n \geq 1$, $R(0) = 0$.

8. Suppose Step 1 of the merge sort algorithm did take a significant amount of time. Assume it takes 0.1 time unit, independent of the value of n.

   (a) Write out a new recurrence relation for $T(n)$ that takes this factor into account.

   (b) Solve for $T(2^r)$, $r \geq 0$.

   (c) Assuming the solution for powers of 2 is a good estimate for all n, compare your result to the solution in the text. As gets large, is there really much difference?

8.5 Generating Functions

This section contains an introduction to the topic of generating functions and how they are used to solve recurrence relations, among other problems. Methods that employ generating functions are based on the concept that you can take a problem involving sequences and translate it into a problem involving generating functions. Once you've solved the new problem, a translation back to sequences gives you a solution of the original problem.

This section covers:

   (1) The definition of a generating function.

   (2) Solution of a recurrence relation using generating functions to identify the skills needed to use generating functions.

   (3) An introduction and/or review of the skills identified in point 2.

   (4) Some applications of generating functions.

8.5.1 Definition

Definition 8.5.1 Generating Function of a Sequence. The generating function of a sequence S with terms $S_0, S_1, S_2, \ldots$, is the infinite sum

$$G(S; z) = \sum_{n=0}^{\infty} S_n z^n = S_0 + S_1 z + S_2 z^2 + S_3 z^3 + \cdots$$

The domain and codomain of generating functions will not be of any concern to us since we will only be performing algebraic operations on them. $\Diamond$

Example 8.5.2 First Examples.

(a) If $S_n = 3^n, n \geq 0$, then

$$G(S; z) = 1 + 3z + 9z^2 + 27z^3 + \cdots$$

$$= \sum_{n=0}^{\infty} 3^n z^n$$

$$= \sum_{n=0}^{\infty} (3z)^n$$

We can get a closed form expression for $G(S; z)$ by observing that $G(S; z) - 3zG(S; z) = 1$. Therefore, $G(S; z) = \frac{1}{1-3z}$.

(b) Finite sequences have generating functions. For example, the sequence of binomial coefficients $\binom{n}{0}, \binom{n}{1}, \ldots, \binom{n}{n}, n \geq 1$ has generating function

$$G\left(\binom{n}{\cdot}; z\right) = \binom{n}{0} + \binom{n}{1}z + \cdots + \binom{n}{n}z^n$$

$$= \sum_{k=0}^{\infty} \binom{n}{k}z^k$$

$$= (1+z)^n$$

by application of the binomial formula.

(c) If $Q(n) = n^2$, $G(Q; z) = \sum_{n=0}^{\infty} n^2 z^n = \sum_{k=0}^{\infty} k^2 z^k$. Note that the index that is used in the summation has no significance. Also, note that the lower limit of the summation could start at 1 since $Q(0) = 0$.

□

8.5.2 Solution of a Recurrence Relation Using Generating Functions

We illustrate the use of generating functions by solving $S(n) - 2S(n-1) - 3S(n-2) = 0$, $n \geq 2$, with $S(0) = 3$ and $S(1) = 1$.

(1) Translate the recurrence relation into an equation about generating functions.

Let $V(n) = S(n) - 2S(n-1) - 3S(n-2)$, $n \geq 2$, with $V(0) = 0$ and $V(1) = 0$. Therefore,

$$G(V; z) = 0 + 0z + \sum_{n=2}^{\infty} (S(n) - 2S(n-1) - 3S(n-2))z^n = 0$$

(2) Solve for the generating function of the unknown sequence, $G(S; z) = \sum_{n=0}^{\infty} S_n z^n$.

$$0 = \sum_{n=2}^{\infty} (S(n) - 2S(n-1) - 3S(n-2))z^n$$

$$= \sum_{n=2}^{\infty} S(n)z^n - 2\left(\sum_{n=2}^{\infty} S(n-1)z^n\right) - 3\left(\sum_{n=2}^{\infty} S(n-2)z^n\right)$$

Close examination of the three sums above shows:

(a)

$$\sum_{n=2}^{\infty} S_n z^n = \sum_{n=0}^{\infty} S_n z^n - S(0) - S(1)z$$

$$= G(S; z) - 3 - z$$

since $S(0) = 3$ and $S(1) = 1$.

(b)

$$\sum_{n=2}^{\infty} S(n-1)z^n = z\left(\sum_{n=2}^{\infty} S(n-1)z^{n-1}\right)$$

$$= z\left(\sum_{n=1}^{\infty} S(n)z^n\right)$$

$$= z\left(\sum_{n=0}^{\infty} S(n)z^n - S(0)\right)$$

$$= z(G(S; z) - 3)$$

(c)

$$\sum_{n=2}^{\infty} S(n-2)z^n = z^2\left(\sum_{n=2}^{\infty} S(n-2)z^{n-2}\right)$$

$$= z^2 G(S; z)$$

Therefore,

$$(G(S; z) - 3 - z) - 2z(G(S; z) - 3) - 3z^2 G(S; z) = 0$$

$$\Rightarrow G(S; z) - 2zG(S; z) - 3z^2 G(S; z) = 3 - 5z$$

$$\Rightarrow G(S; z) = \frac{3 - 5z}{1 - 2z - 3z^2}$$

(3) Determine the sequence whose generating function is the one we got in Step 2.

For our example, we need to know one general fact about the closed form expression of an exponential sequence (a proof will be given later):

$$T(n) = ba^n, n \geq 0 \Leftrightarrow G(T; z) = \frac{b}{1 - az} \qquad (8.5.1)$$

Now, in order to recognize S in our example, we must write our closed form expression for $G(S; z)$ as a sum of terms like $G(T; z)$ above. Note that the denominator of $G(S; z)$ can be factored:

$$G(S; z) = \frac{3 - 5z}{1 - 2z - 3z^2} = \frac{3 - 5z}{(1 - 3z)(1 + z)}$$

If you look at this last expression for $G(S; z)$ closely, you can imagine how it could be the result of addition of two fractions,

$$\frac{3 - 5z}{(1 - 3z)(1 + z)} = \frac{A}{1 - 3z} + \frac{B}{1 + z} \qquad (8.5.2)$$

where A and B are two real numbers that must be determined. Starting on the right of (8.5.2), it should be clear that the sum, for any A and B, would look like the left-hand side. The process of finding values of A and B that make (8.5.2) true is called the **partial fractions decomposition** of the left-hand side:

$$\frac{A}{1 - 3z} + \frac{B}{1 + z} = \frac{A(1 + z)}{(1 - 3z)(1 + z)} + \frac{B(1 - 3z)}{(1 - 3z)(1 + z)}$$
$$= \frac{(A + B) + (A - 3B)z}{(1 - 3z)(1 + z)}$$

Therefore,

$$\left\{ \begin{array}{c} A + B = 3 \\ A - 3B = -5 \end{array} \right\} \Rightarrow \left\{ \begin{array}{c} A = 1 \\ B = 2 \end{array} \right\}$$

and

$$G(S; z) = \frac{1}{1 - 3z} + \frac{2}{1 + z}$$

We can apply (8.5.1) to each term of $G(S; z)$:

- $\frac{1}{1-3z}$ is the generating function for $S_1(n) = 1 \cdot 3^n = 3^n$
- $\frac{2}{1+z}$ is the generating function for $S_2(n) = 2(-1)^n$.

Therefore, $S(n) = 3^n + 2(-1)^n$.

From this example, we see that there are several skills that must be mastered in order to work with generating functions. You must be able to:

(a) Manipulate summation expressions and their indices (in Step 2).

(b) Solve algebraic equations and manipulate algebraic expressions, including partial function decompositions (Steps 2 and 3).

(c) Identify sequences with their generating functions (Steps 1 and 3).

We will concentrate on the last skill first, a proficiency in the other skills is a product of doing as many exercises and reading as many examples as possible.
First, we will identify the operations on sequences and on generating functions.

8.5.3 Operations on Sequences

Definition 8.5.3 Operations on Sequences. Let S and T be sequences of numbers and let c be a real number. Define the sum $S + T$, the scalar product cS, the product ST, the convolution $S * T$, the pop operation $S \uparrow$ (read "S pop"), and the push operation $S \downarrow$ (read "S push") term-wise for $k \geq 0$ by

$$(S + T)(k) = S(k) + T(k) \tag{8.5.3}$$

$$(cS)(k) = cS(k) \tag{8.5.4}$$

$$(S \cdot T)(k) = S(k)T(k) \tag{8.5.5}$$

$$(S * T)(k) = \sum_{j=0}^{k} S(j)T(k - j) \tag{8.5.6}$$

$$(S \uparrow)(k) = S(k + 1) \tag{8.5.7}$$

$$(S \downarrow)(k) = \begin{cases} 0 & \text{if } k = 0 \\ S(k - 1) & \text{if } k > 0 \end{cases} \tag{8.5.8}$$

$\Diamond$

If one imagines a sequence to be a matrix with one row and an infinite number of columns, $S + T$ and cS are exactly as in matrix addition and scalar multiplication. There is no obvious similarity between the other operations and matrix operations.

The pop and push operations can be understood by imagining a sequence to be an infinite stack of numbers with $S(0)$ at the top, $S(1)$ next, etc., as in Figure 8.5.4a. The sequence $S \uparrow$ is obtained by "popping" S(0) from the stack, leaving a stack as in Figure 8.5.4b, with S(1) at the top, S(2) next, etc. The sequence $S \downarrow$ is obtained by placing a zero at the top of the stack, resulting in a stack as in Figure 8.5.4c. Keep these figures in mind when we discuss the pop and push operations.

$S(0)$	$S(1)$	0
$S(1)$	$S(2)$	$S(0)$
$S(2)$	$S(3)$	$S(1)$
$S(3)$	$S(4)$	$S(2)$
$S(4)$	$S(5)$	$S(3)$
$\vdots$	$\vdots$	$\vdots$
(a)	*(b)*	*(c)*

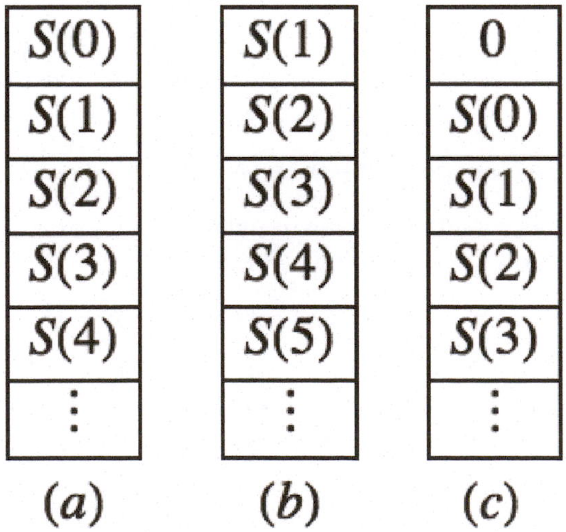

Figure 8.5.4: Stack interpretation of pop and push operation

Example 8.5.5 Some Sequence Operations. If $S(n) = n$, $T(n) = n^2$, $U(n) = 2^n$, and $R(n) = n2^n$:

(a) $(S + T)(n) = n + n^2$

(b) $(U + R)(n) = 2^n + n2^n = (1 + n)2^n$

(c) $(2U)(n) = 2 \cdot 2^n = 2^{n+1}$

(d) $\left(\frac{1}{2}R\right)(n) = \frac{1}{2}n2^n = n2^{n-1}$

(e) $(S \cdot T)(n) = nn^2 = n^3$

(f) $(S * T)(n) = \sum_{j=0}^{n} S(j)T(n - j) = \sum_{j=0}^{n} j(n - j)^2$

$\qquad = \sum_{j=0}^{n} \left(jn^2 - 2nj^2 + j^3\right)$

$\qquad = n^2 \sum_{j=0}^{n} j - 2n \sum_{j=0}^{n} j^2 + \sum_{j=0}^{n} j^3$

$\qquad = n^2 \left(\frac{n(n+1)}{2}\right) - 2n \left(\frac{(2n+1)(n+1)n}{6}\right) + \frac{1}{4}n^2(n + 1)^2$

$\qquad = \frac{n^2(n+1)(n-1)}{12}$

(g) $(U * U)(n) = \sum_{j=0}^{n} U(j)U(n - j)$

$\qquad = \sum_{j=0}^{n} 2^j 2^{n-j}$

$\qquad = (n + 1)2^n$

(h) $(S \uparrow)(n) = n + 1$

(i) $(S \downarrow)(n) = \max(0, n - 1)$

(j) $((S \downarrow) \downarrow)(n) = \max(0, n - 2)$

(k) $(U \downarrow)(n) = \begin{cases} 2^{n-1} & \text{if } n > 0 \\ 0 & \text{if } n = 0 \end{cases}$

(l) $((U \downarrow) \uparrow)(n) = (U \downarrow)(n + 1) = 2^n = U(n)$

(m) $((U \uparrow) \downarrow)(n) = \begin{cases} 0 & \text{if } n = 0 \\ U(n) & \text{if } n > 0 \end{cases}$

□

Note that $(U \downarrow) \uparrow \neq (U \uparrow) \downarrow$.

Definition 8.5.6 Multiple Pop and Push. If S is a sequence of numbers and p a positive integer greater than 1, define

$$S \uparrow p = (S \uparrow (p - 1)) \uparrow \quad \text{if } p \geq 2 \text{ and } S \uparrow 1 = S \uparrow$$

Similarly, define

$$S \downarrow p = (S \downarrow (p - 1)) \downarrow \quad \text{if } p \geq 2 \text{ and } S \downarrow 1 = S \downarrow$$

◊

In general, $(S \uparrow p)(k) = S(k + p)$, and

$$(S \downarrow p)(k) = \begin{cases} 0 & \text{if } k < p \\ S(k - p) & \text{if } k \geq p \end{cases}$$

8.5.4 Operations on Generating Functions

Definition 8.5.7   Operations on Generating Functions.   If $G(z) = \sum_{k=0}^{\infty} a_k z^k$ and $H(z) = \sum_{k=0}^{\infty} b_k z^k$ are generating functions and c is a real number, then the sum $G + H$, scalar product cG, product GH, and monomial product $z^p G$, $p \geq 1$ are generating functions, where

$$(G + H)(z) = \sum_{k=0}^{\infty} (a_k + b_k) z^k \tag{8.5.9}$$

$$(cG)(z) = \sum_{k=0}^{\infty} c a_k z^k \tag{8.5.10}$$

$$(GH)(z) = \sum_{k=0}^{\infty} c z^k \text{ where } c_k = \sum_{j=0}^{k} a_j b_{k-j} \tag{8.5.11}$$

$$(z^p G)(z) = z^p \sum_{k=0}^{\infty} a_k z^k = \sum_{k=0}^{\infty} a_k z^{k+p} = \sum_{n=p}^{\infty} a_{n-p} z^n \tag{8.5.12}$$

The last sum is obtained by substituting $n - p$ for k in the previous sum.

$\diamond$

Example 8.5.8   Some operations on generating functions.   If $D(z) = \sum_{k=0}^{\infty} k z^k$ and $H(z) = \sum_{k=0}^{\infty} 2^k z^k$ then

$$(D + H)(z) = \sum_{k=0}^{\infty} \left(k + 2^k \right) z^k$$

$$(2H)(z) = \sum_{k=0}^{\infty} 2 \cdot 2^k z^k = \sum_{k=0}^{\infty} 2^{k+1} z^k$$

$$(zD)(z) = z \sum_{k=0}^{\infty} k z^k = \sum_{k=0}^{\infty} k z^{k+1}$$

$$= \sum_{k=1}^{\infty} (k - 1) z^k = D(z) - \sum_{k=1}^{\infty} z^k$$

$$(DH)(z) = \sum_{k=0}^{\infty} \left(\sum_{j=0}^{k} j 2^{k-j} \right) z^k$$

$$(HH)(z) = \sum_{k=0}^{\infty} \left(\sum_{j=0}^{k} 2^j 2^{k-j} \right) z^k = \sum_{k=0}^{\infty} (k + 1) 2^k z^k$$

Note: $D(z) = G(S; z)$, and $H(z) = G(U; z)$ from Example 5. $\square$

Now we establish the connection between the operations on sequences and generating functions. Let S and T be sequences and let c be a real number.

$$G(S + T; z) = G(S; z) + G(T; z) \tag{8.5.13}$$

$$G(cS; z) = cG(S; z) \tag{8.5.14}$$

$$G(S * T; z) = G(S; z)G(T; z) \tag{8.5.15}$$

$$G(S \uparrow; z) = (G(S; z) - S(0))/z \tag{8.5.16}$$

$$G(S \downarrow; z) = zG(S; z) \tag{8.5.17}$$

In words, (8.5.13) says that the generating function of the sum of two sequences equals the sum of the generating functions of those sequences. Take the time to write out the other four identities in your own words. From the previous examples, these identities should be fairly obvious, with the possible exception of the last two. We will prove (8.5.16) as part of the next theorem and leave the proof of (8.5.17) to the interested reader. Note that there is no operation on generating functions that is related to sequence multiplication; that is, $G(S \cdot T; z)$ cannot be simplified.

Theorem 8.5.9   Generating functions related to Pop and Push. *If* $p > 1$,

(a) $G(S \uparrow p; z) = \left(G(S; z) - \sum_{k=0}^{p-1} S(k)z^k\right)/z^k$

(b) $G(S \downarrow p; z) = z^p G(S; z)$.

Proof. We prove (a) by induction and leave the proof of (b) to the reader.
   Basis:

$$G(S \uparrow; z) = \sum_{k=0}^{\infty} S(k+1)z^k$$

$$= \sum_{k=1}^{\infty} S(k)z^{k-1}$$

$$= \left(\sum_{k=1}^{\infty} S(k)z^k\right)/z$$

$$= \left(S(0) + \sum_{k=1}^{\infty} S(k)z^k - S(0)\right)/z$$

$$= (G(S; z) - S(0))/z$$

Therefore, part (a) is true for $p = 1$.
   Induction: Suppose that for some $p \geq 1$, the statement in part (a) is true:

$$\begin{aligned}
G(S \uparrow (p+1); z) &= G((S \uparrow p) \uparrow; z) \\
&= (G(S \uparrow p; z) - (S \uparrow p)(0))/z \text{ by the basis} \\
&= \frac{\frac{\left(G(S;z) - \sum_{k=0}^{p-1} S(k)z^k\right)}{z^p} - S(p)}{z}
\end{aligned}$$

by the induction hypothesis. Now write $S(p)$ in the last expression above as $(S(p)z^p)/z^p$ so that it fits into the finite summation:

$$\begin{aligned}
G(S \uparrow (p+1); z) &= \left(\frac{G(S; z) - \sum_{k=0}^{p} S(k)z^k}{z^p}\right)/z \\
&= \left(G(S; z) - \sum_{k=0}^{p} S(k)z^k\right)/z^{p+1}
\end{aligned}$$

Therefore the statement is true for $p + 1$.   ■

8.5.5 Closed Form Expressions for Generating Functions

The most basic tool used to express generating functions in closed form is the closed form expression for the geometric series, which is an expression of the form $a + ar + ar^2 + \cdots$. It can either be terminated or extended infinitely.

Finite Geometric Series:

$$a + ar + ar^2 + \cdots + ar^n = a\left(\frac{1 - r^{n+1}}{1 - r}\right) \qquad (8.5.18)$$

Infinite Geometric Series:

$$a + ar + ar^2 + \cdots = \frac{a}{1 - r} \qquad (8.5.19)$$

Restrictions: a and r represent constants and the right sides of the two equations apply under the following conditions:

(1) r must not equal 1 in the finite case. Note that $a + ar + \cdots ar^n = (n+1)a$ if $r = 1$.

(2) In the infinite case, the absolute value of r must be less than 1.

These restrictions don't come into play with generating functions. We could derive (8.5.18) by noting that if $S(n) = a + ar + \cdots + ar^n$, $n > 0$, then $S(n) = rS(n-1) + a$ (See Exercise 10 of Section 8.3). An alternative derivation was used in Section 8.4. We will take the same steps to derive (8.5.19). Let $x = a + ar + ar^2 + \cdots$. Then

$$rx = ar + ar^2 + \cdots = x - a \Rightarrow x - rx = a \Rightarrow x = \frac{a}{1 - r}$$

Example 8.5.10 Generating Functions involving Geometric Sums.

(a) If $S(n) = 9 \cdot 5^n$, $n \geq 0$, $G(S; z)$ is an infinite geometric series with $a = 9$ and $r = 5z$. Therefore, $G(S; z) = \frac{9}{1 - 5z}$.

(b) If $T(n) = 4$, $n \geq 0$, then $G(T; z) = 4/(1 - z)$.

(c) If $U(n) = 3(-1)^n$, then $G(U; z) = 3/(1 + z)$.

(d) Let $C(n) = S(n) + T(n) + U(n) = 9 \cdot 5^n + 4 + 3(-1)^n$. Then

$$G(C; z) = G(S; z) + G(T; z) + G(U; z)$$

$$= \frac{9}{1 - 5z} + \frac{4}{1 - z} + \frac{3}{1 + z} \qquad .$$

$$= -\frac{14z^2 + 34z - 16}{5z^3 - z^2 - 5z + 1}$$

Given a choice between the last form of $G(C; z)$ and the previous sum of three fractions, we would prefer leaving it as a sum of three functions. As we saw in an earlier example, a partial fractions decomposition of a fraction such as the last expression requires some effort to produce.

(e) If $G(Q; z) = 34/(2 - 3z)$, then Q can be determined by multiplying the numerator and denominator by $1/2$ to obtain $\frac{17}{1 - \frac{3}{2}z}$. We recognize this fraction as the sum of the infinite geometric series with $a = 17$ and $r = \frac{3}{2}z$. Therefore $Q(n) = 17(3/2)^n$.

(f) If $G(A; z) = (1 + z)^3$, then we expand $(1 + z)^3$ to $1 + 3z + 3z^2 + z^3$. Therefore $A(0) = 1$, $A(1) = 3$ $A(2) = 3$, $A(3) = 1$, and, since there are no higher-powered terms, $A(n) = 0$, $n \geq 4$. A more concise way of describing A is $A(k) = \binom{3}{k}$, since $\binom{n}{k}$ is interpreted as 0 of $k > n$.

□

Table 8.5.11 lists some closed form expressions for the generating functions of some common sequences.

Sequence	Generating Function
$S(k) = ba^k$	$G(S; z) = \frac{b}{1-az}$
$S(k) = k$	$G(S; z) = \frac{z}{(1-z)^2}$
$S(k) = bka^k$	$G(S; z) = \frac{abz}{(1-az)^2}$
$S(k) = \frac{1}{k!}$	$G(S; z) = e^z$
$S(k) = \begin{cases} \binom{n}{k} & 0 \leq k \leq n \\ 0 & k > n \end{cases}$	$G(S; z) = (1+z)^n$

Table 8.5.11: Closed Form Expressions of some Generating Functions

Example 8.5.12 Another Complete Solution. Solve $S(k) + 3S(k-1) - 4S(k-2) = 0$, $k \geq 2$, with $S(0) = 3$ and $S(1) = -2$. The solution will be derived using the same steps that were used earlier in this section, with one variation.

(1) Translate to an equation about generating functions. First, we change the index of the recurrence relation by substituting $n + 2$ for k. The result is $S(n+2) + 3S(n+1) - 4S(n) = 0$, $n \geq 0$. Now, if $V(n) = S(n+2) + 3S(n+1) - 4S(n)$, then V is the zero sequence, which has a zero generating function. Furthermore, $V = S \uparrow 2 + 3(S \uparrow) - 4S$. Therefore,

$$\begin{aligned}
0 &= G(V; z) \\
&= G(S \uparrow 2; z) + 3G(S \uparrow; z) - 4G(S; z) \\
&= \frac{G(S; z) - S(0) - S(1)z}{z^2} + 4\frac{(G(S; z) - S(0))}{z} - 4G(S; z)
\end{aligned}$$

(2) We want to now solve the following equation for $G(S; z)$:

$$\frac{G(S; z) - S(0) - S(1)z}{z^2} + 4\frac{(G(S; z) - S(0))}{z} - 4G(S; z) = 0$$

Multiply by z^2 :

$$G(S; z) - 3 + 2z + 3z(G(S; z) - 3) - 4z^2 G(S; z) = 0$$

Expand and collect all terms involving $G(S; z)$ on one side of the equation:

$$\begin{aligned}
G(S; z) + 3zG(S; z) - 4z^2 G(S; z) &= 3 + 7z \\
\left(1 + 3z - 4z^2\right) G(S; z) &= 3 + 7z
\end{aligned}$$

Therefore,

$$G(S; z) = \frac{3 + 7z}{1 + 3z - 4z^2}$$

(3) Determine S from its generating function. $1 + 3z - 4z^2 = (1 + 4z)(1 - z)$ thus a partial fraction decomposition of $G(S; z)$ would be:

$$\frac{A}{1 + 4z} + \frac{B}{1 - z} = \frac{Az - A - 4Bz - B}{(z - 1)(4z + 1)} = \frac{(A + B) + (4B - A)z}{(z - 1)(4z + 1)}$$

Therefore, $A + B = 3$ and $4B - A = 7$. The solution of this set of equations is $A = 1$ and $B = 2$. $G(S; z) = \frac{1}{1+4z} + \frac{2}{1-z}$.

$\frac{1}{1+4z}$ is the generating function of $S_1(n) = (-4)^n$, and
$\frac{2}{1-z}$ is the generating function of $S_2(n) = 2(1)^n = 2$

In conclusion, since $G(S; z) = G(S_1; z) + G(S_2; z)$, $S(n) = 2 + (-4)^n$.

$\square$

Example 8.5.13 An Application to Counting. Let $A = \{a, b, c, d, e\}$ and let A^* be the set of all strings of length zero or more that can be made using each of the elements of A zero or more times. By the generalized rule of products, there are 5^n such strings that have length n, $n \geq 0$, Suppose that X_n is the set of strings of length n with the property that all of the a's and b's precede all of the c's, d's, and e's. Thus $aaabde \in X_6$, but $abcabc \notin X_6$. Let $R(n) = |X_n|$. A closed form expression for R can be obtained by recognizing R as the convolution of two sequences. To illustrate our point, we will consider the calculation of $R(6)$.

Note that if a string belongs to X_6, it starts with k characters from $\{a, b\}$ and is followed by $6 - k$ characters from $\{c, d, e\}$. Let $S(k)$ be the number of strings of a's and b's with length k and let $T(k)$ be the number of strings of c's, d's, and e's with length k. By the generalized rule of products, $S(k) = 2^k$ and $T(k) = 3^k$. Among the strings in X_6 are the ones that start with two a's and b's and end with c's, d's, and e's. There are $S(2)T(4)$ such strings. By the law of addition,

$$|X_6| = R(6) = S(0)T(6) + S(1)T(5) + \cdots + S(5)T(1) + S(6)T(0)$$

Note that the sixth term of R is the sixth term of the convolution of S with T, $S * T$. Think about the general situation for a while and it should be clear that $R = S * T$. Now, our course of action will be to:

(a) Determine the generating functions of S and T,

(b) Multiply $G(S; z)$ and $G(T; z)$ to obtain $G(S * T; z) = G(R; z)\backslash$, and

(c) Determine R on the basis of $G(R; z)$.

(a) $G(S; z) = \sum_{k=0}^{\infty} 2^k z^k = \frac{1}{1-2z}$, and $G(T; z) = \sum_{k=0}^{\infty} 3^k z^k = \frac{1}{1-3z}$

(b) $G(R; z) = G(S; z)G(T; z) = \frac{1}{(1-2z)(1-3z)}$

(c) To recognize R from $G(R; z)$, we must do a partial fractions decomposition:

$$\frac{1}{(1 - 2z)(1 - 3z)} = \frac{A}{1 - 2z} + \frac{B}{1 - 3z} = \frac{-3Az + A - 2Bz + B}{(2z - 1)(3z - 1)} = \frac{(A + B) + (-3A - 2B)z}{(2z - 1)(3z - 1)}$$

Therefore, $A + B = 1$ and $-3A - 2B = 0$. The solution of this pair of equations is $A = -2$ and $B = 3$. Since $G(R; z) = \frac{-2}{1-2z} + \frac{3}{1-3z}$, which is the sum of the generating functions of $-2(2)^k$ and $3(3)^k$, $R(k) = -2(2)^k + 3(3)^k = 3^{k+1} - 2^{k+1}$

For example, $R(6) = 3^7 - 2^7 = 2187 - 128 = 2059$. Naturally, this equals the sum that we get from $(S * T)(6)$. To put this number in perspective, the total number of strings of length 6 with no restrictions is $5^6 = 15625$, and $\frac{2059}{15625} \approx 0.131776$. Therefore approximately 13 percent of the strings of length 6 satisfy the conditions of the problem.

$\square$

8.5.6 Extra for Experts

The remainder of this section is intended for readers who have had, or who intend to take, a course in combinatorics. We do not advise that it be included in a typical course. The method that was used in the previous example is a very powerful one and can be used to solve many problems in combinatorics. We close this section with a general description of the problems that can be solved in this way, followed by some examples.

Consider the situation in which $P_1, P_2, \ldots, P_m$ are m actions that must be taken, each of which results in a well-defined outcome. For each $k = 1, 2, ..., m$ define X_k to be the set of possible outcomes of P_k. We will assume that each outcome can be quantified in some way and that the quantification of the elements of X_k is defined by the function $Q_k : X_k \to \{0, 1, 2, ...\}$. Thus, each outcome has a non-negative integer associated with it. Finally, define a frequency function $F_k : \{0, 1, 2, ...\} \to \{0, 1, 2, ...\}$ such that $F_k(n)$ is the number of elements of X_k that have a quantification of n.

Now, based on these assumptions, we can define the problems that can be solved. If a process P is defined as a sequence of actions $P_1, P_2, \ldots, P_m$ as above, and if the outcome of P, which would be an element of $X_1 \times X_2 \times \cdots \times X_m$, is quantified by

$$Q(a_1, a_2, \ldots, a_m) = \sum_{k=1}^{m} Q_k(a_k)$$

then the frequency function, F, for P is the convolution of the frequency functions for P_1, P_2, ..., P_m, which has a generating function equal to the product of the generating functions of the frequency functions $F_1, F_2, \ldots, F_m$. That is,

$$G(F; z) = G(F_1; z) G(F_2; z) \cdots (F_m; z)$$

Example 8.5.14 Rolling Two Dice. Suppose that you roll a die two times and add up the numbers on the top face for each roll. Since the faces on the die represent the integers 1 through 6, the sum must be between 2 and 12. How many ways can any one of these sums be obtained? Obviously, 2 can be obtained only one way, with two 1's. There are two sequences that yield a sum of 3: 1-2 and 2-1. To obtain all of the frequencies with which the numbers 2 through 12 can be obtained, we set up the situation as follows. For $j = 1, 2$; P_j is the rolling of the die for the jth time. $X_j = \{1, 2, ..., 6\}$ and $Q_j : X_j \to \{0, 1, 2, 3, ...\}$ is defined by $Q_j(x) = x$. Since each number appears on a die exactly once, the frequency function is $F_j(k) = 1$ if $1 \leq k \leq 6$, and $F_j(k) = 0$ otherwise. The process of rolling the die two times is quantified by adding up the $Q_j's$; that is, $Q(a_1, a_2) = Q_1(a_1) + Q_2(a_2)$. The generating function for the frequency function of rolling the die two times is then

$$G(F; z) = G(F_1; z) G(F_2; z)$$
$$= (z^6 + z^5 + z^4 + z^3 + z^2 + z)^2$$
$$= z^{12} + 2z^{11} + 3z^{10} + 4z^9 + 5z^8 + 6z^7 + 5z^6 + 4z^5 + 3z^4 + 2z^3 + z^2$$

Now, to get $F(k)$, just read the coefficient of z^k. For example, the coefficient of z^5 is 4, so there are four ways to roll a total of 5.

To apply this method, the crucial step is to decompose a large process in the proper way so that it fits into the general situation that we've described.
$\square$

Example 8.5.15 Distribution of a Committee. Suppose that an organization is divided into three geographic sections, A, B, and C. Suppose that an executive committee of 11 members must be selected so that no more than 5 members from any one section are on the committee and that Sections A, B, and C must have minimums of 3, 2, and 2 members, respectively, on the committee. Looking only at the number of members from each section on the committee, how many ways can the committee be made up? One example of a valid committee would be 4 A's, 4 B's, and 3 C's.

Let P_A be the action of deciding how many members (not who) from Section A will serve on the committee. $X_A = \{3, 4, 5\}$ and $Q_A(k) = k$. The frequency function, F_A, is defined by $F_A(k) = 1$ if $k \in X_k$, with $F_A(k) = 0$ otherwise. $G(F_A; z)$ is then $z^3 + z^4 + z^5$. Similarly, $G(F_B; z) = z^2 + z^3 + z^4 + z^5 = G(F_C; z)$. Since the committee must have 11 members, our answer will be the coefficient of z^{11} in $G(F_A; z) G(F_B; z) G(F_C; z)$, which is 10.

```
var('z')
expand((z^3+ z^4+z^5)*(z^2+ z^3+ z ^4 + z^5)^2)
```

```
z^15 + 3*z^14 + 6*z^13 + 9*z^12 + 10*z^11 + 9*z^10 + 6*z^9 +
   3*z^8 + z^7
```

$\square$

8.5.7 Exercises for Section 8.5

1.   What sequences have the following generating functions?

    (a) 1

    (b) $\frac{10}{2-z}$

    (c) $1 + z$

    (d) $\frac{3}{1+2z} + \frac{3}{1-3z}$

2.   What sequences have the following generating functions?

    (a) $\frac{1}{1+z}$

    (b) $\frac{1}{4-3z}$

    (c) $\frac{2}{1-z} + \frac{1}{1+z}$

    (d) $\frac{z+2}{z+3}$

3. Find closed form expressions for the generating functions of the following sequences:

   (a) $V(n) = 9^n$

   (b) P, where $P(k) - 6P(k-1) + 5P(k-2) = 0$ for $k \geq 2$, with $P(0) = 2$ and $P(1) = 2$.

   (c) The Fibonacci sequence: $F(k+2) = F(k+1) + F(k)$, $k \geq 0$, with $F(0) = F(1) = 1$.

4. Find closed form expressions for the generating functions of the following sequences:

   (a) $W(n) = \binom{5}{n} 2^n$ for $0 \leq n \leq 5$ and $W(n) = 0$ for $n > 5$.

   (b) Q, where $Q(k) + Q(k-1) - 42Q(k-2) = 0$ for $k \geq 2$, with $Q(0) = 2$ and $Q(1) = 2$.

   (c) G, where $G(k+3) = G(k+2) + G(k+1) + G(k)$ for $k \geq 0$, with $G(0) = G(1) = G(2) = 1$.

5. For each of the following expressions, find the partial fraction decomposition and identify the sequence having the expression as a generating function.

   (a) $\frac{5+2z}{1-4z^2}$

   (b) $\frac{32-22z}{2-3z+z^2}$

   (c) $\frac{6-29z}{1-11z+30z^2}$

6. Find the partial fraction decompositions and identify the sequence having the following expressions:

   (a) $\frac{1}{1-9z^2}$

   (b) $\frac{1+3z}{16-8z+z^2}$

   (c) $\frac{2z}{1-6z-7z^2}$

7. Given that $S(k) = k$ and $T(k) = 10k$, what is the kth term of the generating function of each of the following sequences:

   (a) $S + T$

   (b) $S \uparrow * T$

   (c) $S * T$

   (d) $S \uparrow * S \uparrow$

8. Given that $P(k) = \binom{10}{k}$ and $Q(k) = k!$, what is the kth term of the generating function of each of the following sequences:

   (a) $P * P$

    (b) $P + P\uparrow$

    (c) $P * Q$

    (d) $Q * Q$

9. A game is played by rolling a die five times. For the kth roll, one point is added to your score if you roll a number higher than k. Otherwise, your score is zero for that roll. For example, the sequence of rolls $2, 3, 4, 1, 2$ gives you a total score of three; while a sequence of 1,2,3,4,5 gives you a score of zero. Of the $6^5 = 7776$ possible sequences of rolls, how many give you a score of zero?, of one? ... of five?

10. Suppose that you roll a die ten times in a row and record the square of each number that you roll. How many ways could the sum of the squares of your rolls equal 40? What is the most common outcome?

Chapter 9

Graph Theory

Bipartite

Draw some lines joining dots in set A
To some dots in set B. Then we say
It's **bipartite** if we
Have no "B" joined to "B"
And no "A" joined to "A". That okay?

Chris Howlett, The Omnificent English Dictionary In Limerick Form

This chapter has three principal goals. First, we will identify the basic components of a graph and some of the features that many graphs have. Second, we will discuss some of the questions that are most commonly asked of graphs. Third, we want to make the reader aware of how graphs are used. In Section 9.1, we will discuss these topics in general, and in later sections we will take a closer look at selected topics in graph theory.

Chapter 10 will continue our discussion with an examination of trees, a special type of graph.

9.1 Graphs - General Introduction

9.1.1 Definitions

Recall that we introduced directed graphs in Chapter 6 as a tool to visualize relations on a set. Here is a formal definition.

Definition 9.1.1 Simple Directed Graph. A simple directed graph consists of a nonempty **set of vertices**, V, and a **set of edges**, E, that is a subset of the set $V \times V$. ◇

Note 9.1.2 Some Terminology and Comments. Each edge is an ordered pair of elements from the vertex set. The first entry is the **initial vertex** of the edge and the second entry is the **terminal vertex**. Despite the set terminology in this definition, we often think of a graph as a picture, an aid in visualizing a situation. In Chapter 6, we introduced this concept to help understand relations on sets. Although those relations were principally of a mathematical nature, it remains true that when we see a graph, it tells us how the elements of a set are related to one another. We have chosen not to allow a graph with an empty vertex set, the so-called empty graph. There are both advantages and disadvantages to allowing the empty graph, so you may encounter it in other references.

Example 9.1.3 A Simple Directed Graph. Figure 9.1.4 is an example of a simple directed graph. In set terms, this graph is (V, E), where $V = \{s, a, b\}$ and $E = \{(s, a), (s, b), (a, b), (b, a), (b, b)\}$. Note how each edge is labeled either 0 or 1. There are often reasons for labeling even simple graphs. Some labels are to help make a graph easier to discuss; others are more significant. We will discuss the significance of the labels on this graph later.

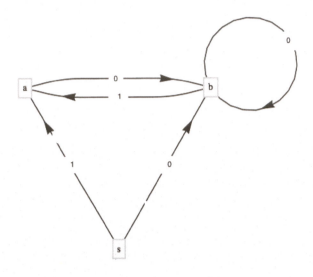

Figure 9.1.4: A directed graph

$\square$

In certain cases there may be a need for more than one edge between two vertices, and we need to expand the class of directed graphs.

Definition 9.1.5 Multigraph. A multigraph is a set of vertices V with a set of edges that can contain more than one edge between the vertices. $\Diamond$

One important point to keep in mind is that if we identify a graph as being a multigraph, it isn't necessary that there are two or more edges between some of the vertices. It is only just *allowed*. In other words, every simple graph is a multigraph. This is analogous to how a rectangle is a more general geometric figure than a square, but a square is still considered a rectangle.

Example 9.1.6 A Multigraph. A common occurrence of a multigraph is a road map. The cities and towns on the map can be thought of as vertices, while the roads are the edges. It is not uncommon to have more than one road connecting two cities. In order to give clear travel directions, we name or number roads so that there is no ambiguity. We use the same method to describe the edges of the multigraph in Figure 9.1.7. There is no question what $e3$ is; however, referring to the edge $(2, 3)$ would be ambiguous.

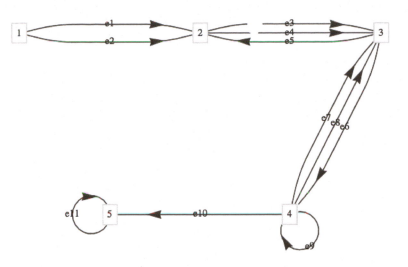

Figure 9.1.7: A directed multigraph

☐

There are cases where the order of the vertices is not significant and so we use a different mathematical model for this situation:

Definition 9.1.8 Undirected Graph. An undirected graph consists of a set V, called a vertex set, and a set E of two-element subsets of V, called the edge set. The two-element subsets are drawn as lines connecting the vertices. It is customary to not allow "self loops" in undirected graphs.   ◇

Example 9.1.9 An Undirected Graph. A network of computers can be described easily using a graph. Figure 9.1.10 describes a network of five computers, a, b, c, d, and e. An edge between any two vertices indicates that direct two-way communication is possible between the two computers. Note that the edges of this graph are not directed. This is due to the fact that the relation that is being displayed is symmetric (i.e., if X can communicate with Y, then Y can communicate with X). Although directed edges could be used here, it would simply clutter the graph.

Figure 9.1.10: Communications Map   **Figure 9.1.11:** Island Road Map

This undirected graph, in set terms, is $V = \{a, b, c, d, e\}$ and $E = \{\{a, b\}, \{a, d\}, \{b, c\}, \{b, d\}, \{c, e\}, \{b, e\}\}$

There are several other situations for which this graph can serve as a model. One of them is to interpret the vertices as cities and the edges as roads, an abstraction of a map such as the one in Figure 9.1.11 . Another interpretation is as an abstraction of the floor plan of a house. See Exercise 9.1.5.11. Vertex a represents the outside of the house; all others represent rooms. Two vertices are connected if there is a door between them.   ☐

Definition 9.1.12 Complete Undirected Graph. A complete undirected graph on n vertices is an undirected graph with the property that each pair of distinct vertices are connected to one another. Such a graph is usually denoted by K_n.   ◊

Example 9.1.13 A Labeled Graph. A flowchart is a common example of a simple graph that requires labels for its vertices and some of its edges. Figure 9.1.14 is one such example that illustrates how many problems are solved.

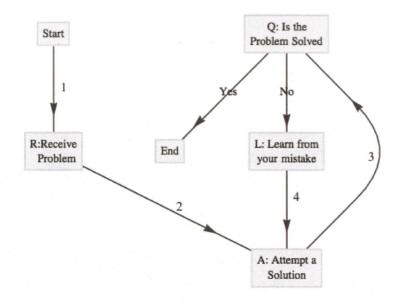

Figure 9.1.14: A flow chart - an example of a labeled graph

At the start of the problem-solving process, we are at the vertex labeled "Start" and at the end (if we are lucky enough to have solved the problem) we will be at the vertex labeled "End." The sequence of vertices that we pass through as we move from "Start" to "End" is called a path. The "Start" vertex is called the initial vertex of the path, while the "End" is called the final, or terminal, vertex. Suppose that the problem is solved after two attempts; then the path that was taken is Start, R, A, Q, L, A, Q, End. An alternate path description would be to list the edges that were used: $1, 2, 3$, No, $4, 3$, Yes. This second method of describing a path has the advantage of being applicable for multigraphs. On the graph in Figure 9.1.7, the vertex list $1, 2, 3, 4, 3$ does not clearly describe a path between 1 and 3, but e_1, e_4, e_6, e_7 is unambiguous. □

Note 9.1.15 A Summary of Path Notation and Terminology. If x and y are two vertices of a graph, then a **path** between x and y describes a motion from x and y along edges of the graph. Vertex x is called the initial vertex of the path and y is called the terminal vertex. A path between x and y can always be described by its edge list, the list of edges that were used: $(e_1, e_2, \ldots, e_n)$, where: (1) the initial vertex of e_1 is x; (2) the terminal vertex of e_i is the initial vertex of e_{i+1}, $i = 1, 2, \ldots, n - 1$; and (3) the terminal vertex of e_n is y. The number of edges in the edge list is the **path length**. A path on a simple graph can also be described by a vertex list. A path of length n will have a list of $n + 1$ vertices $v_0 = x$, $v_1, v_2, \ldots, v_n = y$, where, for $k = 0, 1, 2, \ldots, n - 1$, (v_k, v_{k+1}) is an edge on the graph. A **circuit** is a path that terminates at its initial vertex.

Suppose that a path between two vertices has an edge list $(e_1, e_2, ..., e_n)$. A **subpath** of this graph is any portion of the path described by one or more consecutive edges in the edge list. For example, $(3, \text{No}, 4)$ is a subpath of $(1, 2, 3, \text{No}, 4, 3, \text{Yes})$. Any path is its own subpath; however, we call it an improper subpath of itself. All other nonempty subpaths are called proper subpaths.

A path or circuit is simple if it contains no proper subpath that is a circuit. This is the same as saying that a path or circuit is simple if it does not visit any vertex more than once except for the common initial and terminal vertex in the circuit. In the problem-solving method described in Figure 9.1.14, the path that you take is simple only if you reach a solution on the first try.

9.1.2 Subgraphs

Intuitively, you could probably predict what the term "subgraph" means. A graph contained within a graph, right? But since a graph involves two sets, vertices and edges, does it involve a subset of both of these sets, or just one of them? The answer is it could be either. There are different types of subgraphs. The two that we will define below will meet most of our future needs in discussing the theory of graphs.

Definition 9.1.16 Subgraph. Let $G = (V, E)$ be a graph of any kind: directed, directed multigraph, or undirected. $G' = (V', E')$ is a subgraph of G if $V' \subseteq V$ and $e \in E'$ only if $e \in E$ and the vertices of e are in V'. You create a subgraph of G by removing zero or more vertices and all edges that include the removed vertices and then you possibly remove some other edges.

If the only removed edges are those that include the removed vertices, then we say that G is an **induced subgraph**. Finally, G' is a **spanning subgraph** of G if $V' = V$, or, in other words, no vertices are removed from G, only edges.

$\Diamond$

Example 9.1.17 Some subgraphs. Consider the graph, G, in the top left of Figure 9.1.18. The other three graphs in that figure are all subgraphs of G. The graph in the top right was created by first removing vertex 5 and all edges connecting it. In addition, we have removed the edge $\{1, 4\}$. That removed edge disqualifies the graph from being an induced subgraph. The graphs in the bottom left and right are both spanning subgraphs. The one on the bottom right is a tree, and is referred to as a spanning subtree. Spanning subtrees will be a discussed in Chapter 10.

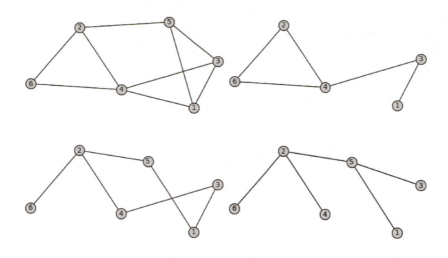

Figure 9.1.18: A few subgraphs

$\square$

One set of subgraphs of any graph is the connected components of a graph. For simplicity, we will define them for undirected graphs. Given a graph $G = (V, E)$, consider the relation "is connected to" on V. We interpret this relation so that each vertex is connected to itself, and any two distinct vertices are related if there is a path along edges of the graph from one to the other. It shouldn't be too difficult convince yourself that this is an equivalence relation on V.

Definition 9.1.19 Connected Component. Given a graph $G = (V, E)$, let C be the relation "is connected to" on V. Then the connected components of G are the induced subgraphs of G each with a vertex set that is an equivalence class with respect to C. $\diamond$

Example 9.1.20 If you ignore the duplicate names of vertices in the four graphs of Figure 9.1.18, and consider the whole figure as one large graph, then there are four connected components in that graph. It's as simple as that! It's harder to describe precisely than to understand the concept. $\square$

From the examples we've seen so far, we can see that although a graph can be defined, in short, as a collection of vertices and edges, an integral part of most graphs is the labeling of the vertices and edges that allows us to interpret the graph as a model for some situation. We continue with a few more examples to illustrate this point.

Example 9.1.21 A Graph as a Model for a Set of Strings. Suppose that you would like to mechanically describe the set of strings of 0's and 1's having no consecutive 1's. One way to visualize a string of this kind is with the graph in Figure 9.1.4. Consider any path starting at vertex s. If the label on each graph is considered to be the output to a printer, then the output will have no consecutive 1's. For example, the path that is described by the vertex list $(s, a, b, b, a, b, b, a, b)$ would result in an output of 10010010. Conversely, any string with no consecutive 1's determines a path starting at s. $\square$

Example 9.1.22 A Tournament Graph. Suppose that four teams compete in a round-robin sporting event; that is, each team meets every other team once, and each game is played until a winner is determined. If the teams are named A, B, C, and D, we can define the relation β on the set of teams by $X\beta Y$ if X

beat Y. For one set of results, the graph of β might look like Figure 9.1.23.

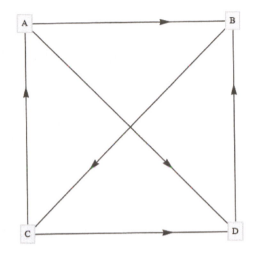

Figure 9.1.23: Round-robin tournament graph with four vertices

□

There are many types of tournaments and they all can be modeled by different types of graphs.

Definition 9.1.24 Tournament Graph.

  (a) A tournament graph is a directed graph with the property that no edge connects a vertex to itself, and between any two vertices there is at most one edge.

  (b) A complete (or round-robin) tournament graph is a tournament graph with the property that between any two distinct vertices there is exactly one edge.

  (c) A single-elimination tournament graph is a tournament graph with the properties that: (i) one vertex (the champion) has no edge terminating at it and at least one edge initiating from it; (ii) every other vertex is the terminal vertex of exactly one edge; and (iii) there is a path from the champion vertex to every other vertex.

◇

Example 9.1.25   Graph of a Single Elimination Tournament. The major league baseball championship is decided with a single-elimination tournament, where each "game" is actually a series of games. From 1969 to 1994, the two divisional champions in the American League (East and West) competed in a series of games. The loser is eliminated and the winner competed against the winner of the National League series (which is decided as in the American League). The tournament graph of the 1983 championship is in Figure 9.1.26

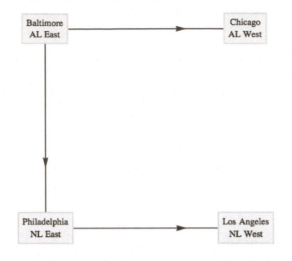

Figure 9.1.26: A single elimination tournament graph

□

9.1.3 Graph Isomorphisms

Next, we establish the relation "is isomorphic to," a form of equality on graphs. The graphs in Figure 9.1.27 obviously share some similarities, such as the number of vertices and the number of edges. It happens that they are even more similar than just that. If the letters a, b, c, and d in the left graph are replaced with the numbers 1,3,4, and 2, respectively, and the vertices are moved around so that they have the same position as the graph on the right, you get the graph on the right.

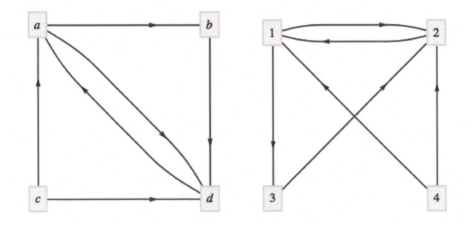

Figure 9.1.27: Isomorphic Graphs

Here is a more precise definition that reflects the fact that the actual positioning (or embedding) of vertices isn't an essential part of a graph.

Definition 9.1.28 Isomorphic Graphs. Two graphs (V, E) and (V', E') are isomorphic if there exists a bijection $f : V \to V'$ such that $(v_i, v_j) \in E$ if and only if $(f(v_i), f(v_j)) \in E'$. For multigraphs, we add that the number of edges connecting v_i to v_j must equal the number of edges from $f(v_i)$ to $f(v_j)$.

◇

The most significant local characteristic of a vertex within a graph is its degree. Collectively, the degrees can partially characterize a graph.

Definition 9.1.29 Degree of a vertex.

(a) Let v be a vertex of an undirected graph. The degree of v, denoted $deg(v)$, is the number of edges that connect v to the other vertices in the graph.

(b) If v is a vertex of a directed graph, then the outdegree of v, denoted $outdeg(v)$, is the number of edges of the graph that initiate at v. The indegree of v, denoted $indeg(v)$, is the number of edges that terminate at v.

◇

Definition 9.1.30 Degree Sequence of a Graph. The degree sequence of an undirected graph is the non-increasing sequence of its vertex degrees.   ◇

Example 9.1.31 Some degrees.

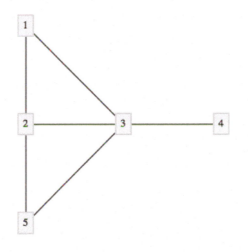

Figure 9.1.32: An undirected graph

(a) The degrees of vertices 1 through 5 in Figure 9.1.32 are 2, 3, 4, 1, and 2, respectively. The degree sequence of the graph is $(4, 3, 2, 2, 1)$.

(b) In a tournament graph, $outdeg(v)$ is the number of wins for v and $indeg(v)$ is the number of losses. In a complete (round-robin) tournament graph with n vertices, $outdeg(v) + indeg(v) = n - 1$ for each vertex.

□

Definition 9.1.33 Graphic Sequence. A finite nonincreasing sequence of integers $d_1, d_2, \ldots, d_n$ is a graphic if there exists an undirected graph with n vertices having the sequence as its degree sequence.   ◇

For example, $4, 2, 1, 1, 1, 1$ is graphic because the degrees of the graph in Figure 9.1.34 match these numbers. There is no connection between the vertex number and its degree in this graph.

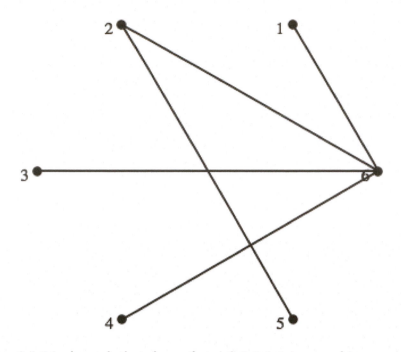

Figure 9.1.34: A graph that shows that $4, 2, 1, 1, 1, 1$ is a graphic sequence.

See [26] for more details on what are also referred to as **graphical degree sequences**, including an algorithm for determining whether or not a sequence is graphic.

9.1.4 Next Steps

The question "Once you have a graph, what do you do with it?" might come to mind. The following list of common questions and comments about graphs is a partial list that will give you an overview of the remainder of the chapter.

(1) How can a graph be represented as a data structure for use on a computer? We will discuss some common data structures that are used to represent graphs in Section 9.2.

(2) Given two vertices in a graph, does there exist a path between them? The existence of a path between any or all pairs of vertices in a graph will be discussed in Section 9.3. A related question is: How many paths of a certain type or length are there between two vertices?

(3) Is there a path (or circuit) that passes through every vertex (or uses every edge) exactly once? Paths of this kind are called traversals. We will discuss traversals in Section 9.4.

(4) Suppose that a cost is associated with the use of each vertex and/or edge in a path. What is the "cheapest" path, circuit, or traversal of a given kind? Problems of this kind will be discussed in Section 9.5.

(5) Given the specifications of a graph, or the graph itself, what is the best way to draw the graph? The desire for neatness alone makes this a reasonable question, but there are other motivations. Another goal might be to avoid having edges of the graph cross one another. This is discussed in Section 9.6.

List 9.1.35

9.1.5 Exercises for Section 9.1

1. What is the significance of the fact that there is a path connecting vertex b with every other vertex in Figure 9.1.10, as it applies to various situations that it models?

2. Draw a graph similar to Figure 9.1.4 that represents the set of strings of 0's and 1's containing no more than two consecutive 1's in any part of the string.

3. Draw a directed graph that models the set of strings of 0's and 1's (zero or more of each) where all of the 1's must appear consecutively.

4. In the NCAA final-four basketball tournament, the East champion plays the West champion, and the champions from the Mideast and Midwest play. The winners of the two games play for the national championship. Draw the eight different single-elimination tournament graphs that could occur.

5. What is the maximum number of edges in an undirected graph with eight vertices?

6. Which of the graphs in Figure 9.1.36 are isomorphic? What is the correspondence between their vertices?

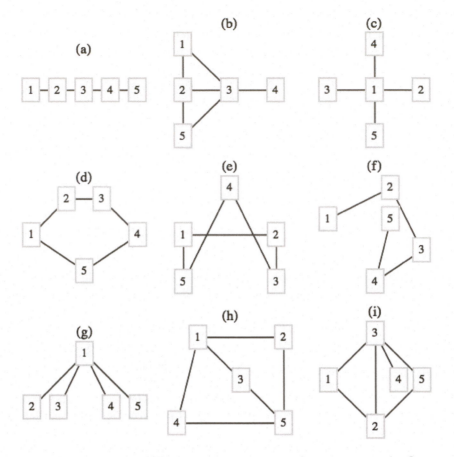

Figure 9.1.36: Which graphs are isomorphic to one another?

7.

    (a) How many edges does a complete tournament graph with n vertices have?

    (b) How many edges does a single-elimination tournament graph with n vertices have?

8.   Draw complete undirected graphs with 1, 2, 3, 4, and 5 vertices. How many edges does a K_n, a complete undirected graph with n vertices, have?

9.   Determine whether the following sequences are graphic. Explain your logic.

    (a) $(6, 5, 4, 3, 2, 1, 0)$

    (b) $(2, 2, 2, 2, 2, 2)$

    (c) $(3, 2, 2, 2, 2, 2)$

    (d) $(5, 3, 3, 3, 3, 3)$

    (e) $(1, 1, 1, 1, 1, 1)$

    (f) $(5, 5, 4, 3, 2, 1)$

10.

    (a) Based on observations you might have made in exercise 9, describe as many characteristics as you can about graphic sequences of length n.

(b) Consider the two graphs in Figure 9.1.37. Notice that they have the same degree sequences, $(2, 2, 2, 2, 2, 2)$. Explain why the two graphs are not isomorphic.

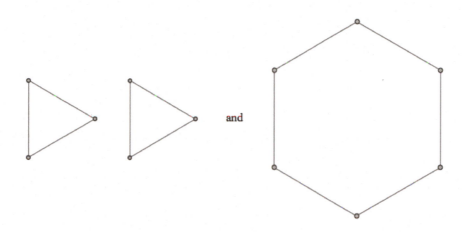

Figure 9.1.37: Two graphs with the same degree sequences

11. Draw a plan for the rooms of a house so that Figure 9.1.10 models connectedness of the rooms. That is, (a, b) is an edge if and only if a door connects rooms a and b.

12. How many subgraphs are there of a K_n, $n \geq 1$. How many of them are spanning graphs?

9.2 Data Structures for Graphs

In this section, we will describe data structures that are commonly used to represent graphs. In addition we will introduce the basic syntax for graphs in Sage.

9.2.1 Basic Data Stuctures

> Assume that we have a graph with n vertices that can be indexed by the integers $1, 2, \ldots, n$. Here are three different data structures that can be employed to represent graphs.
>
> (a) Adjacency Matrix: As we saw in Chapter 6, the information about edges in a graph can be summarized with an adjacency matrix, G, where $G_{ij} = 1$ if and only if vertex i is connected to vertex j in the graph. Note that this is the same as the adjacency matrix for a relation.
>
> (b) Edge Dictionary: For each vertex in our graph, we maintain a list of edges that initiate at that vertex. If G represents the graph's edge information, then we denote by G_i the list of vertices that are terminal vertices of edges initiating at

vertex i. The exact syntax that would be used can vary. We will use Sage/Python syntax in our examples.

(c) Edge List: Note that in creating either of the first two data structures, we would presume that a list of edges for the graph exists. A simple way to represent the edges is to maintain this list of ordered pairs, or two element sets, depending on whether the graph is intended to be directed or undirected. We will not work with this data stucture here, other than in the first example.

List 9.2.1

Example 9.2.2 A Very Small Example. We consider the representation of the following graph:

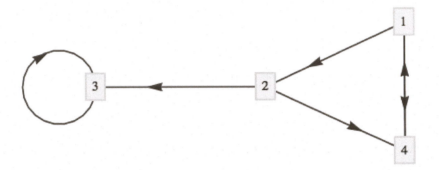

Figure 9.2.3: Graph for a Very Small Example

The adjacency matrix that represents the graph would be

$$G = \begin{pmatrix} 0 & 1 & 0 & 1 \\ 0 & 0 & 1 & 1 \\ 0 & 0 & 1 & 0 \\ 1 & 0 & 0 & 0 \end{pmatrix}.$$

The same graph could be represented with the edge dictionary

`{1:[2,4],2:[3,4],3:[3],4:[1]}`.

Notice the general form of each item in the dictionary: `vertex:[list of vertices]`.

Finally, a list of edges `[(1,2),(1,4),(2,3),(2,4),(3,3),(4,1)]` also describes the same graph.   □

A natural question to ask is: Which data structure should be used in a given situation? For small graphs, it really doesn't make much difference. For larger matrices the edge count would be a consideration. If n is large and the number of edges is relatively small, it might use less memory to maintain an edge dictionary or list of edges instead of building an $n \times n$ matrix. Some software for working with graphs will make the decision for you.

Example 9.2.4 NCAA Basketball. Consider the tournament graph representing a NCAA Division 1 men's (or women's) college basketball season in the United States. There are approximately 350 teams in Division 1. Suppose we constructed the graph with an edge from team A to team B if A beat B at least once in the season; and we label the edge with the number of wins. Since the average team plays around 30 games in a season, most of which will be against other Division I teams, we could expect around $\frac{30 \cdot 350}{2} = 5,250$ edges in the graph. This would be somewhat reduced by games with lower division teams and cases where two or more wins over the same team produces one edge. Since 5,250 is much smaller than $350^2 = 122,500$ entries in an adjacency matrix, an edge dictionary or edge list would be more compact than an adjacency matrix. Even if we were use software to create an adjacency matrix, many programs will identify the fact that a matrix such as the one in this example would be "sparse" and would leave data in list form and use sparse array methods to work with it.   □

9.2.2 Sage Graphs

The most common way to define a graph in Sage is to use an edge dictionary. Here is how the graph in Example 9.2.2 is generated and then displayed. Notice that we simply wrap the function `DiGraph()` around the same dictionary expression we identified earlier.

```
G1 = DiGraph( {1 : [4, 2], 2 : [3, 4], 3 : [3], 4 : [1]})
G1.show()
```

You can get the adjacency matrix of a graph with the `adjacency_matrix` method.

```
G1.adjacency_matrix()
```

```
[0 1 0 1]
[0 0 1 1]
[0 0 1 0]
[1 0 0 0]
```

You can also define a graph based on its adjacency matrix.

```
M = Matrix([[0,1,0,0,0],[0,0,1,0,0],[0,0,0,1,0],
                   [0,0,0,0,1],[1,0,0,0,0]])
DiGraph(M).show()
```

```
[0 1 0 1]
[0 0 1 1]
[0 0 1 0]
[1 0 0 0]
```

The edge list of any directed graph can be easily retrieved. If you replace `edges` with `edge_iterator`, you can iterate through the edge list. The third coordinate of the items in the edge is the label of the edge, which is `None` in this case.

```
DiGraph(M).edges()
```

```
[(0, 1, None), (1, 2, None), (2, 3, None), (3, 4, None), (4,
    0, None)]
```

Replacing the wrapper `DiGraph()` with `Graph()` creates an undirected graph.

```
G2 = Graph( {1 : [4, 2], 2 : [3, 4], 3 : [3], 4 : [1]})
G2.show()
```

There are many special graphs and graph families that are available in Sage through the `graphs` module. They are referenced with the prefix `graphs.` followed by the name and zero or more paramenters inside parentheses. Here are a couple of them, first a complete graph with five vertices.

```
graphs.CompleteGraph(5).show()
```

Here is a wheel graph, named for an obvious pattern of vertices and edges. We assign a name to it first and then show the graph without labeling the vertices.

```
w=graphs.WheelGraph(20)
w.show(vertex_labels=false)
```

There are dozens of graph methods, one of which determines the degree sequence of a graph. In this case, it's the wheel graph above.

```
w.degree_sequence()
```

```
[19, 3, 3, 3, 3, 3, 3, 3, 3, 3, 3, 3, 3, 3, 3, 3, 3, 3, 3, 3]
```

The degree sequence method is defined within the graphs module, but the prefix `graphs.` is not needed because the value of w inherits the graphs methods.

9.2.3 Exercises for Section 9.2

1. Estimate the number of vertices and edges in each of the following graphs. Would the graph be considered sparse, so that an adjacency matrix would be inefficient?

   (a) Vertices: Cities of the world that are served by at least one airline. Edges: Pairs of cities that are connected by a regular direct flight.

   (b) Vertices: ASCII characters. Edges: connect characters that differ in their binary code by exactly two bits.

   (c) Vertices: All English words. Edges: An edge connects word x to word y if x is a prefix of y.

2. Each edge of a graph is colored with one of the four colors red, blue, yellow, or green. How could you represent the edges in this graph using a variation of the adjacency matrix structure?

3. Directed graphs $G_1, \ldots, G_6$, each with vertex set $\{1, 2, 3, 4, 5\}$ are represented by the matrices below. Which graphs are isomorphic to one another?

$$G_1: \begin{pmatrix} 0 & 1 & 0 & 0 & 0 \\ 0 & 0 & 1 & 0 & 0 \\ 0 & 0 & 0 & 1 & 0 \\ 0 & 0 & 0 & 0 & 1 \\ 1 & 0 & 0 & 0 & 0 \end{pmatrix} \quad G_2: \begin{pmatrix} 0 & 0 & 0 & 0 & 0 \\ 0 & 0 & 1 & 0 & 0 \\ 0 & 0 & 0 & 0 & 0 \\ 1 & 1 & 1 & 0 & 1 \\ 0 & 0 & 0 & 0 & 0 \end{pmatrix} \quad G_3: \begin{pmatrix} 0 & 0 & 0 & 0 & 0 \\ 1 & 0 & 0 & 0 & 1 \\ 0 & 1 & 0 & 0 & 0 \\ 0 & 0 & 1 & 0 & 0 \\ 0 & 0 & 1 & 0 & 0 \end{pmatrix}$$

$$G_4: \begin{pmatrix} 0 & 1 & 1 & 1 & 1 \\ 0 & 0 & 0 & 0 & 0 \\ 0 & 0 & 0 & 0 & 0 \\ 0 & 0 & 1 & 0 & 0 \\ 0 & 0 & 0 & 0 & 0 \end{pmatrix} \quad G_5: \begin{pmatrix} 0 & 0 & 0 & 0 & 1 \\ 0 & 0 & 0 & 0 & 0 \\ 0 & 1 & 0 & 1 & 0 \\ 0 & 0 & 0 & 0 & 1 \\ 0 & 0 & 1 & 0 & 0 \end{pmatrix} \quad G_6: \begin{pmatrix} 0 & 0 & 0 & 1 & 0 \\ 0 & 0 & 0 & 0 & 0 \\ 1 & 1 & 0 & 0 & 0 \\ 0 & 0 & 1 & 0 & 0 \\ 0 & 0 & 0 & 1 & 0 \end{pmatrix}$$

4. The following Sage command verifies that the wheel graph with four vertices is isomorphic to the complete graph with four vertices.

```
graphs.WheelGraph(4).is_isomorphic(graphs.CompleteGraph(4))
```

```
True
```

A list of all graphs in this the `graphs` database is available via tab completion. Type "graphs." and then hit the tab key to see which graphs are available. This can be done using the Sage application or SageMath-Cloud, but not sage cells. Find some other pairs of isomorphic graphs in the database.

9.3 Connectivity

This section is devoted to a question that, when posed in relation to the graphs that we have examined, seems trivial. That question is: Given two vertices, s and t, of a graph, is there a path from s to t? If $s = t$, this question is interpreted as asking whether there is a circuit of positive length starting at s. Of course, for the graphs we have seen up to now, this question can be answered after a brief examination.

9.3.1 Preliminaries

There are two situations under which a question of this kind is nontrivial. One is where the graph is very large and an "examination" of the graph could take a considerable amount of time. Anyone who has tried to solve a maze may have run into a similar problem. The second interesting situation is when we want to pose the question to a machine. If only the information on the edges between the vertices is part of the data structure for the graph, how can you put that information together to determine whether two vertices can be connected by a path?

Note 9.3.1 Connectivity Terminology. Let v and w be vertices of a directed graph. Vertex v is connected to vertex w if there is a path from v to w. Two vertices are strongly connected if they are connected in both directions to one another. A graph is connected if, for each pair of distinct vertices, v and w, v is connected to w or w is connected to v. A graph is strongly connected if every pair of its vertices is strongly connected. For an undirected graph, in which edges can be used in either direction, the notions of strongly connected and connected are the same.

Theorem 9.3.2 Maximal Path Theorem. *If a graph has n vertices and vertex u is connected to vertex w, then there exists a path from u to w of length no more than n.*

Proof. (Indirect): Suppose u is connected to w, but the shortest path from u to w has length m, where $m > n$. A vertex list for a path of length m will have $m + 1$ vertices. This path can be represented as $(v_0, v_1, \ldots, v_m)$, where $v_0 = u$ and $v_m = w$. Note that since there are only n vertices in the graph and m vertices are listed in the path after v_0, we can apply the pigeonhole principle and be assured that there must be some duplication in the last m vertices of the vertex list, which represents a circuit in the path. This means that our path of minimum length can be reduced, which is a contradiction. ∎

9.3.2 Adjacency Matrix Method

Algorithm 9.3.3 Adjacency Matrix Method. *Suppose that the information about edges in a graph is stored in an adjacency matrix, G. The relation, r, that G defines is vrw if there is an edge connecting v to w. Recall that the composition of r with itself, r^2, is defined by vr^2w if there exists a vertex y such that vry and yrw; that is, v is connected to w by a path of length 2. We could prove by induction that the relation r^k, $k \geq 1$, is defined by vr^kw if and only if there is a path of length k from v to w. Since the transitive closure, r^+, is the union of r, r^2, r^3, ..., we can answer our connectivity question by determining the transitive closure of r, which can be done most easily by keeping our relation in matrix form. Theorem 9.3.2 is significant in our calculations because it tells us that we need only go as far as G^n to determine the matrix of the transitive closure.*

The main advantage of the adjacency matrix method is that the transitive closure matrix can answer all questions about the existence of paths between any vertices. If G^+ is the matrix of the transitive closure, v_i is connected to v_j if and only if $(G^+)_{ij} = 1$. A directed graph is connected if $(G^+)_{ij} = 1$ or $(G^+)_{ji} = 1$ for each $i \neq j$. A directed graph is strongly connected if its transitive closure matrix has no zeros.

A disadvantage of the adjacency matrix method is that the transitive closure matrix tells us whether a path exists, but not what the path is. The next algorithm solve this problem

9.3.3 Breadth-First Search

We will describe the Breadth-First Search Algorithm first with an example.

The football team at Mediocre State University (MSU) has had a bad year, 2 wins and 9 losses. Thirty days after the end of the football season, the university trustees are meeting to decide whether to rehire the head coach; things look bad for him. However, on the day of the meeting, the coach issues the following press release with results from the past year:

The Mediocre State University football team compared favorably with national champion Enormous State University this season.

- Mediocre State defeated Local A and M.

- Local A and M defeated City College.

- City College defeated Corn State U.

- ... (25 results later)

- Tough Tech defeated Enormous State University (ESU).

...and ESU went on to win the national championship!

List 9.3.4

The trustees were so impressed that they rehired the coach with a raise! How did the coach come up with such a list?

In reality, such lists exist occasionally and have appeared in newspapers from time to time. Of course they really don't prove anything since each team that defeated MSU in our example above can produce a similar, shorter chain of results. Since college football records are readily available, the coach could have found this list by trial and error. All that he needed to start with was that his team won at least one game. Since ESU lost one game, there was some hope of producing the chain.

The problem of finding this list is equivalent to finding a path in the tournament graph for last year's football season that initiates at MSU and ends at ESU. Such a graph is far from complete and is likely to be represented using edge lists. To make the coach's problem interesting, let's imagine that only the winner of any game remembers the result of the game. The coach's problem has now taken on the flavor of a maze. To reach ESU, he must communicate with the various teams along the path. One way that the coach could have discovered his list in time is by sending the following messages to the coaches of the two teams that MSU defeated during the season:

Note 9.3.5 When this example was first written, we commented that ties should be ignored. Most recent NCAA rules call for a tiebreaker in college football and so ties are no longer an issue. Email was also not common and we described the process in terms of letter, not email messages. Another change is that the coach could also have asked the MSU math department to use Mathematica or Sage to find the path!

Dear Football Coach:

    Please follow these directions exactly.

(1) If you are the coach at ESU, contact the coach at MSU now and tell him who sent you this message.

(2) If you are not the coach at ESU and this is the first message of this type that you have received, then:

- Remember from whom you received this message.

- Forward a copy of this message, signed by you, to each of the coaches whose teams you defeated during the past year.

- Ignore this message if you have received one like it already.

    Signed,
        Coach of MSU

List 9.3.6

From the conditions of this message, it should be clear that if everyone cooperates and if coaches participate within a day of receiving the message:

(1) If a path of length n exists from MSU to ESU, then the coach will know about it in n days.

(2) By making a series of phone calls, the coach can construct a path that he wants by first calling the coach who defeated

ESU (the person who sent ESU's coach that message). This coach will know who sent him a letter, and so on. Therefore, the vertex list of the desired path is constructed in reverse order.

(3) If a total of M football games were played, no more than M messages will be sent out.

(4) If a day passes without any message being sent out, no path from MSU to ESU exists.

(5) This method could be extended to construct a list of all teams that a given team can be connected to. Simply imagine a series of letters like the one above sent by each football coach and targeted at every other coach.

List 9.3.7

The general problem of finding a path between two vertices in a graph, if one exists, can be solved exactly as we solved the problem above. The following algorithm, commonly called a breadth-first search, uses a stack.

Stacks. A stack is a fundamental data structure in computer science. A common analogy used to describe stacks is of a stack of plates. If you put a plate on the top of a stack and then want to use a plate, it's natural to use that top plate. So the last plate in is the first plate out. "Last in, first out" is the short description of the rule for stacks. This is contrast with a queue which uses a "First in, first out" rule.

Algorithm 9.3.8 Breadth-first Search. *A broadcasting algorithm for finding a path between vertex i and vertex j of a graph having n vertices. Each item V_k of a list $V = \{V_1, V_2, \ldots, V_n\}$, consist of a Boolean field $V_k.found$ and an integer field $V_k.from$. The sets D_1, D_2, $\ldots$, called depth sets, have the property that if $k \in D_r$, then the shortest path from vertex i to vertex k is of length r. In Step 5, a stack is used to put the vertex list for the path from the vertex i to vertex j in the proper order.*

(1) *Set the value $V_k.found$ equal to False, $k = 1, 2, \ldots, n$*

(2) $r = 0$

(3) $D_0 = \{i\}$

(4) *while ($\neg V_j.found$) and ($D_r \neq \emptyset$)*

- $D_{r+1} = \emptyset$
- *for each k in D_r:*
    for each edge (k,t):
        If $V_t.found == False$:
        $V_t.found = True$
        $V_t.from = k$
        $D_{r+1} = D_{r+1} \cup \{t\}$
- $r = r + 1$

(5) *if $V_j.found$:*

- $S = EmptyStack$
- $k = j$
- $while\ V_k.from \neq i:$
  $Push\ k\ onto\ S$
  $k = V_k.from$

- This algorithm will produce one path from vertex i to vertex j, if one exists, and that path will be as short as possible. If more than one path of this length exists, then the one that is produced depends on the order in which the edges are examined and the order in which the elements of D_r are examined in Step 4.

- The condition $D_r \neq \emptyset$ is analogous to the condition that no mail is sent in a given stage of the process, in which case MSU cannot be connected to ESU.

- This algorithm can be easily revised to find paths to all vertices that can be reached from vertex i. Step 5 would be put off until a specific path to a vertex is needed since the information in V contains an efficient list of all paths. The algorithm can also be extended further to find paths between any two vertices.

List 9.3.9

Example 9.3.10   A simple example.   Consider the graph below. The existence of a path from vertex 2 to vertex 3 is not difficult to determine by examination. After a few seconds, you should be able to find two paths of length four. Algorithm 9.3.8 will produce one of them.

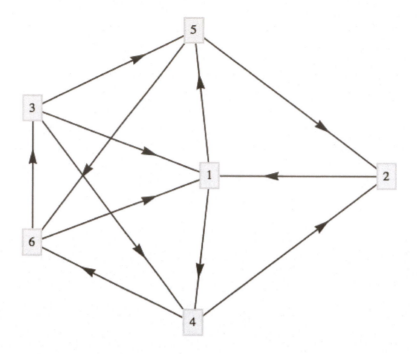

Figure 9.3.11: A simple example of breadth-first search

Suppose that the edges from each vertex are sorted in ascending order by terminal vertex. For example, the edges from vertex 3 would be in the order $(3, 1), (3, 4), (3, 5)$. In addition, assume that in the body of Step 4 of the algorithm, the elements of D_r are used in ascending order. Then at the end of Step 4, the value of V will be

k	1	2	3	4	5	6
V_k.found	T	T	T	T	T	T
V_k.from	2	4	6	1	1	4
Depthset	1	3	4	2	2	3

Therefore, the path $(2, 1, 4, 6, 3)$ is produced by the algorithm. Note that if we wanted a path from 2 to 5, the information in V produces the path $(2, 1, 5)$ since V_k.from $= 1$ and V_1.from $= 2$. A shortest circuit that initiates at vertex 2 is also available by noting that V_2.from $= 4$, V_4.from $= 1$, and V_1.from $= 2$; thus the circuit $(2, 1, 4, 2)$ is the output of the algorithm. □

9.3.4 SageMath Note - Graph Searching

The following sequence of Sage cells illustrates how searching can be done in graphs.

Generate a random undirected graph with 18 vertices. For each pair of vertices, an edge is included between them with probability 0.2. Since there are $\binom{18}{2} = 153$ potential edges, we expect that there will be approximately $0.2 \cdot 153 \approx 31$ edges. The random number generation is seeded first so that the result will always be the same in spite of the random graph function. Changing or removing that first line will let you experiment with different graphs.

```
set_random_seed(2002)
```

```
Gr=graphs.RandomGNP(18,0.2)
Gr.show()
```

Count the number of edges. In this case the number is a bit less than expected.

```
len(Gr.edges(labels=False))
```

27

Find a shortest path from vertex 0 to vertex 8.

```
Gr.shortest_path(0, 8)
```

[0, 7, 3, 8]

Generate a list of vertices that would be reached in a breadth-first search. The expression `Gr.breadth_first_search(0)` creates an iterator that is convenient for programming. Wrapping `list( )` around the expression shows the order in which the vertices are visited with the depth set indicated in the second coordinates.

```
list(Gr.breadth_first_search(0,report_distance='True'))
```

[(0, 0),(7, 1),(14, 1),(15, 1),(16, 2),(2, 2),(3, 2),(13,
    2),(17, 2),
 (4, 2),(5, 2),(10, 2),(6, 2),(11, 2),(8, 3),(1, 3),(9,
    3),(12, 3)]

Generate a list of vertices that would be reached in a depth-first search. In this type of search you travel in one direction away from the starting point until no further new vertices can be reached. We will discuss this search later.

```
list(Gr.depth_first_search(0))
```

[0, 15, 11, 10, 14, 5, 13, 7, 3, 8, 9, 12, 6, 16, 1, 2, 17, 4]

9.3.5 Exercises for Section 9.3

1. Apply Algorithm 9.3.8 to find a path from 5 to 1 in Figure 9.3.11. What would be the final value of V? Assume that the terminal vertices in edge lists and elements of the depth sets are put into ascending order, as we assumed in Example 9.3.10.

2. Apply Algorithm 9.3.8 to find a path from d to c in the road graph in Example 9.1.9 using the edge list in that example. Assume that the elements of the depth sets are put into ascending order.

3. In a simple undirected graph with no self-loops, what is the maximum number of edges you can have, keeping the graph unconnected? What is the minimum number of edges that will assure that the graph is connected?

4. Use a broadcasting algorithm to determine the shortest path from vertex a to vertex i in the graphs shown in the Figure 9.3.12 below. List the depth sets and the stack that is created.

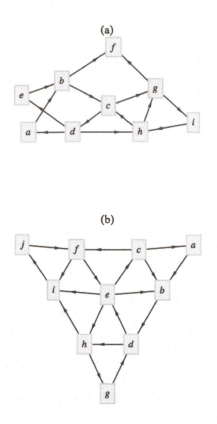

Figure 9.3.12: Shortest paths from *a* to *i*?

5.   Prove (by induction on k) that if the relation r on vertices of a graph is defined by vrw if there is an edge connecting v to w, then r^k, $k \geq 1$, is defined by vr^kw if there is a path of length k from v to w.

9.4 Traversals: Eulerian and Hamiltonian Graphs

The subject of graph traversals has a long history. In fact, the solution by Leonhard Euler (Switzerland, 1707-83) of the Koenigsberg Bridge Problem is considered by many to represent the birth of graph theory.

9.4.1 Eulerian Graphs

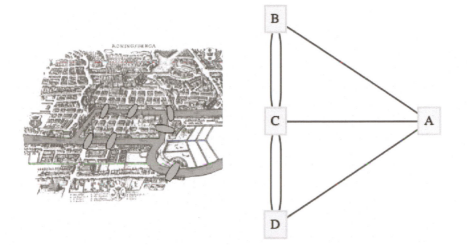

Figure 9.4.1: A map of Koenigsberg, **Figure 9.4.2:** A multigraph for the
circa 1735   bridges of Koenigsberg

A map of the Prussian city of Koenigsberg (circa 1735) in Figure 1 shows
that there were seven bridges connecting the four land masses that made up
the city. The legend of this problem states that the citizens of Koenigsberg
searched in vain for a walking tour that passed over each bridge exactly once.
No one could design such a tour and the search was abruptly abandoned with
the publication of Euler's Theorem.

Theorem 9.4.3 Euler's Theorem: Koenigsberg Case. *No walking tour
of Koenigsberg can be designed so that each bridge is used exactly once.*

Proof. The map of Koenigsberg can be represented as an undirected multigraph, as in Figure 9.4.2. The four land masses are the vertices and each edge
represents a bridge.

The desired tour is then a path that uses each edge once and only once.
Since the path can start and end at two different vertices, there are two remaining vertices that must be intermediate vertices in the path. If x is an
intermediate vertex, then every time that you visit x, you must use two of its
incident edges, one to enter and one to exit. Therefore, there must be an even
number of edges connecting x to the other vertices. Since every vertex in the
Koenigsberg graph has an odd number of edges, no tour of the type that is
desired is possible.   ∎

As is typical of most mathematicians, Euler wasn't satisfied with solving
only the Koenigsberg problem. His original theorem, which is paraphrased below, concerned the existence of paths and circuits like those sought in Koenigsberg. These paths and circuits have become associated with Euler's name.

Definition 9.4.4 Eulerian Paths, Circuits, Graphs. An Eulerian path
through a graph is a path whose edge list contains each edge of the graph
exactly once. If the path is a circuit, then it is called an Eulerian circuit. A
Eulerian graph is a graph that possesses a Eulerian circuit.   ◇

Example 9.4.5 An Eulerian Graph. Without tracing any paths, we can
be sure that the graph below has an Eulerian circuit because all vertices have
an even degree. This follows from the following theorem.

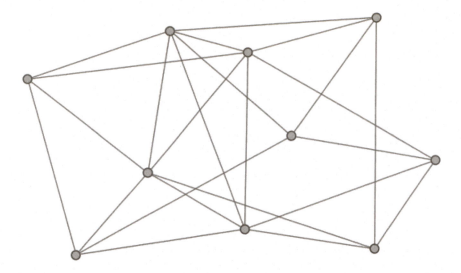

Figure 9.4.6: An Eulerian graph

□

Theorem 9.4.7 Euler's Theorem: General Case. *An undirected graph has an Eulerian path if and only if it is connected and has either zero or two vertices with an odd degree. If no vertex has an odd degree, then the graph is Eulerian.*

Proof. It can be proven by induction that the number of vertices in an undirected graph that have an odd degree must be even. We will leave the proof of this fact to the reader as an exercise. The necessity of having either zero or two vertices of odd degree is clear from the proof of the Koenigsberg case of this theorem. Therefore, we will concentrate on proving that this condition is sufficient to ensure that a graph has an Eulerian path. Let k be the number of vertices with odd degree.

Phase 1. If $k = 0$, start at any vertex, v_0, and travel along any path, not using any edge twice. Since each vertex has an even degree, this path can always be continued past each vertex that you reach except v_0. The result is a circuit that includes v_0. If $k = 2$, let v_0 be either one of the vertices of odd degree. Trace any path starting at v_0 using up edges until you can go no further, as in the $k = 0$ case. This time, the path that you obtain must end at the other vertex of odd degree that we will call v_1. At the end of Phase 1, we have an initial path that may or may not be Eulerian. If it is not Eulerian, Phase 2 can be repeated until all of the edges have been used. Since the number of unused edges is decreased in any use of Phase 2, an Eulerian path must be obtained in a finite number of steps.

Phase 2. As we enter this phase, we have constructed a path that uses a proper subset of the edges in our graph. We will refer to this path as the current path. Let V be the vertices of our graph, E the edges, and E_u the edges that have been used in the current path. Consider the graph $G' = (V, E - E_u)$. Note that every vertex in G' has an even degree. Select any edge, e, from G'. Let v_a and v_b be the vertices that e connects. Trace a new path starting at v_a whose first edge is e. We can be sure that at least one vertex of the new path is also in the current path since (V, E) is connected. Starting at v_a, there exists a path in (V, E) to any vertex in the current path. At some point along this path, which we can consider the start of the new path, we will have intersected the current path. Since the degree of each vertex in G' is even, any path that we start at v_a can be continued until it is a circuit. Now, we simply augment the current path with this circuit. As we travel along the current path, the first time that we intersect the new path, we travel along it (see Figure 9.4.8). Once we complete the circuit that is the new path, we resume the traversal of the current path.

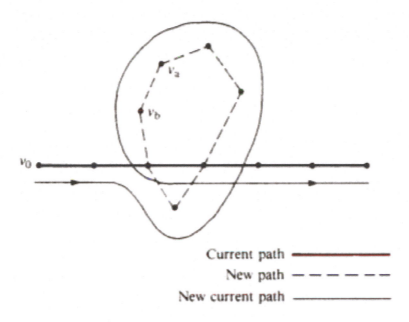

Current path ————————

New path — — — — — —

New current path ——————

Figure 9.4.8: Path Augmentation Plan

If the result of this phase is an Eulerian path, then we are finished; otherwise, repeat this phase.   ∎

Example 9.4.9 Complete Eulerian Graphs. The complete undirected graphs K_{2n+1}, $n = 1, 2, 3, \ldots$ .., are Eulerian. If $n \geq 1$, then K_{2n} is not Eulerian.   □

9.4.2 Hamiltonian Graphs

To search for a path that uses every vertex of a graph exactly once seems to be a natural next problem after you have considered Eulerian graphs. The Irish mathematician Sir William Rowan Hamilton (1805-65) is given credit for first defining such paths. He is also credited with discovering the quaternions, for which he was honored by the Irish government with a postage stamp in 2004.

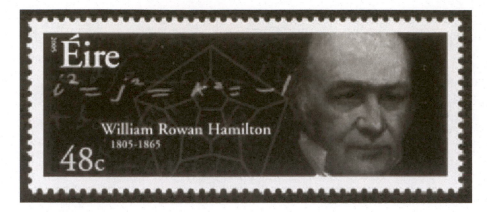

Figure 9.4.10: Irish stamp honoring Sir William Rowan Hamilton

Definition 9.4.11 Hamiltonian Path, Circuit, and Graphs. A Hamiltonian path through a graph is a path whose vertex list contains each vertex of the graph exactly once, except if the path is a circuit, in which case the initial vertex appears a second time as the terminal vertex. If the path is a circuit, then it is called a Hamiltonian circuit. A Hamiltonian graph is a graph that possesses a Hamiltonian circuit. ◊

Example 9.4.12 The Original Hamiltonian Graph. Figure 9.4.14 shows a graph that is Hamiltonian. In fact, it is the graph that Hamilton used as an example to pose the question of existence of Hamiltonian paths in 1859. In its original form, the puzzle that was posed to readers was called "Around the World." The vertices were labeled with names of major cities of the world and the object was to complete a tour of these cities. The graph is also referred to as the dodecahedron graph, where vertices correspond with the corners of a dodecahedron and the edges are the edges of the solid that connect the corners.

 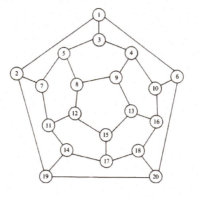

Figure 9.4.13: A Dodecahedron **Figure 9.4.14:** The Dodecahedron Graph

☐

Problem 9.4.15 Unfortunately, a simple condition doesn't exist that characterizes a Hamiltonian graph. An obvious necessary condition is that the graph be connected; however, there is a connected undirected graph with four vertices that is not Hamiltonian. Can you draw such a graph? ☐

Note 9.4.16 What Is Possible and What Is Impossible? The search for a Hamiltonian path in a graph is typical of many simple-sounding problems in graph theory that have proven to be very difficult to solve. Although there are simple algorithms for conducting the search, they are impractical for large problems because they take such a long time to complete as graph size increases. Currently, every algorithm to search for a Hamiltonian path in a graph takes a time that grows at a rate that is greater than any polynomial as a function of the number of vertices. Rates of this type are called "**super-polynomial**." That is, if $T(n)$ is the time it takes to search a graph of n vertices, and $p(n)$ is any polynomal, then $T(n) > p(n)$ for all but possibly a finite number of positive values for n.

It is an unproven but widely held belief that no faster algorithm exists to search for Hamiltonian paths in general graphs. To sum up, the problem of determining whether a graph is Hamiltonian is theoretically possible; however, for large graphs we consider it a practical impossibility. Many of the problems we will discuss in the next section, particularly the Traveling Salesman Problem, are thought to be impossible in the same sense.

Definition 9.4.17 The n-cube. Let $n \geq 1$, and let B^n be the set of strings of 0's and 1's with length n. The n-cube is the undirected graph with a vertex for each string in B^n and an edge connecting each pair of strings that differ in exactly one position. The n-cube is normally denoted Q_n. $\diamond$

The n-cube is among the graphs that are defined within the `graphs` package of SageMath and is created with the expression `graphs.CubeGraph(n)`.

```
graphs.CubeGraph(4).show(layout="spring")
```

Example 9.4.18 Analog-to-digital Conversion and the Gray Code. A common problem encountered in engineering is that of analog-to-digital (a-d) conversion, where the reading on a dial, for example, must be converted to a numerical value. In order for this conversion to be done reliably and quickly, one must solve an interesting problem in graph theory. Before this problem is posed, we will make the connection between a-d conversion and the graph problem using a simple example. Suppose a dial can be turned in any direction, and that the positions will be converted to one of the numbers zero through seven as depicted in Figure 9.4.19. The angles from 0 to 360 are divided into eight equal parts, and each part is assigned a number starting with 0 and increasing clockwise. If the dial points in any of these sectors the conversion is to the number of that sector. If the dial is on the boundary, then we will be satisfied with the conversion to either of the numbers in the bordering sectors. This conversion can be thought of as giving an approximate angle of the dial, for if the dial is in sector k, then the angle that the dial makes with east is approximately $45k°$.

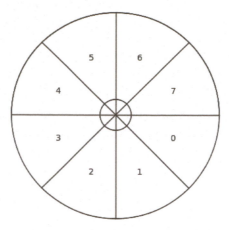

Figure 9.4.19: Analog-Digital Dial

Now that the desired conversion has been identified, we will describe a "solution" that has one major error in it, and then identify how this problem can be rectified. All digital computers represent numbers in binary form, as a sequence of 0's and 1's called bits, short for binary digits. The binary representations of numbers 0 through 7 are:

$$0 = 000_{two} = 0 \cdot 4 + 0 \cdot 2 + 0 \cdot 1$$
$$1 = 001_{two} = 0 \cdot 4 + 0 \cdot 2 + 1 \cdot 1$$
$$2 = 010_{two} = 0 \cdot 4 + 1 \cdot 2 + 0 \cdot 1$$
$$3 = 011_{two} = 0 \cdot 4 + 1 \cdot 2 + 1 \cdot 1$$
$$4 = 100_{two} = 1 \cdot 4 + 0 \cdot 2 + 0 \cdot 1$$
$$5 = 101_{two} = 1 \cdot 4 + 0 \cdot 2 + 1 \cdot 1$$
$$6 = 110_{two} = 1 \cdot 4 + 1 \cdot 2 + 0 \cdot 1$$
$$7 = 111_{two} = 1 \cdot 4 + 1 \cdot 2 + 1 \cdot 1$$

The way that we could send those bits to a computer is by coating parts of the back of the dial with a metallic substance, as in Figure 9.4.20. For each of the three concentric circles on the dial there is a small magnet. If a magnet lies under a part of the dial that has been coated with metal, then it will turn a switch ON, whereas the switch stays OFF when no metal is detected above a magnet. Notice how every ON/OFF combination of the three switches is possible given the way the back of the dial is coated.

If the dial is placed so that the magnets are in the middle of a sector, we expect this method to work well. There is a problem on certain boundaries, however. If the dial is turned so that the magnets are between sectors three and four, for example, then it is unclear what the result will be. This is due to the fact that each magnet will have only a fraction of the required metal above it to turn its switch ON. Due to expected irregularities in the coating of the dial, we can be safe in saying that for each switch either ON or OFF could be the result, and so if the dial is between sectors three and four, any number could be indicated. This problem does not occur between every sector. For example, between sectors 0 and 1, there is only one switch that cannot be predicted. No matter what the outcome is for the units switch in this case, the indicated sector must be either 0 or 1. This consistent with the original objective that a positioning of the dial on a boundary of two sectors should produce the number of either sector.

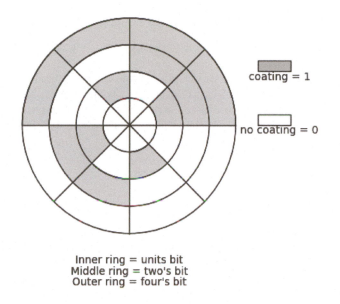

coating = 1

no coating = 0

Inner ring = units bit
Middle ring = two's bit
Outer ring = four's bit

Figure 9.4.20: Coating scheme for the Analog-Digital Dial

Is there a way to coat the sectors on the back of the dial so that each of the eight patterns corresponding to the numbers 0 to 7 appears once, and so that between any two adjacent sectors there is only one switch that will have a questionable setting? What we are describing here is a Hamiltonian circuit of the (Figure 9.4.21). If one can draw a path along the edges in the 3-cube that starts at any vertex, passes through every other vertex once, and returns to the start, then that sequence of bit patterns can be used to coat the back of the dial so that between every sector there is only one questionable switch. Such a path is not difficult to find, as we will see below.

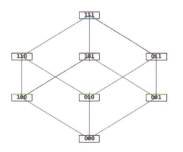

Figure 9.4.21: The 3-cube

Many A-D conversion problems require many more sectors and switches than this example, and the same kinds of problems can occur. The solution would be to find a path within a much larger yet similar graph. For example, there might be 1,024 sectors with 10 switches, resulting in a graph with 1,024. Fortunately, our solution will apply to the n-cube for any positive value of n.

A Hamiltonian circuit of the n-cube can be described recursively. The circuit itself, called the Gray Code, is not the only Hamiltonian circuit of the n-cube, but it is the easiest to describe. The standard way to write the Gray Code is as a column of strings, where the last string is followed by the first string to complete the circuit.

Basis for the Gray Code ($n = 1$): The Gray Code for the 1-cube is $G_1 =$

$\begin{pmatrix} 0 \\ 1 \end{pmatrix}$. Note that the edge between 0 and 1 is used twice in this circuit. That doesn't violate any rules for Hamiltonian circuits, but can only happen if a graph has two vertices.

Recursive definition of the Gray Code: Given the Gray Code for the n-cube, $n \geq 1$, then G_{n+1} is obtained by (1) listing G_n with each string prefixed with 0, and then (2) reversing the list of strings in G_n with each string prefixed with 1. Symbolically, the recursion can be expressed as follows, where G_n^r is the reverse of list G_n.

$$G_{n+1} = \begin{pmatrix} 0G_n \\ 1G_n^r \end{pmatrix}$$

The Gray Codes for the 2-cube and 3-cube are

$$G_2 = \begin{pmatrix} 00 \\ 01 \\ 11 \\ 10 \end{pmatrix} \text{ and } G_3 = \begin{pmatrix} 000 \\ 001 \\ 011 \\ 010 \\ 110 \\ 111 \\ 101 \\ 100 \end{pmatrix}$$

One question might come to mind at this point. If the coatings of the dial no longer in the sequence from 0 to 7, how would you interpret the patterns that are on the back of the dial as numbers from 0 to 7? In Chapter 14 we will see that if the Gray Code is used, this "decoding" is quite easy. □

Example 9.4.22 Applications of the Gray Code. One application of the Gray code was discussed in the Introduction to this book. Another application is in statistics. In a statistical analysis, there is often a variable that depends on several factors, but exactly which factors are significant may not be obvious. For each subset of factors, there would be certain quantities to be calculated. One such quantity is the multiple correlation coefficient for a subset. If the correlation coefficient for a given subset, A, is known, then the value for any subset that is obtained by either deleting or adding an element to A can be obtained quickly. To calculate the correlation coefficient for each set, we simply travel along G_n, where n is the number of factors being studied. The first vertex will always be the string of 0's, which represents the empty set. For each vertex that you visit, the set that it corresponds to contains the kth factor if the kth character is a 1. □

The 3-cube and its generalization, the n-cube, play a role in the design of a multiprocessor called a hypercube. A multiprocessor is a computer that consists of several independent processors that can operate simultaneously and are connected to one another by a network of connections. In a hypercube with $M = 2^n$ processors, the processors are numbered 0 to $M - 1$. Two processors are connected if their binary representations differ in exactly one bit. The hypercube has proven to be the best possible network for certain problems requiring the use of a "supercomputer."

9.4.3 Exercises for Section 9.4

1.   Locate a map of New York City and draw a graph that represents its land masses, bridges and tunnels. Is there an Eulerian path through New York? You can do the same with any other city that has at least two land masses.

2.   Which of the drawings in Figure 9.4.23 can be drawn without removing your pencil from the paper and without drawing any line twice?

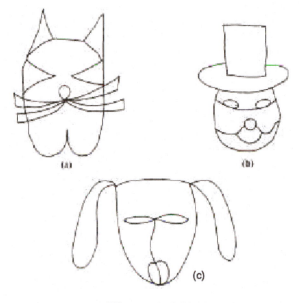

Figure 9.4.23

3.   Write out the Gray Code for the 4-cube.

4.   Find a Hamiltonian circuit for the dodecahedron graph in Figure 9.4.14.

5.   The Euler Construction Company has been contracted to construct an extra bridge in Koenigsberg so that an Eulerian path through the town exists. Can this be done, and if so, where should the bridge be built?

6.   Consider the graphs in Figure 9.4.24. Determine which of the graphs have an Eulerian path, and find an Eulerian path for the graphs that have one.

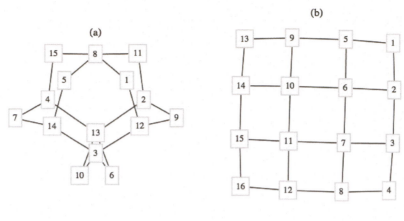

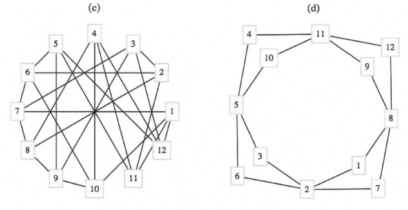

Figure 9.4.24: Graphs for exercise 6

7.   Formulate Euler's theorem for directed graphs.

8.   Prove that the number of vertices in an undirected graph with odd degree must be even.

   Hint.   Prove by induction on the number of edges.

9.

   (a) Under what conditions will a round-robin tournament graph be Eulerian?

   (b) Prove that every round-robin tournament graph is Hamiltonian.

10. For what values of n is the n-cube Eulerian?

9.5 Graph Optimization

The common thread that connects all of the problems in this section is the desire to optimize (maximize or minimize) a quantity that is associated with a graph. We will concentrate most of our attention on two of these problems, the Traveling Salesman Problem and the Maximum Flow Problem. At the close of this section, we will discuss some other common optimization problems.

9.5.1 Weighted Graphs

Definition 9.5.1 Weighted Graph. A weighted graph, (V, E, w), is a graph (V, E) together with a weight function $w : E \rightarrow \mathbb{R}$. If $e \in E$, $w(e)$ is the weight on edge e. $\diamond$

As you will see in our examples, $w(e)$ is often a cost associated with the edge e; therefore, most weights will be positive.

Example 9.5.2 A Distance Graph. Let V be the set of six capital cities in New England: Boston, Augusta, Hartford, Providence, Concord, and Montpelier. Let E be $\{\{a, b\} \in V \times V \mid a \neq b\}$; that is, (V, E) is a complete unordered graph. An example of a weight function on this graph is $w(c_1, c_2) =$ the distance, in miles, from c_1 to c_2.

Many road maps define distance functions as in the following table.

--	Augusta	Boston	Concord	Hartford	Montpelier	Providence
Augusta, ME	--	165	148	266	190	208
Boston, MA	165	--	75	103	192	43
Concord, NH	148	75	--	142	117	109
Hartford, CT	266	103	142	--	204	70
Montpelier, VT	190	192	117	204	--	223
Providence, RI	208	43	109	70	223	--

Table 9.5.3: Distances between capital cities in New England

9.5.2 The Traveling Salesman Problem

The Traveling Salesman Problem is, given a weighted graph, to find a circuit $(e_1, e_2, \ldots, e_n)$ that visits every vertex at least once and minimizes the sum of the weights, $\sum_{i=1}^{n} w(e_i)$. Any such circuit is called an optimal path.

Some statements of the Traveling Salesman Problem require that the circuit be Hamiltonian. In many applications, the graph in question will be complete and this restriction presents no problem. If the weight on each edge is constant, for example, $w(e) = 1$, then an optimal path would be any Hamiltonian circuit.

Example 9.5.4 The problem of a Boston salesman. The Traveling Salesman Problem gets its name from the situation of a salesman who wants to minimize the number of miles that he travels in visiting his customers. For example, if a salesman from Boston must visit the other capital cities of New England, then the problem is to find a circuit in the weighted graph of Example 9.5.2. Note that distance and cost are clearly related in this case. In addition, tolls and traffic congestion might also be taken into account. $\square$

The search for an efficient algorithm that solves the Traveling Salesman has occupied researchers for years. If the graph in question is complete, there are $(n-1)!$ different circuits. As n gets large, it is impossible to check every possible circuit. The most efficient algorithms for solving the Traveling Salesman Problem take an amount of time that is proportional to $n2^n$. Since this quantity grows so quickly, we can't expect to have the time to solve the Traveling Salesman Problem for large values of n. Most of the useful algorithms that have been developed have to be heuristic; that is, they find a circuit that should be close to the optimal one. One such algorithm is the "closest neighbor" algorithm, one of the earliest attempts at solving the Traveling Salesman

Problem. The general idea behind this algorithm is, starting at any vertex, to visit the closest neighbor to the starting point. At each vertex, the next vertex that is visited is the closest one that has not been reached. This shortsighted approach typifies heuristic algorithms called greedy algorithms, which attempt to solve a minimization (maximization) problem by minimizing (maximizing) the quantity associated with only the first step.

Algorithm 9.5.5 The Closest Neighbor Algorithm. *Let $G = (V, E, w)$ be a complete weighted graph with $|V| = n$. The closest neighbor circuit through G starting at v_1 is $(v_1, v_2, \ldots, v_n)$, defined by the steps:*

(1) $V_1 = V - \{v_1\}$.

(2) For $k = 2$ to $n - 1$

    (a) $v_k = $ the closest vertex in V_{k-1} to v_{k-1}:

$$w(v_{k-1}, v_k) = \min(w(v_{k-1}, v) \mid v \in V_{k-1})$$

    In case of a tie for closest, v_k may be chosen arbitrarily.

    (b) $V_k = V_{k-1} - \{v_k\}$

(3) $v_n = $ the only element of V_n

The cost of the closest neighbor circuit is $\sum_{k=1}^{n-1} w(v_k, v_{k+1}) + w(v_n, v_1)$

Example 9.5.6 A small example. The closest neighbor circuit starting at A in Figure 9.5.7 is $(1, 3, 2, 4, 1)$, with a cost of 29. The optimal path is $(1, 2, 3, 4, 1)$, with a cost of 27.

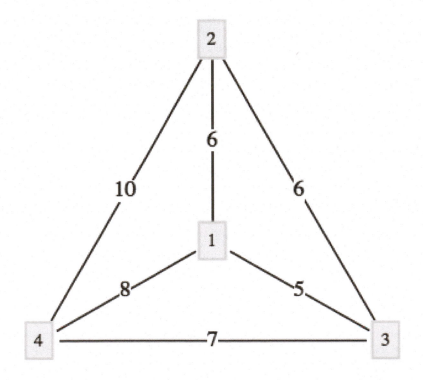

Figure 9.5.7: A small example

□

Although the closest neighbor circuit is often not optimal, we may be satisfied if it is close to optimal. If C_{opt} and C_{cn} are the costs of optimal and closest neighbor circuits in a graph, then it is always the case that $C_{opt} \leq C_{cn}$ or $\frac{C_{cn}}{C_{opt}} \geq 1$. We can assess how good the closest neighbor algorithm is by determining how small the quantity $\frac{C_{cn}}{C_{opt}}$ gets. If it is always near 1, then the algorithm is good. However, if there are graphs for which it is large, then the algorithm may be discarded. Note that in Example 9.5.6, $\frac{C_{cn}}{C_{opt}} = \frac{29}{27} \approx 1.074$. A 7 percent increase in cost may or may not be considered significant, depending on the situation.

Example 9.5.8 The One-way Street. A salesman must make stops at vertices A, B, and C, which are all on the same one-way street. The graph in Figure 9.5.9 is weighted by the function $w(i,j)$ equal to the time it takes to drive from vertex i to vertex j.

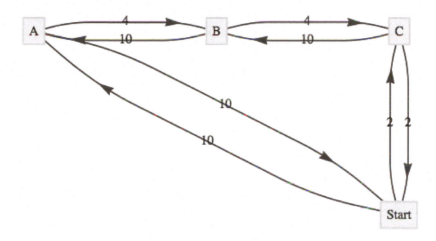

Figure 9.5.9: Traveling a one-way street

Note that if j is down the one-way street from i, then $w(i,j) < w(j,i)$. The values of C_{opt}, and C_{cn} are 20 and 32, respectively. Verify that C_{cn} is 32 by using the closest neighbor algorithm. The value of $\frac{C_{cn}}{C_{opt}} = 1.6$ is significant in this case since our salesman would spend 60 percent more time on the road if he used the closest neighbor algorithm. □

A more general result relating to the closest neighbor algorithm presumes that the graph in question is complete and that the weight function satisfies the conditions

- $w(x,v) = w(y,x)$ for all x, y in the vertex set, and

- $w(x,y) + w(y,z) \geq w(x,z)$ for all x, y, z in the vertex set.

The first condition is called the symmetry condition and the second is the triangle inequality.

Theorem 9.5.10 *If (V, E, w) is a complete weighted graph that satisfies the symmetry and triangle inequality conditions, then*

$$\frac{C_{cn}}{C_{opt}} \leq \frac{\lceil \log_2(2n) \rceil}{2}$$

Observation 9.5.11 If $|V| = 8$, then this theorem says that C_{cn} can be no larger than twice the size of C_{opt}; however, it doesn't say that the closest neighbor circuit will necessarily be that far from an optimal circuit. The quantity $\frac{\lceil \log_2(2n) \rceil}{2}$ is called an upper bound for the ratio $\frac{C_{cn}}{C_{opt}}$. It tells us only that things can't be any worse than the upper bound. Certainly, there are many graphs with eight vertices such that the optimal and closest neighbor circuits are the same. What is left unstated in this theorem is whether there are graphs for which the quantities are equal. If there are such graphs, we say that the upper bound is sharp.

The value of $\frac{C_{cn}}{C_{opt}}$ in Example Example 9.5.8 is 1.6, which is greater than $\frac{\lceil \log_2(2 \cdot 4) \rceil}{2} = 1.5$; however, the weight function in this example does not satisfy the conditions of the theorem.

Example 9.5.12 The Unit Square Problem. Suppose a robot is programmed to weld joints on square metal plates. Each plate must be welded at prescribed points on the square. To minimize the time it takes to complete the job, the total distance that a robot's arm moves should be minimized. Let $d(P, Q)$ be the distance between P and Q. Assume that before each plate can be welded, the arm must be positioned at a certain point P_0 . Given a list of n points, we want to put them in order so that

$$d\left(P_0, P_1\right) + d\left(P_1, P_2\right) + \cdots + d\left(P_{n-1}, P_n\right) + d\left(P_n, P_0\right)$$

is as small as possible.   □

The type of problem that is outlined in the example above is of such importance that it is one of the most studied version of the Traveling Salesman Problem. What follows is the usual statement of the problem. Let $[0, 1] = \{x \in \mathbb{R} \mid 0 \leq x \leq 1\}$, and let $S = [0, 1]^2$, the unit square. Given n pairs of real numbers $(x_1, y_1), (x_2, y_2), \ldots, (x_n, y_n)$ in S that represent the n vertices of a K_n, find a circuit of the graph that minimizes the sum of the distances traveled in traversing the circuit.

Since the problem calls for a circuit, it doesn't matter which vertex we start at; assume that we will start at (x_1, y_1). Once the problem is solved, we can always change our starting position. A function can most efficiently describe a circuit in this problem. Every bijection $f : \{1, ..., n\} \to \{1, ..., n\}$ with $f(1) = 1$ describes a circuit

$$(x_1, y_1), \left(x_{f(2)}, y_{f(2)}\right), \ldots, \left(x_{f(n)}, y_{f(n)}\right)$$

There are $(n-1)!$ such bijections. Since a circuit and its reversal have the same associated cost, there are $\frac{(n-1)!}{2}$ cases to consider. An examination of all possible cases is not feasible for large values of n.

One popular heuristic algorithm is the strip algorithm:

Heuristic 9.5.13 The Strip Algorithm. *Given n points in the unit square:*
  Phase 1:

(1) Divide the square into $\left\lceil \sqrt{n/2} \right\rceil$ vertical strips, as in Figure 9.5.14. Let d be the width of each strip. If a point lies on a boundary between two strips, consider it part of the left-hand strip.

(2) Starting from the left, find the first strip that contains one of the points. Locate the starting point by selecting the first point that is encountered in that strip as you travel from bottom to top. We will assume that the first point is (x_1, y_1)

(3) Alternate traveling up and down the strips that contain vertices until all

of the vertices have been reached.

(4) Return to the starting point.

Phase 2:

(1) Shift all strips d/2 units to the right (creating a small strip on the left).

(2) Repeat Steps 1.2 through 1.4 of Phase 1 with the new strips.

When the two phases are complete, choose the shorter of the two circuits obtained.

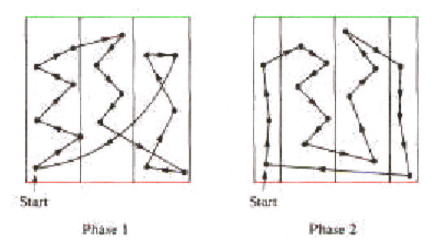

Figure 9.5.14: The Strip Algorithm

Step Item 3 may need a bit more explanation. How do you travel up or down a strip? In most cases, the vertices in a strip will be vertically distributed so that the order in which they are visited is obvious. In some cases, however, the order might not be clear, as in the third strip in Phase I of Figure 9.5.14. Within a strip, the order in which you visit the points (if you are going up the strip) is determined thusly: (x_i, y_i) precedes (x_j, y_j) if $y_i < y_j$ or if $y_i = y_j$ and $x_i < x_j$. In traveling down a strip, replace $y_i < y_j$ with $y_i > y_j$.

The selection of $\left\lceil \sqrt{n/2} \right\rceil$ strips was made in a 1959 paper by Beardwood, Halton, and Hammersley. It balances the problems that arise if the number of strips is too small or too large. If the square is divided into too few strips, some strips may be packed with vertices so that visiting them would require excessive horizontal motion. If too many strips are used, excessive vertical motion tends to be the result. An update on what is known about this algorithm is contained in [38].

Since the construction of a circuit in the square consists of sorting the given points, it should come as no surprise that the strip algorithm requires a time that is roughly a multiple of $n \log n$ time units when n points are to be visited.

The worst case that has been encountered with this algorithm is one in which the circuit obtained has a total distance of approximately $\sqrt{2n}$ (see Sopowit et al.).

9.5.3 Networks and the Maximum Flow Problem

Definition 9.5.15 Network. A network is a simple weighted directed graph that contains two distinguished vertices called the source and the sink with the properties that the indegree of the source and outdegree of the sink are both zero, and source is connected to sink. The weight function on a network is the capacity function, which has positive weights.   ◇

An example of a real situation that can be represented by a network is a city's water system. A reservoir would be the source, while a distribution point in the city to all of the users would be the sink. The system of pumps and pipes that carries the water from source to sink makes up the remaining network. We can assume that the water that passes through a pipe in one minute is controlled by a pump and the maximum rate is determined by the size of the pipe and the strength of the pump. This maximum rate of flow through a pipe is called its capacity and is the information that the weight function of a network contains.

Example 9.5.16 A City Water System. Consider the system that is illustrated in Figure 9.5.17. The numbers that appear next to each pipe indicate the capacity of that pipe in thousands of gallons per minute. This map can be drawn in the form of a network, as in Figure 9.5.18.

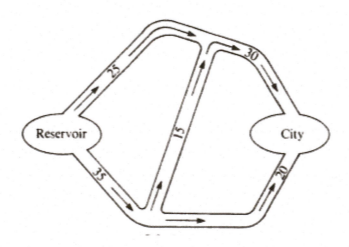

Figure 9.5.17: City Water System

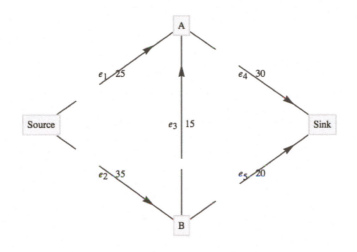

Figure 9.5.18: Flow Diagram for a City's Water Network

Although the material passing through this network is water, networks can also represent the flow of other materials, such as automobiles, electricity, bits, telephone calls, or patients in a health system. ☐

Problem 9.5.19   The Maximum Flow Problem. The Maximum Flow Problem is derived from the objective of moving the maximum amount of water or other material from the source to the sink. To measure this amount, we define a flow as a function $f : E \to \mathbb{R}$ such that (1) the flow of material through any edge is nonnegative and no larger than its capacity: $0 \leq f(e) \leq w(e)$, for all $e \in E$; and (2) for each vertex other than the source and sink, the total amount of material that is directed into a vertex is equal to the total amount that is directed out:

$$\sum_{(x,v)\in E} f(x,v) \quad = \quad \sum_{(v,y)\in E} f(v,y) \qquad (9.5.1)$$
$$\text{Flow into } v \quad = \quad \text{Flow out of } v$$

The summation on the left of (9.5.1) represents the sum of the flows through each edge in E that has v as a terminal vertex. The right-hand side indicates that you should add all of the flows through edges that initiate at v. ☐

Theorem 9.5.20   Flow out of Source equals Flow in Sink. *If f is a flow, then* $\qquad \sum_{(source,v)\in E} f(source, v) = \sum_{(v,sink)\in E} f(v, sink)$

Proof. Subtract the right-hand side of (9.5.1) from the left-hand side. The result is:

$$\text{Flow into } v - \text{ Flow out of } v = 0$$

Now sum up these differences for each vertex in $V' = V - \{source, sink\}$. The result is

$$\sum_{v\in V'} \left(\sum_{(x,v)\in E} f(x,v) - \sum_{(v,y)\in E} f(v,y) \right) = 0 \qquad (9.5.2)$$

Now observe that if an edge connects two vertices in V', its flow appears as both a positive and a negative term in (9.5.2). This means that the only positive terms that are not cancelled out are the flows into the sink. In addition, the only negative terms that remain are the flows out of the source. Therefore,

$$\sum_{(v,\text{sink})\in E} f(v, \text{sink}) - \sum_{(\text{source},v)\in E} f(\text{source}, v) = 0$$

∎

Definition 9.5.21 The Value of a Flow. The two values **flow into the sink** and **flow out of the source** were proved to be equal in Theorem 9.5.20 and this common value is called the **value of the flow**. It is denoted by $V(f)$. The value of a flow represents the amount of material that passes through the network with that flow. ◇

Since the Maximum Flow Problem consists of maximizing the amount of material that passes through a given network, it is equivalent to finding a flow with the largest possible value. Any such flow is called a **maximal flow**.

For the network in Figure 9.5.18, one flow is f_1, defined by $f_1(e_1) = 25$, $f_1(e_2) = 20$, $f_1(e_3) = 0$, $f_1(e_4) = 25$, and $f_1(e_5) = 20$. The value of f_1, $V(f_1)$, is 45. Since the total flow into the sink can be no larger than 50 ($w(e_4) + w(e_5) = 30 + 20$), we can tell that f_1 is not very far from the solution. Can you improve on f_1 at all? The sum of the capacities into the sink can't always be obtained by a flow. The same is true for the sum of the capacities out of the source. In this case, the sum of the capacities out of the source is 60, which obviously can't be reached in this network.

A solution of the Maximum Flow Problem for this network is the maximal flow f_2, where $f_2(e_1) = 25$, $f_2(e_2) = 25$, $f_2(e_3) = 5$, $f_2(e_4) = 30$, and $f_2(e_5) = 20$, with $V(f_2) = 50$. This solution is not unique. In fact, there is an infinite number of maximal flows for this problem.

There have been several algorithms developed to solve the Maximal Flow Problem. One of these is the Ford and Fulkerson Algorithm (FFA). The FFA consists of repeatedly finding paths in a network called flow augmenting paths until no improvement can be made in the flow that has been obtained.

Definition 9.5.22 Flow Augmenting Path. Given a flow f in a network (V, E), a flow augmenting path with respect to f is a simple path from the source to the sink using edges both in their forward and their reverse directions such that for each edge e in the path, $w(e) - f(e) > 0$ if e is used in its forward direction and $f(e) > 0$ if e is used in the reverse direction. ◇

Example 9.5.23 Augmenting City Water Flow. For f_1 in Figure 9.5.18, a flow augmenting path would be (e_2, e_3, e_4) since $w(e_2) - f_1(e_2) = 15$, $w(e_3) - f_1(e_3) = 5$, and $w(e_4) - f_1(e_4) = 5$.

These positive differences represent unused capacities, and the smallest value represents the amount of flow that can be added to each edge in the path. Note that by adding 5 to each edge in our path, we obtain f_2, which is maximal. If an edge with a positive flow is used in its reverse direction, it is contributing a movement of material that is counterproductive to the objective of maximizing flow. This is why the algorithm directs us to decrease the flow through that edge. □

Algorithm 9.5.24 The Ford and Fulkerson Algorithm.

(1) Define the flow function f_0 by $f_0(e) = 0$ for each edge $e \in E$.

(2) $i = 0$.

(3) Repeat:

(a) If possible, find a flow augmenting path with respect to f_i.

(b) *If a flow augmenting path exists, then:*

(i) *Determine*

$$d = \min\{\{w(e) - f_i(e) \mid e \text{ is used in the forward direction}\},$$
$$\{f_i(e) \mid e \text{ is used in the reverse direction}\}\}$$

(ii) *Define f_{i+1} by*

$$
\begin{array}{ll}
f_{i+1}(e) = f_i(e) & \text{if } e \text{ is not part of the flow augmenting path} \\
f_{i+1}(e) = f_i(e) + d & \text{if } e \text{ is used in the forward direction} \\
f_{i+1}(e) = f_i(e) - d & \text{if } e \text{ is used in the reverse direction}
\end{array}
$$

(iii) $i = i + 1$

until no flow augmenting path exists.

(4) *Terminate with a maximal flow f_i*

---

(1) It should be clear that every flow augmenting path leads to a flow of increased value and that none of the capacities of the network can be violated.

(2) The depth-first search should be used to find flow augmenting paths since it is far more efficient than the breadth-first search in this situation. The depth-first search differs from the breadth-first algorithm in that you sequentially visit vertices until you reach a "dead end" and then backtrack.

(3) There have been networks discovered for which the FFA does not terminate in a finite number of steps. These examples all have irrational capacities. It has been proven that if all capacities are positive integers, the FFA terminates in a finite number of steps. See Ford and Fulkerson, Even, or Berge for details.

(4) When you use the FFA to solve the Maximum Flow Problem by hand it is convenient to label each edge of the network with the fraction $\frac{f_i(e)}{w(e)}$.

List 9.5.25

Algorithm 9.5.26 Depth-First Search for a Flow Augmenting Path.
This is a depth-first search for the Sink Initiating at the Source. Let E' be the set of directed edges that can be used in producing a flow augmenting path. Add to the network a vertex called start and the edge (start, source).

(1) $S =$ *vertex set of the network.*

(2) $p =$ *source*   *Move p along the edge (start, source)*

(3) *while p is not equal to start or sink:*

(a) if an edge in E' exists that takes you from p to another vertex in S:

then set p to be that next vertex and delete the edge from E'

else reassign p to be the vertex that p was reached from (i.e., backtrack).

(4) if p = start:

then no flow augmenting path exists.

else p = sink and you have found a flow augmenting path.

Example 9.5.27 A flow augmenting path going against the flow.
Consider the network in Figure 9.5.28, where the current flow, f, is indicated
by a labeling of the edges.

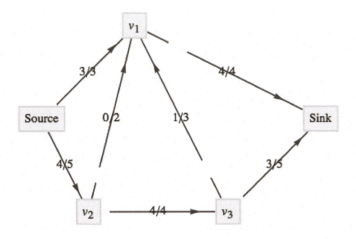

Figure 9.5.28: Current Flow

The path $(Source, v_2, v_1, v_3, Sink)$ is a flow augmenting path that allows
us to increase the flow by one unit. Note that (v_1, v_3) is used in the reverse
direction, which is allowed because $f(v_1, v_3) > 0$. The value of the new flow
that we obtain is 8. This flow must be maximal since the capacities out of the
source add up to 8. This maximal flow is defined by Figure 9.5.29.

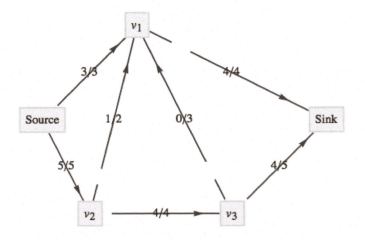

Figure 9.5.29: Updated Flow

□

9.5.4 Other Graph Optimization Problems

(1) The Minimum Spanning Tree Problem: Given a weighted graph, (V, E, w), find a subset E' of E with the properties that (V, E') is connected and the sum of the weights of edges in E' is as small as possible. We will discuss this problem in Chapter 10.

(2) The Minimum Matching Problem: Given an undirected weighted graph, (K, E, w), with an even number of vertices, pair up the vertices so that each pair is connected by an edge and the sum of these edges is as small as possible. A unit square version of this problem has been studied extensively. See [38] for details on what is known about this version of the problem.

(3) The Graph Center Problem: Given a connected, undirected, weighted graph, find a vertex (called a center) in the graph with the property that the distance from the center to every other vertex is as small as possible. "As small as possible" is normally interpreted as minimizing the maximum distance from the center to a vertex.

9.5.5 Exercises for Section 9.5

1. Find the closest neighbor circuit through the six capitals of New England starting at Boston. If you start at a different city, will you get a different circuit?

2. Is the estimate in Theorem 9.5.10 sharp for $n = 3$? For $n = 4$?

3. Given the following sets of points in the unit square, find the shortest circuit that visits all the points and find the circuit that is obtained with the strip algorithm.

    (a) $\{(0.1k, 0.1k) : k = 0, 1, 2, ..., 10\}$

    (b) $\{(0.1, 0.3), (0.3, 0.8), (0.5, 0.3), (0.7, 0.9), (0.9, 0.1)\}$

    (c) $\{(0.0, 0.5), (0.5, 0.0), (0.5, 1.0), (1.0, 0.5)\}$

    (d) $\{(0, 0), (0.2, 0.6), (0.4, 0.1), (0.6, 0.8), (0.7, 0.5)\}$

4. For $n = 4, 5,$ and 6, locate n points in the unit square for which the strip algorithm works poorly.

5. Consider the network whose maximum capacities are shown on the following graph.

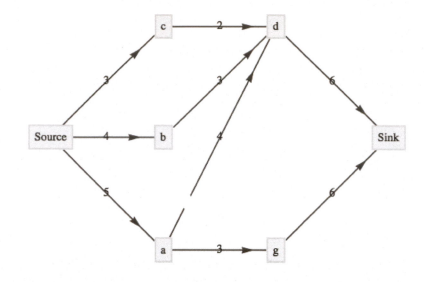

Figure 9.5.30

(a) A function f is partially defined on the edges of this network by: $f(\text{Source}, c) = 2$, $f(\text{Source}, b) = 2$, $f(\text{Source}, a) = 2$, and $f(a, d) = 1$. Define f on the rest of the other edges so that f is a flow. What is the value of f ?

(b) Find a flow augmenting path with respect to f for this network. What is the value of the augmented flow?

(c) Is the augmented flow a maximum flow? Explain.

6. Given the following network with capacity function c and flow function f, find a maximal flow function. The labels on the edges of the network are of the form $f(e)/c(e)$, where $c(e)$ is the capacity of edge e and $f(e)$ is the used capacity for flow f.

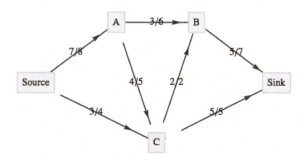

Figure 9.5.31

7. Find maximal flows for the following networks.

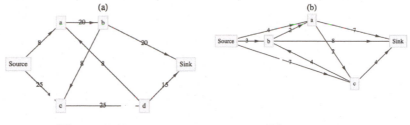

Figure 9.5.32   Figure 9.5.33

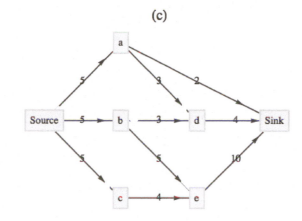

Figure 9.5.34

8.

(a) Find two maximal flows for the network in Figure 9.5.28 other than the one found in the text.

(b) Describe the set of all maximal flows for the same network.

(c) Prove that if a network has two maximal flows, then it has an infinite number of maximal flows.

9. Discuss reasons that the closest neighbor algorithm is not used in the unit square version of the Traveling Salesman Problem.

   Hint. Count the number of comparisons of distances that must be done.

10. Explore the possibility of solving the Traveling Salesman Problem in the "unit box": $[0,1]^3$.

11. Devise a "closest neighbor" algorithm for matching points in the unit square.

9.6 Planarity and Colorings

The topics in this section are related to how graphs are drawn.

Planarity: Can a given graph be drawn in a plane so that no edges intersect? Certainly, it is natural to avoid intersections, but up to now we haven't gone out of our way to do so.

Colorings: Suppose that each vertex in an undirected graph is to be colored so that no two vertices that are connected by an edge have the same color. How many colors are needed? This question is motivated by the problem of drawing a map so that no two bordering countries are colored the same. A similar question can be asked for coloring edges.

9.6.1 Planar Graphs

Definition 9.6.1 Planar Graph/Plane Graph. A graph is planar if it can be drawn in a plane so that no edges cross. If a graph is drawn so that no edges intersect, it is a plane graph, and such a drawing is a planar embedding of the graph.   ◇

Example 9.6.2 A Planar Graph. The graph in Figure 9.6.3(a) is planar but not a plane graph. The same graph is drawn as a plane graph in Figure 9.6.3(b).

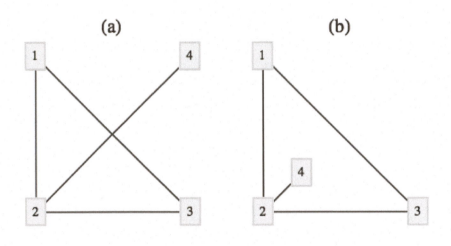

Figure 9.6.3: A Planar Graph

□

(a) In discussing planarity, we need only consider simple undirected graphs with no self-loops. All other graphs can be treated as such since all of the edges that relate any two vertices can be considered as one "package" that clearly can be drawn in a plane.

(b) Can you think of a graph that is not planar? How would you prove that it isn't planar? Proving the nonexistence of something is usually more difficult than proving its existence. This case is no exception. Intuitively, we would expect that sparse graphs would be planar and dense graphs would be nonplanar. Theorem 9.6.10 will verify that dense graphs are indeed nonplanar.

(c) The topic of planarity is a result of trying to restrict a graph to two dimensions. Is there an analogous topic for three dimensions? What graphs can be drawn in one dimension?

Definition 9.6.4 Path Graph. A path graph of length n, denoted P_n, is an undirected graph with $n + 1$ vertices $v_0, v_1, \ldots, v_n$ having n edges $\{v_i, v_{i+1}\}$, $i = 0, 1, \ldots, n - 1$.   ◇

Observation 9.6.5 Graphs in other dimensions. If a graph has only a finite number of vertices, it can always be drawn in three dimensions with no edge crossings. Is this also true for all graphs with an infinite number of vertices? The only "one-dimensional" graphs are graphs consisting of a single vertix, and path graphs, as shown in Figure 9.6.6.

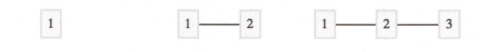

Figure 9.6.6: One dimensional graphs

A discussion of planarity is not complete without mentioning the famous Three Utilities Puzzle. The object of the puzzle is to supply three houses, A, B, and C, with the three utilities, gas, electric, and water. The constraint that makes this puzzle impossible to solve is that no utility lines may intersect. There is no planar embedding of the graph in Figure 9.6.7, which is commonly denoted $K_{3,3}$. This graph is one of two fundamental nonplanar graphs. The Kuratowski Reduction Theorem states that if a graph is nonplanar then "contains" either a $K_{3,3}$ or a K_5. Containment is in the sense that if you start with a nonplanar graph you can always perform a sequence of edge deletions and contractions (shrinking an edge so that the two vertices connecting it coincide) to produce one of the two graphs.

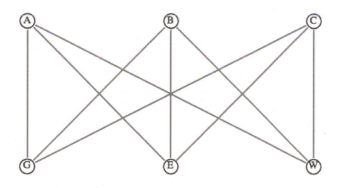

Figure 9.6.7: The Three Utilities Puzzle

A planar graph divides the plane into one or more regions. Two points on the plane lie in the same region if you can draw a curve connecting the two points that does not pass through an edge. One of these regions will be of infinite area. Each point on the plane is either a vertex, a point on an edge, or a point in a region. A remarkable fact about the geography of planar graphs is the following theorem that is attributed to Euler.

Activity 9.6.1

(a) Experiment: Jot down a graph right now and count the number of vertices, regions, and edges that you have. If $v + r - e$ is not 2, then your graph is either nonplanar or not connected.

Theorem 9.6.8 Euler's Formula. *If $G = (V, E)$ is a connected planar graph with r regions, v vertices and e edges, then*

$$v + r - e = 2 \tag{9.6.1}$$

Proof. We prove Euler's Formula by Induction on e, for $e \geq 0$.

Basis: If $e = 0$, then G must be a graph with one vertex, $v = 1$; and there is one infinite region, $r = 1$. Therefore, $v + r - e = 1 + 1 - 0 = 2$, and the basis is true.

Induction: Suppose that G has k edges, $k \geq 1$, and that all connected planar graphs with less than k edges satisfy (9.6.1). Select any edge that is part of the boundary of the infinite region and call it e_1. Let G' be the graph obtained from G by deleting e_1. Figure 9.6.9 illustrates the two different possibilities we need to consider: either G' is connected or it has two connected components, G_1 and G_2.

Case 1:

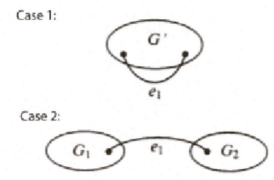

Case 2:

Figure 9.6.9: Two cases in the proof of Euler's Formula

If G' is connected, the induction hypothesis can be applied to it. If G' has v' vertices, r' edges and e' edges, then $v' + r' - e' = 2$ and in terms of the corresponding numbers for G,

$v' = v$ No vertices were removed to form G'

$r' = r - 1$ One region of G was merged with the infinite region when e_1 was removed

$e' = k - 1$ We assumed that G had k edges.

For the case where G' is connected,

$$\begin{aligned}
v + r - e &= v + r - k \\
&= v' + (r' + 1) - (e' + 1) \\
&= v' + r' - e' \\
&= 2
\end{aligned}$$

If G' is not connected, it must consist of two connected components, G_1 and G_2, since we started with a connected graph, G. We can apply the induction hypothesis to each of the two components to complete the proof. We leave it to the students to do this, with the reminder that in counting regions, G_1 and G_2 will share the same infinite region. ∎

Theorem 9.6.10 A Bound on Edges of a Planar Graph. *If $G = (V, E)$ is a connected planar graph with v vertices, $v \geq 3$, and e edges, then*

$$e \leq 3v - 6 \tag{9.6.2}$$

Proof. (Outline of a Proof)

(a) Let r be the number of regions in G. For each region, count the number of edges that comprise its border. The sum of these counts must be at least $3r$. Recall that we are working with simple graphs here, so a region made by two edges connecting the same two vertices is not possible.

(b) Based on (a), infer that the number of edges in G must be at least $\frac{3r}{2}$.

(c) $e \geq \frac{3r}{2} \Rightarrow r \leq \frac{2e}{3}$

(d) Substitute $\frac{2e}{3}$ for r in Euler's Formula to obtain an inequality that is equivalent to (9.6.2)

∎

Remark 9.6.11 One implication of (9.6.2) is that the number of edges in a connected planar graph will never be larger than three times its number of vertices (as long as it has at least three vertices). Since the maximum number of edges in a graph with v vertices is a quadratic function of v, as v increases, planar graphs are more and more sparse.

The following theorem will be useful as we turn to graph coloring.

Theorem 9.6.12 A Vertex of Degree Five. *If G is a connected planar graph, then it has a vertex with degree 5 or less.*

Proof. (by contradiction): We can assume that G has at least seven vertices, for otherwise the degree of any vertex is at most 5. Suppose that G is a connected planar graph and each vertex has a degree of 6 or more. Then, since each edge contributes to the degree of two vertices, $e \geq \frac{6v}{2} = 3v$. However, Theorem 9.6.10 states that the $e \leq 3v - 6 < 3v$, which is a contradiction. ∎

9.6.2 Graph Coloring

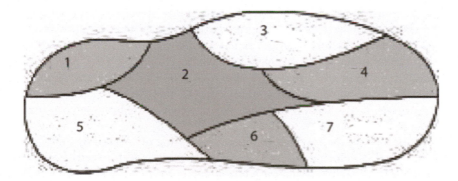

Figure 9.6.13: A 3-coloring of Euler Island

The map of Euler Island in Figure 9.6.13 shows that there are seven towns on the island. Suppose that a cartographer must produce a colored map in which no two towns that share a boundary have the same color. To keep costs down, she wants to minimize the number of different colors that appear on the map. How many colors are sufficient? For Euler Island, the answer is three. Although it might not be obvious, this is a graph problem. We can represent the map with a graph, where the vertices are countries and an edge between two vertices indicates that the two corresponding countries share a boundary of positive length. This problem motivates a more general problem.

Definition 9.6.14 Graph Coloring. Given an undirected graph $G = (V, E)$, find a "coloring function" f from V into a set of colors H such that $(v_i, v_j) \in E \Rightarrow f(v_i) \neq f(v_j)$ and H has the smallest possible cardinality. The cardinality of H is called the chromatic number of G, $\chi(G)$. ◇

- A coloring function onto an n-element set is called an n-coloring.

- In terms of this general problem, the chromatic number of the graph of Euler Island is three. To see that no more than three colors are needed, we need only display a 3-coloring: $f(1) = f(4) = f(6) = $ blue, $f(2) = $ red, and $f(3) = f(5) = f(7) = $ white. This coloring is not unique. The next smallest set of colors would be of two colors, and you should be able to convince yourself that no 2-coloring exists for this graph.

In the mid-nineteenth century, it became clear that the typical planar graph had a chromatic number of no more than 4. At that point, mathematicians attacked the Four-Color Conjecture, which is that if G is any planar graph, then its chromatic number is no more than 4. Although the conjecture is quite easy to state, it took over 100 years, until 1976, to prove the conjecture in the affirmative.

Theorem 9.6.15 The Four-Color Theorem. *If G is a planar graph, then* $\chi(G) \leq 4$.

A proof of the Four-Color Theorem is beyond the scope of this text, but we can prove a theorem that is only 25 percent inferior.

Theorem 9.6.16 The Five-Color Theorem. *If G is a planar graph, then* $\chi(G) \leq 5$.

Proof. The number 5 is not a sharp upper bound for $\chi(G)$ because of the Four-Color Theorem.

This is a proof by Induction on the Number of Vertices in the Graph.

Basis: Clearly, a graph with one vertex has a chromatic number of 1.

Induction: Assume that all planar graphs with $n - 1$ vertices have a chromatic number of 5 or less. Let G be a planar graph. By Theorem 9.6.12, there exists a vertex v with $\deg v \leq 5$. Let $G - v$ be the planar graph obtained by deleting v and all edges that connect v to other vertices in G. By the induction hypothesis, $G - v$ has a 5-coloring. Assume that the colors used are red, white, blue, green, and yellow.

If $\deg v < 5$, then we can produce a 5-coloring of G by selecting a color that is not used in coloring the vertices that are connected to v with an edge in G.

If $\deg v = 5$, then we can use the same approach if the five vertices that are adjacent to v are not all colored differently. We are now left with the possibility that v_1, v_2, v_3, v_4, and v_5 are all connected to v by an edge and they are all colored differently. Assume that they are colored red, white blue, yellow, and green, respectively, as in Figure 9.6.17.

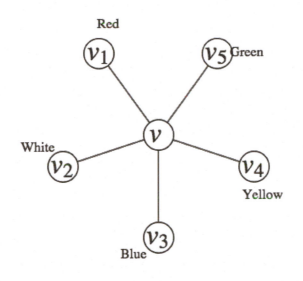

Figure 9.6.17

Starting at v_1 in $G - v$, suppose we try to construct a path v_3 that passes through only red and blue vertices. This can either be accomplished or it can't be accomplished. If it can't be done, consider all paths that start at v_1, and go through only red and blue vertices. If we exchange the colors of the vertices in these paths, including v_1 we still have a 5-coloring of $G - v$. Since v_1 is now blue, we can color the central vertex, v, red.

Finally, suppose that v_1 is connected to v_3 using only red and blue vertices. Then a path from v_1 to v_3 by using red and blue vertices followed by the edges (v_3, v) and (v, v_1) completes a circuit that either encloses v_2 or encloses v_4 and v_5. Therefore, no path from v_2 to v_4 exists using only white and yellow vertices. We can then repeat the same process as in the previous paragraph with v_2 and v_4, which will allow us to color v white. ■

Definition 9.6.18 Bipartite Graph. A bipartite graph is a graph that has a 2-coloring. Equivalently, a graph is bipartite if its vertices can be partitioned into two nonempty subsets so that no edge connects vertices from the same subset. ◇

Example 9.6.19 A Few Examples.

(a) The graph of the Three Utilities Puzzle is bipartite. The vertices are partitioned into the utilities and the homes. Of course a 2-coloring of the graph is to color the utilities red and the homes blue.

(b) For $n \geq 1$, the n-cube is bipartite. A coloring would be to color all strings with an even number of 1's red and the strings with an odd number of 1's blue. By the definition of the n-cube, two strings that have the same color couldn't be connected since they would need to differ in at least two positions.

(c) Let V be a set of 64 vertices, one for each square on a chess board. We can index the elements of V by v_{ij} = the square on the row i, column j. Connect vertices in V according to whether or not you can move a knight from one square to another. Using our indexing of V,

$(v_{ij}, v_{kl}) \in E$ if and only if $\begin{array}{l} |i-k| + |j-l| = 3 \\ \text{and } |i-k| \cdot |j-l| = 2 \end{array}$ (V, E) is a bipartite graph. The usual coloring of a chessboard is valid 2-coloring.

$\square$

How can you recognize whether a graph is bipartite? Unlike planarity, there is a nice equivalent condition for a graph to be bipartite.

Theorem 9.6.20 No Odd Circuits in a Bipartite Graph. *An undirected graph is bipartite if and only if it has no circuit of odd length.*

Proof. ($\Rightarrow$) Let $G = (V, E)$ be a bipartite graph that is partitioned into two sets, R(ed) and B(lue) that define a 2-coloring. Consider any circuit in V. If we specify a direction in the circuit and define f on the vertices of the circuit by

$$f(u) = \text{the next vertex in the circuit after } v$$

Note that f is a bijection. Hence the number of red vertices in the circuit equals the number of blue vertices, and so the length of the circuit must be even.

($\Longleftarrow$) Assume that G has no circuit of odd length. For each component of G, select any vertex w and color it red. Then for every other vertex v in the component, find the path of shortest distance from w to v. If the length of the path is odd, color v blue, and if it is even, color v red. We claim that this method defines a 2-coloring of G. Suppose that it does not define a 2-coloring. Then let v_a and v_b be two vertices with identical colors that are connected with an edge. By the way that we colored G, neither v_a nor v_b could equal w. We can now construct a circuit with an odd length in G. First, we start at w and follow the shortest path to v_a . Then follow the edge (v_a, v_b), and finally, follow the reverse of a shortest path from w to v_b. Since v_a and v_b have the same color, the first and third segments of this circuit have lengths that are both odd or even, and the sum of their lengths must be even. The addition of the single edge (v_a, v_b) shows us that this circuit has an odd length. This contradicts our premise. $\blacksquare$

9.6.3 Exercises for Section 9.6

1.   Apply Theorem 9.6.12 to prove that once n gets to a certain size, a K_n is nonplanar. What is the largest complete planar graph?

2.   Can you apply Theorem 9.6.12 to prove that the Three Utilities Puzzle can't be solved?

3.   What are the chromatic numbers of the following graphs?

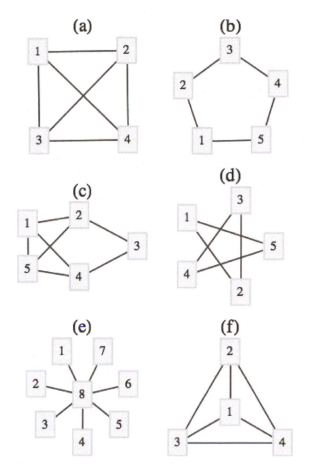

Figure 9.6.21: What are the chromatic numbers?

4. Prove that if an undirected graph has a subgraph that is a K_3 it then its chromatic number is at least 3.

5. What is $\chi(K_n)$, $n \geq 1$?

6. What is the chromatic number of the United States?

7. Complete the proof of Euler's Formula.

8. Use the outline of a proof of Theorem 9.6.10 to write a complete proof. Be sure to point out where the premise $v \geq 3$ is essential.

9. Let $G = (V, E)$ with $|V| \geq 11$, and let U be the set of all undirected edges between distinct vertices in V. Prove that either G or $G' = (V, E^c)$ is nonplanar.

10. Design an algorithm to determine whether a graph is bipartite.

11. Prove that a bipartite graph with an odd number of vertices greater than or equal to 3 has no Hamiltonian circuit.

12. Prove that any graph with a finite number of vertices can be drawn in three dimensions so that no edges intersect.

13. Suppose you had to color the edges of an undirected graph so that for each vertex, the edges that it is connected to have different colors. How can this problem be transformed into a vertex coloring problem?

14.

   (a) Suppose the edges of a K_6 are colored either red or blue. Prove that there will be either a "red K_3" (a subset of the vertex set with three vertices connected by red edges) or a "blue K_3" or both.

(b) Suppose six people are selected at random. Prove that either there exists a subset of three of them with the property that any two people in the subset can communicate in a common language, or there exist three people, no two of whom can communicate in a common language.

15. Let d be a positive integer, and let $a_1, a_2, \ldots a_d$ be positive integers greater than or equal to two. The **mesh graph** $M(a_1, a_2, \ldots, a_d)$ has vertices of the form $x = (x_1, x_2, \ldots, x_d)$ where $1 \le x_i \le a_i$. Two vertices x and y are adjacent if and only if $\sum_{i=1}^{d} |x_i - y_i| = 1$. In other words, two adjacent vertices must differ in only one coordinate and by a difference of 1.

(a) What is the chromatic number of $M(a_1, a_2, \ldots, a_d)$?

(b) For what pairs (a_1, a_2) does $M(a_1, a_2)$ have a Hamiltonian circuit?

(c) For what triples (a_1, a_2, a_3) does $M(a_1, a_2, a_3)$ have a Hamiltonian circuit?

9.6.4 Further Reading

[1]   Wilson, R., *Four Colors Suffice - How the Map Problem Was Solved* Princeton, NJ: Princeton U. Press, 2013.

Chapter 10

Trees

In this chapter we will study the class of graphs called trees. Trees are frequently used in both mathematics and the sciences. Our solution of Example 2.1.1 is one simple instance. Since they are often used to illustrate or prove other concepts, a poor understanding of trees can be a serious handicap. For this reason, our ultimate goals are to: (1) define the various common types of trees, (2) identify some basic properties of trees, and (3) discuss some of the common applications of trees.

10.1 What Is a Tree?

10.1.1 Definition

What distinguishes trees from other types of graphs is the absence of certain paths called cycles. Recall that a path is a sequence of consecutive edges in a graph, and a circuit is a path that begins and ends at the same vertex.

Definition 10.1.1 Cycle. A cycle is a circuit whose edge list contains no duplicates. It is customary to use C_n to denote a cycle with n edges. ◇

The simplest example of a cycle in an undirected graph is a pair of vertices with two edges connecting them. Since trees are cycle-free, we can rule out all multigraphs from consideration as trees.

Trees can either be undirected or directed graphs. We will concentrate on the undirected variety in this chapter.

Definition 10.1.2 Tree. An undirected graph is a tree if it is connected and contains no cycles or self-loops. ◇

Example 10.1.3 Some trees and non-trees.

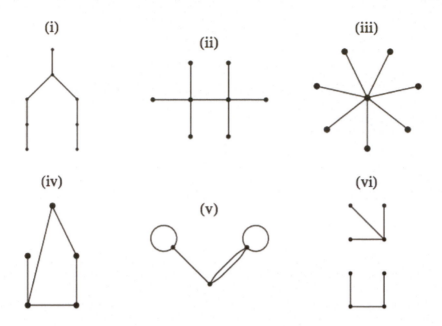

Figure 10.1.4: Some trees and some non-trees

(a) Graphs i, ii and iii in Figure 10.1.4 are all trees, while graphs iv, v, and vi are not trees.

(b) A K_2 is a tree. However, if $n \geq 3$, a K_n is not a tree.

(c) In a loose sense, a botanical tree is a mathematical tree. There are usually no cycles in the branch structure of a botanical tree.

(d) The structures of some chemical compounds are modeled by a tree. For example, butane Figure 10.1.5 consists of four carbon atoms and ten hydrogen atoms, where an edge between two atoms represents a bond between them. A bond is a force that keeps two atoms together. The same set of atoms can be linked together in a different tree structure to give us the compound isobutane Figure 10.1.6. There are some compounds whose graphs are not trees. One example is benzene Figure 10.1.7.

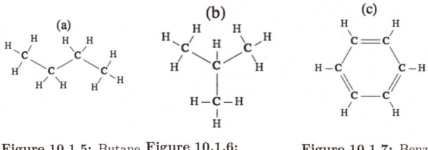

Figure 10.1.5: Butane **Figure 10.1.6:**   **Figure 10.1.7:** Benzene
                    Isobutane

□

One type of graph that is not a tree, but is closely related, is a forest.

Definition 10.1.8 Forest. A forest is an undirected graph whose components are all trees. ◇

Example 10.1.9 A forest. The top half of Figure 10.1.4 can be viewed as a forest of three trees. Graph (vi) in this figure is also a forest. □

10.1.2 Conditions for a graph to be a tree

We will now examine several conditions that are equivalent to the one that defines a tree. The following theorem will be used as a tool in proving that the conditions are equivalent.

Lemma 10.1.10 *Let $G = (V, E)$ be an undirected graph with no self-loops, and let $v_a, v_b \in V$. If two different simple paths exist between v_a and v_b, then there exists a cycle in G.*

Proof. Let $p_1 = (e_1, e_2, \ldots, e_m)$ and $p_2 = (f_1, f_2, \ldots, f_n)$ be two different simple paths from v_a to v_b. The first step we will take is to delete from p_1 and p_2 the initial edges that are identical. That is, if $e_1 = f_1$, $e_2 = f_2, \ldots, e_j = f_j$, and $e_{j+1} \neq f_{j+1}$ delete the first j edges of both paths. Once this is done, both paths start at the same vertex, call it v_c, and both still end at v_b. Now we construct a cycle by starting at v_c and following what is left of p_1 until we first meet what is left of p_2. If this first meeting occurs at vertex v_d, then the remainder of the cycle is completed by following the portion of the reverse of p_2 that starts at v_d and ends at v_c. ∎

Theorem 10.1.11 Equivalent Conditions for a Graph to be a Tree. *Let $G = (V, E)$ be an undirected graph with no self-loops and $|V| = n$. The following are all equivalent:*

(1) G is a tree.

(2) For each pair of distinct vertices in V, there exists a unique simple path between them.

(3) G is connected, and if $e \in E$, then $(V, E - \{e\})$ is disconnected.

(4) G contains no cycles, but by adding one edge, you create a cycle.

(5) G is connected and $|E| = n - 1$.

Proof. Proof Strategy. Most of this theorem can be proven by proving the following chain of implications: $(1) \Rightarrow (2)$, $(2) \Rightarrow (3)$, $(3) \Rightarrow (4)$, and $(4) \Rightarrow (1)$. Once these implications have been demonstrated, the transitive closure of $\Rightarrow$ on $1, 2, 3, 4$ establishes the equivalence of the first four conditions. The proof that Statement 5 is equivalent to the first four can be done by induction, which we will leave to the reader.

$(1) \Rightarrow (2)$ (Indirect). Assume that G is a tree and that there exists a pair of vertices between which there is either no path or there are at least two distinct paths. Both of these possibilities contradict the premise that G is a tree. If no path exists, G is disconnected, and if two paths exist, a cycle can be obtained by Theorem 10.1.11.

$(2) \Rightarrow (3)$. We now use Statement 2 as a premise. Since each pair of vertices in V are connected by exactly one path, G is connected. Now if we select any edge e in E, it connects two vertices, v_1 and v_2. By (2), there is no simple path connecting v_1 to v_2 other than e. Therefore, no path at all can exist between v_1 and v_2 in $(V, E - \{e\})$. Hence $(V, E - \{e\})$ is disconnected.

$(3) \Rightarrow (4)$. Now we will assume that Statement 3 is true. We must show that G has no cycles and that adding an edge to G creates a cycle. We will use an indirect proof for this part. Since (4) is a conjunction, by DeMorgan's Law its negation is a disjunction and we must consider two cases. First, suppose that G has a cycle. Then the deletion of any edge in the cycle keeps the graph connected, which contradicts (3). The second case is that the addition of an edge to G does not create a cycle. Then there are two distinct paths between the vertices that the new edge connects. By Lemma 10.1.10, a cycle can then be created, which is a contradiction.

(4) $\Rightarrow$ (1) Assume that G contains no cycles and that the addition of an edge creates a cycle. All that we need to prove to verify that G is a tree is that G is connected. If it is not connected, then select any two vertices that are not connected. If we add an edge to connect them, the fact that a cycle is created implies that a second path between the two vertices can be found which is in the original graph, which is a contradiction. ∎

The usual definition of a directed tree is based on whether the associated undirected graph, which is created by "erasing" its directional arrows, is a tree. In Section 10.3 we will introduce the rooted tree, which is a special type of directed tree.

10.1.3 Exercises for Section 10.1

1. Given the following vertex sets, draw all possible undirected trees that connect them.

    (a) $V_a = \{\text{right}, \text{left}\}$

    (b) $V_b = \{+, -, 0\}$

    (c) $V_c = \{\text{north}, \text{south}, \text{east}, \text{west}\}$.

2. Are all trees planar? If they are, can you explain why? If they are not, you should be able to find a nonplanar tree.

3. Prove that if G is a simple undirected graph with no self-loops, then G is a tree if and only if G is connected and $|E| = |V| - 1$.

    Hint. Use induction on $|E|$.

4.

    (a) Prove that if $G = (V, E)$ is a tree and $e \in E$, then $(V, E - \{e\})$ is a forest of two trees.

    (b) Prove that if (V_1, E_1) and (V_2, E_2) are disjoint trees and e is an edge that connects a vertex in V_1 to a vertex in V_2, then $(V_1 \cup V_2, E_1 \cup E_2 \cup \{e\})$ is a tree.

5.

    (a) Prove that any tree with at least two vertices has at least two vertices of degree 1.

    (b) Prove that if a tree has n vertices, $n \geq 4$, and is not a path graph, P_n, then it has at least three vertices of degree 1.

10.2 Spanning Trees

10.2.1 Motivation

The topic of spanning trees is motivated by a graph-optimization problem.

A graph of Atlantis University (Figure 10.2.1) shows that there are four campuses in the system. A new secure communications system is being installed and the objective is to allow for communication between any two campuses; to achieve this objective, the university must buy direct lines between certain pairs of campuses. Let G be the graph with a vertex for each campus and an edge for each direct line. Total communication is equivalent to G being a connected graph. This is due to the fact that two campuses can communicate over any number of lines. To minimize costs, the university wants to buy a minimum number of lines.

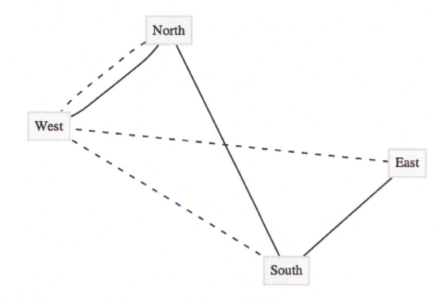

Figure 10.2.1: Atlantis University Graph

The solutions to this problem are all trees. Any graph that satisfies the requirements of the university must be connected, and if a cycle does exist, any line in the cycle can be deleted, reducing the cost. Each of the sixteen trees that can be drawn to connect the vertices North, South, East, and West (see Exercise 10.1.3.1) solves the problem as it is stated. Note that in each case, three direct lines must be purchased. There are two considerations that can help reduce the number of solutions that would be considered.

- Objective 1: Given that the cost of each line depends on certain factors, such as the distance between the campuses, select a tree whose cost is as low as possible.

- Objective 2: Suppose that communication over multiple lines is noisier as the number of lines increases. Select a tree with the property that the maximum number of lines that any pair of campuses must use to communicate with is as small as possible.

Typically, these objectives are not compatible; that is, you cannot always simultaneously achieve these objectives. In the case of the Atlantis university system, the solution with respect to Objective 1 is indicated with solid lines in Figure 10.2.1. There are four solutions to the problem with respect to Objective 2: any tree in which one campus is directly connected to the other three. One solution with respect to Objective 2 is indicated with dotted lines in Figure 10.2.1. After satisfying the conditions of Objective 2, it would seem reasonable to select the cheapest of the four trees.

10.2.2 Definition

Definition 10.2.2 Spanning Tree. Let $G = (V, E)$ be a connected undirected graph. A spanning tree for G is a spanning subgraph 9.1.16 of G that is a tree. $\diamond$

Note 10.2.3

(a) If (V, E') is a spanning tree, $|E'| = |V| - 1$.

(b) The significance of a spanning tree is that it is a minimal spanning set. A smaller set would not span the graph, while a larger set would have a cycle, which has an edge that is superfluous.

For the remainder of this section, we will discuss two of the many topics that relate to spanning trees. The first is the problem of finding Minimal Spanning Trees, which addresses Objective 1 above. The second is the problem of finding Minimum Diameter Spanning Trees, which addresses Objective 2.

Definition 10.2.4 Minimal Spanning Tree. Given a weighted connected undirected graph $G = (V, E, w)$, a minimal spanning tree is a spanning tree (V, E') for which $\sum_{e \in E'} w(e)$ is as small as possible. $\diamond$

10.2.3 Prim's Algorithm

Unlike many of the graph-optimization problems that we've examined, a solution to this problem can be obtained efficiently. It is a situation in which a greedy algorithm works.

Definition 10.2.5 Bridge. Let $G = (V, E)$ be an undirected graph and let $\{L, R\}$ be a partition of V. A bridge between L and R is an edge in E that connects a vertex in L to a vertex in R. $\diamond$

Theorem 10.2.6 *Let $G = (V, E, w)$ be a weighted connected undirected graph. Let V be partitioned into two sets L and R. If e^* is a bridge of least weight between L and R, then there exists a minimal spanning tree for G that includes e^*.*

Proof. Suppose that no minimal spanning tree including e^* exists. Let $T = (V, E')$ be a minimal spanning tree. If we add e^* to T, a cycle is created, and this cycle must contain another bridge, e, between L and R. Since $w(e^*) \leq w(e)$, we can delete e and the new tree, which includes e^* must also be a minimal spanning tree. ∎

Example 10.2.7 Some Bridges. The bridges between the vertex sets $\{a, b, c\}$ and $\{d, e\}$ in Figure 10.2.8 are the edges $\{b, d\}$ and $\{c, e\}$. According to the theorem above, a minimal spanning tree that includes $\{b, d\}$ exists. By examination, you should be able to see that this is true. Is it true that only the bridges of minimal weight can be part of a minimal spanning tree?

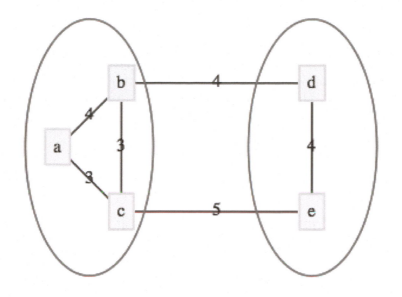

Figure 10.2.8: Bridges between two sets

☐

Theorem 10.2.6 essentially tells us that a minimal spanning tree can be constructed recursively by continually adding minimally weighted bridges to a set of edges.

Algorithm 10.2.9 Prim's Algorithm. *Let $G = (V, E, w)$ be a connected, weighted, undirected graph, and let v_0 be an arbitrary vertex in V. The following steps lead to a minimal spanning tree for G. L and R will be sets of vertices and E' is a set of edges.*

(1) *(Initialize) $L = V - \{v_0\}$; $R = \{v_0\}$; $E' = \emptyset$.*

(2) *(Build the tree) While $L \neq \emptyset$:*

   (1) *Find $e^* = \{v_L, v_R\}$, a bridge of minimum weight between L and R.*
   (2) *$R = R \cup \{v_L\}$; $L = L - \{v_L\}$; $E' = E' \cup \{e^*\}$*

(3) *Terminate with a minimal spanning tree (V, E').*

Note 10.2.10

(a) If more than one minimal spanning tree exists, then the one that is obtained depends on v_0 and the means by which e^* is selected in Step 2.

(b) Warning: If two minimally weighted bridges exist between L and R, do not try to speed up the algorithm by adding both of them to E'.

(c) That Algorithm 10.2.9 yields a minimal spanning tree can be proven by induction with the use of Theorem 10.2.6.

(d) If it is not known whether G is connected, Algorithm 10.2.9 can be revised to handle this possibility. The key change (in Step 2.1) would be to determine whether any bridge at all exists between L and R. The condition of the while loop in Step 2 must also be changed somewhat.

Example 10.2.11 An Small Example. Consider the graph in Figure 10.2.12. If we apply Prim's Algorithm starting at a, we obtain the following edge list in the order given: $\{a, f\}, \{f, e\}, \{e, c\}, \{c, d\}, \{f, b\}, \{b, g\}$. The total of the weights of these edges is 20. The method that we have used (in Step 2.1) to select a bridge when more than one minimally weighted bridge exists is to order all bridges alphabetically by the vertex in L and then, if further ties exist, by the vertex in R. The first vertex in that order is selected in Step 2.1 of the algorithm.

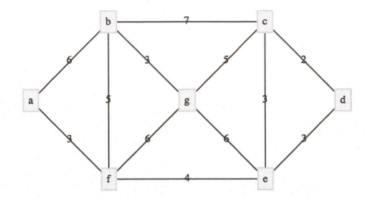

Figure 10.2.12: A small weighted graph

□

Definition 10.2.13 Minimum Diameter Spanning Tree. Given a connected undirected graph $G = (V, E)$, find a spanning tree $T = (V, E')$ of G such that the longest path in T is as short as possible. ◊

Example 10.2.14 The Case for Complete Graphs. The Minimum Diameter Spanning Tree Problem is trivial to solve in a K_n. Select any vertex v_0 and construct the spanning tree whose edge set is the set of edges that connect v_0 to the other vertices in the K_n. Figure 10.2.15 illustrates a solution for $n = 5$.

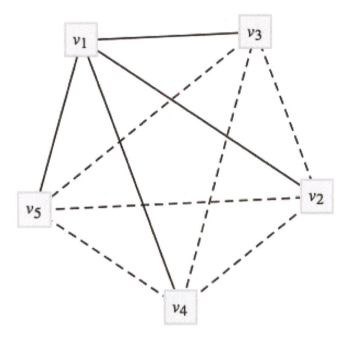

Figure 10.2.15: Minimum diameter spanning tree for K_5

For incomplete graphs, a two-stage algorithm is needed. In short, the first step is to locate a "center" of the graph. The maximum distance from a center to any other vertex is as small as possible. Once a center is located, a breadth-first search of the graph is used to construct the spanning tree.

10.2.4 Exercises for Section 10.2

1. Suppose that after Atlantis University's phone system is in place, a fifth campus is established and that a transmission line can be bought to connect the new campus to any old campus. Is this larger system the most economical one possible with respect to Objective 1? Can you always satisfy Objective 2?

2. Construct a minimal spanning tree for the capital cities in New England (see Table 9.5.3).

3. Show that the answer to the question posed in Example 10.2.7 is "no."

4. Find a minimal spanning tree for the following graphs.

(a)

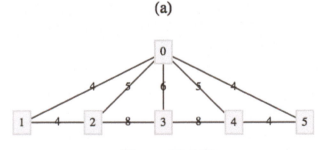

Figure 10.2.16

(b)

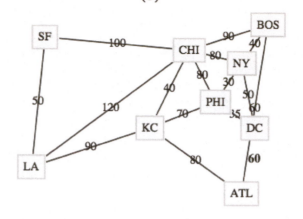

Figure 10.2.17

(c)

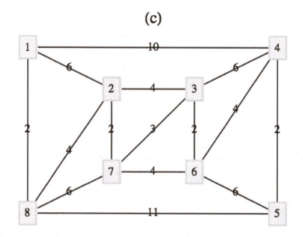

Figure 10.2.18

5.   Find a minimum diameter spanning tree for the following graphs.

(a)

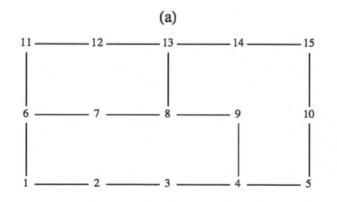

Figure 10.2.19

(b)

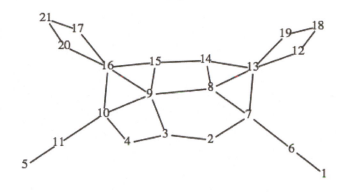

Figure 10.2.20

6. In each of the following parts justify your answer with either a proof or a counterexample.

(a) Suppose a weighted undirected graph had distinct edge weights. Is it possible that no minimal spanning tree includes the edge of minimal weight?

(b) Suppose a weighted undirected graph had distinct edge weights. Is it possible that every minimal spanning tree includes the edge of maximal weight? If true, under what conditions would it happen?

10.3 Rooted Trees

In the next two sections, we will discuss rooted trees. Our primary foci will be on general rooted trees and on a special case, ordered binary trees.

10.3.1 Definition and Terminology

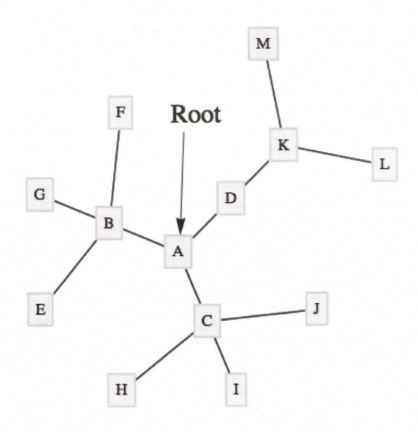

Figure 10.3.1: A Rooted Tree

What differentiates rooted trees from undirected trees is that a rooted tree contains a distinguished vertex, called the root. Consider the tree in Figure 10.3.1. Vertex A has been designated the root of the tree. If we choose any other vertex in the tree, such as M, we know that there is a unique path from A to M. The vertices on this path, (A, D, K, M), are described in genealogical terms:

- M is a child of K (so is L)

- K is M's parent.

- A, D, and K are M's ancestors.

- D, K, and M are descendants of A.

These genealogical relationships are often easier to visualize if the tree is rewritten so that children are positioned below their parents, as in Figure 10.3.3.

With this format, it is easy to see that each vertex in the tree can be thought of as the root of a tree that contains, in addition to itself, all of its descendants. For example, D is the root of a tree that contains D, K, L, and M. Furthermore, K is the root of a tree that contains K, L, and M. Finally, L and M are roots of trees that contain only themselves. From this observation, we can give a formal definition of a rooted tree.

List 10.3.2

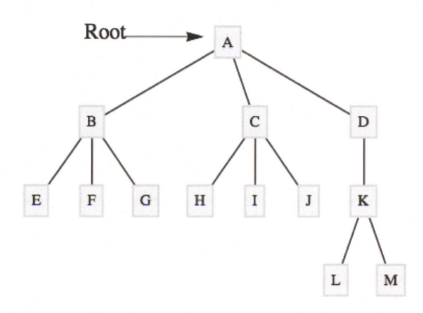

Figure 10.3.3: A Rooted Tree, redrawn

Definition 10.3.4 Rooted Tree.

(a) Basis: A tree with no vertices is a rooted tree (the empty tree).

(b) A single vertex with no children is a rooted tree.

(c) Recursion: Let $T_1, T_2, \ldots, T_r$, $r \geq 1$, be disjoint rooted trees with roots v_1, v_2, ..., v_r, respectively, and let v_0 be a vertex that does not belong to any of these trees. Then a rooted tree, rooted at v_0, is obtained by making v_0 the parent of the vertices v_1, v_2,..., and v_r. We call $T_1, T_2, \ldots, T_r$, subtrees of the larger tree.

$\Diamond$

The **level of a vertex** of a rooted tree is the number of edges that separate the vertex from the root. The level of the root is zero. The depth of a tree is the maximum level of the vertices in the tree. The depth of a tree in Figure 10.3.3 is three, which is the level of the vertices L and M. The vertices E, F, G, H, I, J, and K have level two. B, C, and D are at level one and A has level zero.

Example 10.3.5 A Decision Tree. Figure 2.1.2 is a rooted tree with Start as the root. It is an example of what is called a decision tree. □

Example 10.3.6 Tree Structure of Data. One of the keys to working with large amounts of information is to organize it in a consistent, logical way. A **data structure** is a scheme for organizing data. A simple example of a data structure might be the information a college admissions department might keep on their applicants. Items might look something like this:

$ApplicantItem = (FirstName, MiddleInitial, LastName, StreetAddress,$
$\qquad City, State, Zip, HomePhone, CellPhone, EmailAddress,$
$\qquad HighSchool, Major, ApplicationPaid, MathSAT, VerbalSAT,$
$\qquad Recommendation1, Recommendation2, Recommendation3)$

This structure is called a "flat file".

A spreadsheet can be used to arrange data in this way. Although a "flat file" structure is often adequate, there are advantages to clustering some the information. For example the applicant information might be broken into four parts: name, contact information, high school, and application data:

$ApplicantItem = ((FirstName, MiddleInitial, LastName),$
$\qquad ((StreetAddress, City, State, Zip),$
$\qquad (HomePhone, CellPhone), EmailAddress),$
$\qquad HighSchool,$
$\qquad (Major, ApplicationPaid, (MathSAT, VerbalSAT),$
$\qquad (Recommendation1, Recommendation2, Recommendation3))$

The first item in each ApplicantItem is a list $(FirstName, MiddleInitial, LastName)$, with each item in that list being a single field of the original flat file. The third item is simply the single high school item from the flat file. The application data is a list and one of its items, is itself a list with the recommendation data for each recommendation the applicant has.

The organization of this data can be visualized with a rooted tree such as the one in Figure 10.3.7.

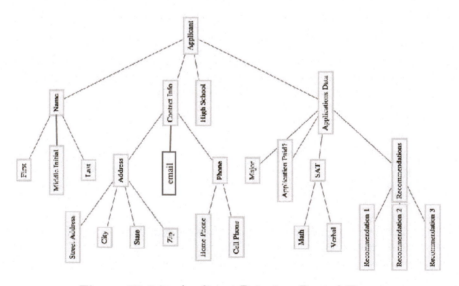

Figure 10.3.7: Applicant Data in a Rooted Tree

In general, you can represent a data item, T, as a rooted tree with T as the root and a subtree for each field. Those fields that are more than just one item are roots of further subtrees, while individual items have no further children in the tree. $\square$

10.3.2 Kruskal's Algorithm

An alternate algorithm for constructing a minimal spanning tree uses a forest of rooted trees. First we will describe the algorithm in its simplest terms. Afterward, we will describe how rooted trees are used to implement the algorithm. Finally, we will demonstrate the SageMath implementation of the algorithm. In all versions of this algorithm, assume that $G = (V, E, w)$ is a weighted undirected graph with $|V| = m$ and $|E| = n$.

Algorithm 10.3.8 Kruskal's Algorithm - Informal Version.

(1) *Sort the edges of G in ascending order according to weight. That is,*

$$i \leq j \Leftrightarrow w(e_j) \leq w(e_j).$$

(2) *Go down the list from Step 1 and add edges to a set (initially empty) of edges so that the set does not form a cycle. When an edge that would create a cycle is encountered, ignore it. Continue examining edges until either $m - 1$ edges have been selected or you have come to the end of the edge list. If $m - 1$ edges are selected, these edges make up a minimal spanning tree for G. If fewer than $m - 1$ edges are selected, G is not connected.*

Step 1 can be accomplished using one of any number of standard sorting routines. Using the most efficient sorting routine, the time required to perform this step is proportional to $n \log n$. The second step of the algorithm, also of $n \log n$ time complexity, is the one that uses a forest of rooted trees to test for whether an edge should be added to the spanning set.

Algorithm 10.3.9 Kruskal's Algorithm.

(1) *Sort the edges of G in ascending order according to weight. That is,*

$$i \leq j \Leftrightarrow w(e_j) \leq w(e_j).$$

(2) (1) *Initialize each vertex in V to be the root of its own rooted tree.*

    (2) *Go down the list of edges until either a spanning tree is completed or the edge list has been exhausted. For each edge $e = \{v_1, v_2\}$, we can determine whether e can be added to the spanning set without forming a cycle by determining whether the root of $v_1's$ tree is equal to the root of $v_2's$ tree. If the two roots are equal, then ignore e. If the roots are different, then we can add e to the spanning set. In addition, we merge the trees that v_1 and v_2 belong to. This is accomplished by either making $v_1's$ root the parent of $v_2's$ root or vice versa.*

Note 10.3.10

(a) Since we start the Kruskal's algorithm with m trees and each addition of an edge decreases the number of trees by one, we end the algorithm with one rooted tree, provided a spanning tree exists.

(b) The rooted tree that we develop in the algorithm is not the spanning tree itself.

10.3.3 SageMath Note - Implementation of Kruskal's Algorithm

Kruskal's algorithm has been implemented in Sage. We illustrate how the spanning tree for a weighted graph in can be generated. First, we create such a graph

   We will create a graph using a list of triples of the form (vertex, vertex, label). The *weighted* method tells Sage to consider the labels as weights.

```
edges=[(1, 2, 4), (2, 8, 4), (3, 8, 4), (4, 7, 5), (6, 8,
    5), (1, 3, 6), (1, 7, 6), (4, 5, 6), (5, 10, 9), (2, 10,
    7), (4, 6, 7), (2, 4, 8), (1,
8, 9), (1, 9, 9), (5, 6, 9), (1, 10, 10), (2, 9, 10), (4, 9,
    10), (5, 9, 10), (6, 9, 10)]
G=Graph(edges)
G.weighted(True)
G.graphplot(edge_labels=True,save_pos=True).show()
```

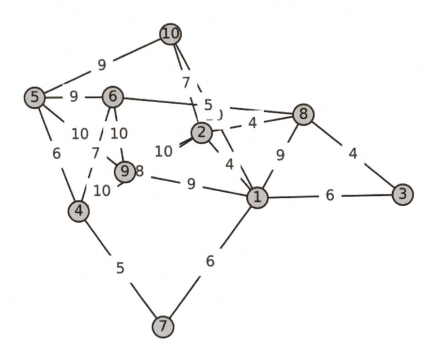

Figure 10.3.11: Weighed graph, SageMath output

Next, we load the kruskal function and use it to generate the list of edges in a spanning tree of G.

```
from sage.graphs.spanning_tree import kruskal
E = kruskal(G, check=True);E
```

```
[(1, 2, 4), (1, 7, 6), (1, 9, 9), (2, 8, 4), (2, 10, 7), (3,
    8, 4), (4, 5, 6), (4, 7, 5), (6, 8, 5)]
```

To see the resulting tree with the same embedding as G, we generate a graph from the spanning tree edges. Next, we set the positions of the vertices to be the same as in the graph. Finally, we plot the tree.

```
T=Graph(E)
```

```
T.set_pos(G.get_pos())
T.graphplot(edge_labels=True).show()
```

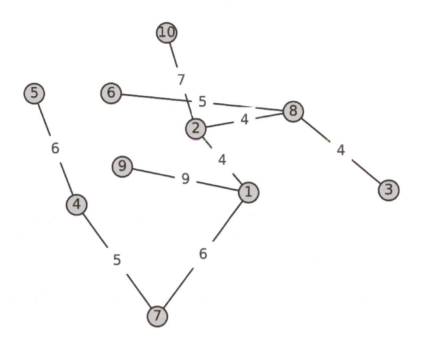

Figure 10.3.12: Spanning tree, SageMath output

10.3.4 Exercises for Section 10.3

1. Suppose that an undirected tree has diameter d and that you would like to select a vertex of the tree as a root so that the resulting rooted tree has the smallest depth possible. How would such a root be selected and what would be the depth of the tree (in terms of d)?

2. Use Kruskal's algorithm to find a minimal spanning tree for the following graphs. In addition to the spanning tree, find the final rooted tree in the algorithm. When you merge two trees in the algorithm, make the root with the lower number the root of the new tree.

(a)

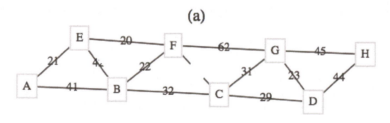

Figure 10.3.13

(b)

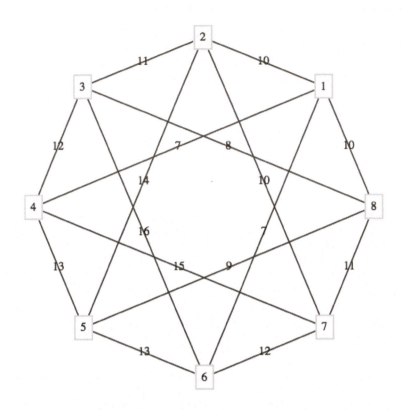

Figure 10.3.14

3. Suppose that information on buildings is arranged in records with five fields: the name of the building, its location, its owner, its height, and its floor space. The location and owner fields are records that include all of the information that you would expect, such as street, city, and state, together with the owner's name (first, middle, last) in the owner field. Draw a rooted tree to describe this type of record

4. Step through Kruskel's Algorthm by hand to verify that the example of a minimal spanning tree using Sage in Subsection 10.3.3 is correct.

10.4 Binary Trees

10.4.1 Definition of a binary tree

An **ordered rooted tree** is a rooted tree whose subtrees are put into a definite order and are, themselves, ordered rooted trees. An empty tree and a single vertex with no descendants (no subtrees) are ordered rooted trees.

Example 10.4.1 Distinct Ordered Rooted Trees. The trees in Figure 10.4.2 are identical rooted trees, with root 1, but as ordered trees, they are different.

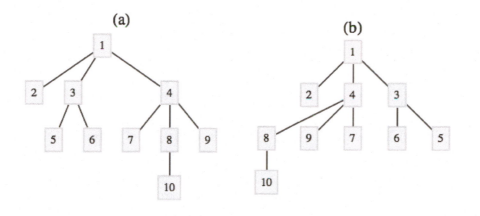

Figure 10.4.2: Two different ordered rooted trees

□

If a tree rooted at v has p subtrees, we would refer to them as the first, second,..., p^{th} subtrees. There is a subtle difference between certain ordered trees and binary trees, which we define next.

Definition 10.4.3 Binary Tree.

(1) A tree consisting of no vertices (the empty tree) is a binary tree

(2) A vertex together with two subtrees that are both binary trees is a binary tree. The subtrees are called the left and right subtrees of the binary tree.

◇

The difference between binary trees and ordered trees is that every vertex of a binary tree has exactly two subtrees (one or both of which may be empty), while a vertex of an ordered tree may have any number of subtrees. But there is another significant difference between the two types of stuctures. The two trees in Figure 10.4.4 would be considered identical as ordered trees. However, they are different binary trees. Tree (a) has an empty right subtree and Tree (b) has an empty left subtree.

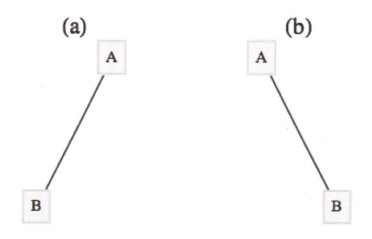

Figure 10.4.4: Two different binary trees

(a) A vertex of a binary tree with two empty subtrees is called a **leaf**. All other vertices are called *internal vertices*.

(b) The number of leaves in a binary tree can vary from one up to roughly half the number of vertices in the tree (see Exercise 4 of this section).

(c) The maximum number of vertices at level k of a binary tree is 2^k, $k \geq 0$ (see Exercise 6 of this section).

(d) A **full binary tree** is a tree for which each vertex has either zero or two empty subtrees. In other words, each vertex has either two or zero children. See Exercise 10.4.6.7 of this section for a general fact about full binary trees.

List 10.4.5

10.4.2 Traversals of Binary Trees

The traversal of a binary tree consists of visiting each vertex of the tree in some prescribed order. Unlike graph traversals, the consecutive vertices that are visited are not always connected with an edge. The most common binary tree traversals are differentiated by the order in which the root and its subtrees are visited. The three traversals are best described recursively and are:

Preorder Traversal: (1) Visit the root of the tree.

   (2) Preorder traverse the left subtree.

   (3) Preorder traverse the right subtree.

Inorder Traversal: (1) Inorder traverse the left subtree.

   (2) Visit the root of the tree.

   (3) Inorder traverse the right subtree.

Postorder Traversal: (1) Postorder traverse the left subtree.

   (2) Postorder traverse the right subtree.

   (3) Visit the root of the tree.

Any traversal of an empty tree consists of doing nothing.

Example 10.4.6 Traversal Examples. For the tree in Figure 10.4.7, the orders in which the vertices are visited are:

- A-B-D-E-C-F-G, for the preorder traversal.

- D-B-E-A-F-C-G, for the inorder traversal.

- D-E-B-F-G-C-A, for the postorder traversal.

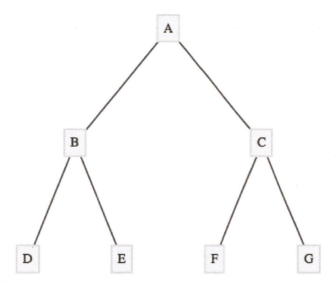

Figure 10.4.7: A Complete Binary Tree to Level 2

□

Binary Tree Sort. Given a collection of integers (or other objects than can be ordered), one technique for sorting is a binary tree sort. If the integers are $a_1, a_2, \ldots, a_n$, $n \geq 1$, we first execute the following algorithm that creates a binary tree:

Algorithm 10.4.8 Binary Sort Tree Creation.

(1) Insert a_1 into the root of the tree.

(2) For $k := 2$ to n // insert a_k into the tree

> *(a) $r = a_1$*
>
> *(b) inserted = false*
>
> *(c) while not(inserted):*
>> *if $a_k < r$:*
>>> *if r has a left child:*
>>>> *r = left child of r*
>>>
>>> *else:*
>>>> *make a_k the left child of r*
>>>> *inserted = true*
>>
>> *else:*
>>> *if r has a right child:*
>>>> *r = right child of r*
>>>
>>> *else:*
>>>> *make a_k the right child of r*
>>>> *inserted = true*

If the integers to be sorted are 25, 17, 9, 20, 33, 13, and 30, then the tree that is created is the one in Figure 10.4.9. The inorder traversal of this tree is 9, 13, 17, 20, 25, 30, 33, the integers in ascending order. In general, the inorder traversal of the tree that is constructed in the algorithm above will produce a sorted list. The preorder and postorder traversals of the tree have no meaning here.

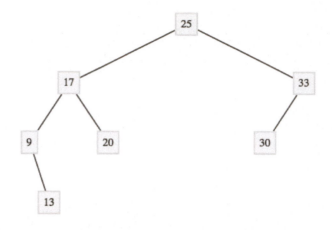

Figure 10.4.9: A Binary Sorting Tree

10.4.3 Expression Trees

A convenient way to visualize an algebraic expression is by its expression tree. Consider the expression

$$X = a * b - c/d + e.$$

Since it is customary to put a precedence on multiplication/divisions, X is evaluated as $((a * b) - (c/d)) + e$. Consecutive multiplication/divisions or addition/subtractions are evaluated from left to right. We can analyze X further by noting that it is the sum of two simpler expressions $(a * b) - (c/d)$ and e. The first of these expressions can be broken down further into the difference of the expressions $a * b$ and c/d. When we decompose any expression into (left expression)operation(right expression), the expression tree of that expression is the binary tree whose root contains the operation and whose left and right subtrees are the trees of the left and right expressions, respectively. Additionally, a simple variable or a number has an expression tree that is a single vertex containing the variable or number. The evolution of the expression tree for expression X appears in Figure 10.4.10.

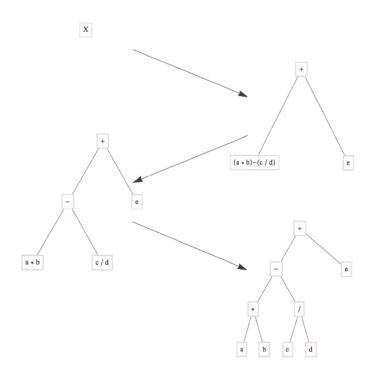

Figure 10.4.10: Building an Expression Tree

Example 10.4.11 Some Expression Trees.

(a) If we intend to apply the addition and subtraction operations in X first, we would parenthesize the expression to $a*(b-c)/(d+e)$. Its expression tree appears in Figure 10.4.12(a).

(b) The expression trees for $a^2 - b^2$ and for $(a+b)*(a-b)$ appear in Figure 10.4.12(b) and Figure 10.4.12(c).

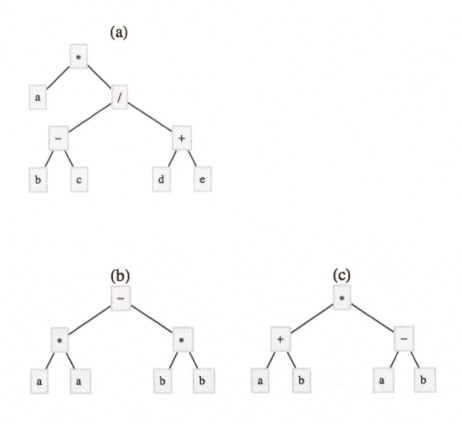

Figure 10.4.12: Expression Tree Examples

The three traversals of an operation tree are all significant. A binary operation applied to a pair of numbers can be written in three ways. One is the familiar infix form, such as $a + b$ for the sum of a and b. Another form is prefix, in which the same sum is written $+ab$. The final form is postfix, in which the sum is written $ab+$. Algebraic expressions involving the four standard arithmetic operations $(+, -, *, \text{and}/)$ in prefix and postfix form are defined as follows:

Prefix (a) A variable or number is a prefix expression

    (b) Any operation followed by a pair of prefix expressions is a prefix expression.

Postfix (a) A variable or number is a postfix expression

    (b) Any pair of postfix expressions followed by an operation is a postfix expression.

List 10.4.13

The connection between traversals of an expression tree and these forms is simple:

(a) The preorder traversal of an expression tree will result in the prefix form of the expression.

(b) The postorder traversal of an expression tree will result in the postfix form of the expression.

(c) The inorder traversal of an operation tree will not, in general, yield the proper infix form of the expression. If an expression requires parentheses in infix form, an inorder traversal of its expression tree has the effect of removing the parentheses.

Example 10.4.14 Traversing an Expression Tree. The preorder traversal of the tree in Figure 10.4.10 is $+ - *ab/cde$, which is the prefix version of expression X. The postorder traversal is $ab * cd/ - e+$. Note that since the original form of X needed no parentheses, the inorder traversal, $a*b-c/d+e$, is the correct infix version. □

10.4.4 Counting Binary Trees

We close this section with a formula for the number of different binary trees with n vertices. The formula is derived using generating functions. Although the complete details are beyond the scope of this text, we will supply an overview of the derivation in order to illustrate how generating functions are used in advanced combinatorics.

Let $B(n)$ be the number of different binary trees of size n (n vertices), $n \geq 0$. By our definition of a binary tree, $B(0) = 1$. Now consider any positive integer $n + 1$, $n \geq 0$. A binary tree of size $n + 1$ has two subtrees, the sizes of which add up to n. The possibilities can be broken down into $n + 1$ cases:

Case 0: Left subtree has size 0; right subtree has size n.

Case 1: Left subtree has size 1; right subtree has size $n - 1$.

$\vdots$

Case k: Left subtree has size k; right subtree has size $n - k$.

$\vdots$

Case n: Left subtree has size n; right subtree has size 0.

In the general Case k, we can count the number of possibilities by multiplying the number of ways that the left subtree can be filled, $B(k)$, by the number of ways that the right subtree can be filled. $B(n - k)$. Since the sum of these products equals $B(n + 1)$, we obtain the recurrence relation for $n \geq 0$:

$$B(n + 1) = B(0)B(n) + B(1)B(n - 1) + \cdots + B(n)B(0)$$

$$= \sum_{k=0}^{n} B(k)B(n - k)$$

Now take the generating function of both sides of this recurrence relation:

$$\sum_{n=0}^{\infty} B(n + 1)z^n = \sum_{n=0}^{\infty} \left(\sum_{k=0}^{n} B(k)B(n - k) \right) z^n \qquad (10.4.1)$$

or

$$G(B \uparrow; z) = G(B * B; z) = G(B; z)^2 \qquad (10.4.2)$$

Recall that $G(B \uparrow; z) = \frac{G(B;z)-B(0)}{z} = \frac{G(B;z)-1}{z}$ If we abbreviate $G(B;z)$ to G, we get

$$\frac{G-1}{z} = G^2 \Rightarrow zG^2 - G + 1 = 0$$

Using the quadratic equation we find two solutions:

$$G_1 = \frac{1 + \sqrt{1-4z}}{2z} \text{ and} \tag{10.4.3}$$

$$G_2 = \frac{1 - \sqrt{1-4z}}{2z} \tag{10.4.4}$$

The gap in our deviation occurs here since we don't presume calculus. If we expand G_1 as an extended power series, we find

$$G_1 = \frac{1 + \sqrt{1-4z}}{2z} = \frac{1}{z} - 1 - z - 2z^2 - 5z^3 - 14z^4 - 42z^5 + \cdots \tag{10.4.5}$$

The coefficients after the first one are all negative and there is singularity at 0 because of the $\frac{1}{z}$ term. However if we do the same with G_2 we get

$$G_2 = \frac{1 - \sqrt{1-4z}}{2z} = 1 + z + 2z^2 + 5z^3 + 14z^4 + 42z^5 + \cdots \tag{10.4.6}$$

Further analysis leads to a closed form expression for $B(n)$, which is

$$B(n) = \frac{1}{n+1} \binom{2n}{n}$$

This sequence of numbers is often called the **Catalan numbers**. For more information on the Catalan numbers, see the entry A000108 in The On-Line Encyclopedia of Integer Sequences.

10.4.5 SageMath Note - Power Series

It may be of interest to note how the extended power series expansions of G_1 and G_2 are determined using Sage. In Sage, one has the capability of being very specific about how algebraic expressions should be interpreted by specifying the underlying ring. This can make working with various algebraic expressions a bit more confusing to the beginner. Here is how to get a Laurent expansion for G_1 above.

```
R.<z>=PowerSeriesRing(ZZ,'z')
G1=(1+sqrt(1-4*z))/(2*z)
G1
```

```
z^-1 - 1 - z - 2*z^2 - 5*z^3 - 14*z^4 - 42*z^5 - 132*z^6
   - 429*z^7 - 1430*z^8 - 4862*z^9 - 16796*z^10 - 58786*z^11
     - 208012*z^12 - 742900*z^13 - 2674440*z^14 - 9694845*z^15
       - 35357670*z^16 - 129644790*z^17 - 477638700*z^18 + O(z^19)
```

The first Sage expression above declares a structure called a **ring** that contains power series. We are not using that whole structure, just a specific element, G1. So the important thing about this first input is that it establishes z as being a variable associated with power series over the integers. When the second expression defines the value of G1 in terms of z, it is automatically converted to a power series.

The expansion of G_2 uses identical code, and it's coefficients are the values of $B(n)$.

```
R.<z>=PowerSeriesRing(ZZ,'z')
G2=(1-sqrt(1-4*z))/(2*z)
G2
```

```
1 + z + 2*z^2 + 5*z^3 + 14*z^4 + 42*z^5 + 132*z^6 + 429*z^7
 + 1430*z^8 + 4862*z^9 + 16796*z^10 + 58786*z^11 + 208012*z^12
  + 742900*z^13 + 2674440*z^14 + 9694845*z^15 + 35357670*z^16
  + 129644790*z^17 + 477638700*z^18 + O(z^19)
```

In Chapter 16 we will introduce rings and will be able to take further advantage of Sage's capabilities in this area.

10.4.6 Exercises for Section 10.4

1. Draw the expression trees for the following expressions:

   (a) $a(b + c)$

   (b) $ab + c$

   (c) $ab + ac$

   (d) $bb - 4ac$

   (e) $((a_3 x + a_2) x + a_1) x + a_0$

2. Draw the expression trees for

   (a) $\frac{x^2 - 1}{x - 1}$

   (b) $xy + xz + yz$

3. Write out the preorder, inorder, and postorder traversals of the trees in Exercise 1 above.

4. Verify the formula for $B(n)$, $0 \leq n \leq 3$ by drawing all binary trees with three or fewer vertices.

5.

   (a) Draw a binary tree with seven vertices and only one leaf.

   (b) Draw a binary tree with seven vertices and as many leaves as possible.

6. Prove that the maximum number of vertices at level k of a binary tree is 2^k and that a tree with that many vertices at level k must have $2^{k+1} - 1$ vertices.

7. Prove that if T is a full binary tree, then the number of leaves of T is one more than the number of internal vertices (non-leaves).

Chapter 11

Group Theory and Applications

This is a stub for Chapter 15 that includes material that is to be included in Part 1 of Applied Discrete Structures.

11.1 Bijections on a three element set

Suppose that $A = \{1, 2, 3\}$. There are $3! = 6$ different permutations on A. We will call the set of all 6 permutations S_3. They are listed in the following table. The matrix form for describing a function on a finite set is to list the domain across the top row and the image of each element directly below it. For example $r_1(1) = 2$.

$$
i = \begin{pmatrix} 1 & 2 & 3 \\ 1 & 2 & 3 \end{pmatrix} \quad
r_1 = \begin{pmatrix} 1 & 2 & 3 \\ 2 & 3 & 1 \end{pmatrix} \quad
r_2 = \begin{pmatrix} 1 & 2 & 3 \\ 3 & 1 & 2 \end{pmatrix}
$$

$$
f_1 = \begin{pmatrix} 1 & 2 & 3 \\ 1 & 3 & 2 \end{pmatrix} \quad
f_2 = \begin{pmatrix} 1 & 2 & 3 \\ 3 & 2 & 1 \end{pmatrix} \quad
f_3 = \begin{pmatrix} 1 & 2 & 3 \\ 2 & 1 & 3 \end{pmatrix}
$$

Table 11.1.1: Elements of S_3

Appendix A

Algorithms

Computer programs, bicycle assembly instructions, knitting instructions, and recipes all have several things in common. They all tell us how to do something; and the usual format is as a list of steps or instructions. In addition, they are usually prefaced with a description of the raw materials that are needed (the input) to produce the end result (the output). We use the term algorithm to describe such lists of instructions. We assume that the reader may be unfamiliar with algorithms, so the first section of this appendix will introduce some of the components of the algorithms that appear in this book. Since we would like our algorithms to become computer programs in many cases, the notation will resemble a computer language such as Python or Sage; but our notation will be slightly less formal. In some cases we will also translate the pseudocode to Sage. Our goal will be to give mathematically correct descriptions of how to accomplish certain tasks. To this end, the second section of this appendix is an introduction to the Invariant Relation Theorem, which is a mechanism for algorithm verification that is related to Mathematical Induction

A.1 An Introduction to Algorithms

Most of the algorithms in this book will contain a combination of three kinds of steps: the assignment step, the conditional step, and the loop.

A.1.1 Assignments

In order to assign a value to a variable, we use an assignment step, which takes the form:

$$\text{Variable} = \text{Expression to be computed}$$

The equals sign in most languages is used for assignment but some languages may use variations such as := or a left pointing arrow. Logical equality, which produces a boolean result and would be used in conditional or looping steps, is most commonly expressed with a double-equals, ==.

An example of an assignment is $k = n - 1$ which tells us to subtract 1 from the value of n and assign that value to variable k. During the execution of an algorithm, a variable may take on only one value at a time. Another example of an assignment is $k = k - 1$. This is an instruction to subtract one from the value of k and then reassign that value to k.

A.1.2 Conditional steps

Frequently there are steps that must be performed in an algorithm if and only if a certain condition is met. The conditional or "if ... then" step is then employed. For example, suppose that in step 2 of an algorithm we want to assure that the values of variables x and y satisfy the condition x <= y. The following step would accomplish this objective.

```
2. If x > y:
        2.1 t = x
        2.2 x = y
        2.3 y = t
```

Listing A.1.1

Steps 2.1 through 2.3 would be bypassed if the condition x > y were false before step 2.

One slight variation is the "if ... then ... else" step, which allows us to prescribe a step to be taken if the condition is false. For example, if you wanted to exercise today, you might look out the window and execute the following algorithm.

```
1. If it is cold or raining:
                exercise indoors
        else:
                go outside and run
2. Rest
```

Listing A.1.2

A.1.3 Loops

The conditional step tells us to do something once if a logical condition is true. A loop tells us to repeat one or more steps, called the body of the loop, while the logical condition is true. Before every execution of the body, the condition is tested. The following flow diagram serves to illustrate the steps in a While loop.

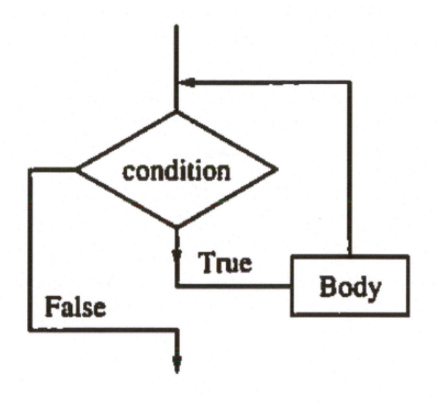

Figure A.1.3: Flow diagram for a while loop

Suppose you wanted to solve the equation $f(x) = 0$. The following initial assignment and loop could be employed.

```
1. c = your first guess
2. While f(c) != 0:
            c = another guess
```

Listing A.1.4

Caution: One must always guard against the possibility that the condition of a While loop will never become false. Such "infinite loops" are the bane of beginning programmers. The loop above could very well be such a situation, particularly if the equation has no solution, or if the variable takes on real values

In cases where consecutive integer values are to be assigned to a variable, a different loop construction, a *For loop*, is often employed. For example, suppose we wanted to assign variable k each of the integer values from m to n and for each of these values perform some undefined steps. We could accomplish this with a While loop:

```
1. k := m
2. While k <= n:
        2.1 execute some steps
        2.2 k = k + l
```

Listing A.1.5

Alternatively, we can perform these steps with a For loop.

```
For k = m to n:
        execute some   steps
```

Listing A.1.6

For loops such as this one have the advantage of being shorter than the equivalent While loop. The While loop construction has the advantage of being able to handle more different situations than the For loop.

A.1.4 Exercises for Part 1 of the Algorithms Appendix

1. What are the inputs and outputs of the algorithms listed in the first sentence of this section?

2. What is wrong with this algorithm?

```
Input: a and b, integers
Output: the value of c will be a - b
(1) c = 0
(2) While a > b:
                (2.1) a := a - l
                (2.2) c := c + l
```

Listing A.1.7

3. Describe, in words, what the following algorithm does:

```
Input: k, a positive integer
Output: s = ?
(1) s = 0
(2) While k > 0:
    (2.1) s = s + k
    (2.2) k = k - 1
```

Listing A.1.8

4. Write While loops to replace the For loops in the following partial algorithms:

(a) (1) S = 0

   (2) for k = 1 to 5: S = S + k^2

(b) The floor of a number is the greatest integer less than or equal to that number.

   (1) m = a positive integer greater than 1

   (2) B = floor(sqrt(m))

(3) for i = 2 to B: if i divides evenly into m, jump to step 5

(4) print "m is a prime" and exit

(5) print "m is composite" and exit

5. Describe in words what the following algorithm does:

```
Input: n, a positive integer
Output: k?
(1) f= 0
(2) k=n
(3) While k is even:
          (3.1) f = f+ 1
          (3.2) k = k div 2
```

Listing A.1.9

6. Fix the algorithm in Exercise 2.

A.2 The Invariant Relation Theorem

A.2.1 Two Exponentiation Algorthms

Consider the following algorithm implemented in Sage to compute $a^m \bmod n$, given an arbitrary integer a, non-negative exponent m, and a modulus n, $n \geq 0$. The default sample evaluation computes $2^5 \bmod 7 = 32 \bmod 7 = 4$, but you can edit the final line for other inputs.

```
def slow_exp(a,m,n):
        b=1
        k=m
        while k>0:
                b=(b*a)%n  # % is integer remainder (mod)
                    operation
                k-=1
        return b

slow_exp(2,5,7)
```

It should be fairly clear that this algorithm will successfully compute $a^m \pmod n$ since it mimics the basic definition of exponentiation. However, this algorithm is highly inefficient. The algorithm that is most commonly used for the task of exponentiation is the following one, also implemented in Sage.

```
def fast_exp(a,m,n):
        t=a
        b=1
        k=m
        while k>0:
                if k%2==1: b=(b*t)%n
                t=(t^2)%n
                k=k//2  # // is the integer quotient
                    operation
        return b
```

```
fast_exp(2,5,7)
```

The only difficulty with the "fast algorithm" is that it might not be so obvious that it always works. When implemented, it can be verified by example, but an even more rigorous verification can be done using the Invariant Relation Theorem. Before stating the theorem, we define some terminology.

A.2.2 Proving the correctness of the fast algorithm

Definition A.2.1 Pre and Post Values. Given a variable x, the pre value of x, denoted $\grave{x}$, is the value before an iteration of a loop. The post value, denoted $\acute{x}$, is the value after the iteration. $\diamond$

Example A.2.2 Pre and post values in the fast exponentiation algorithm. In the fast exponentiation algorithm, the relationships between the pre and post values of the three variables are as follows.

$$\acute{b} \equiv \grave{b}\grave{t}^{\grave{k}\,mod\,2}(mod\,n)$$

$$\acute{t} \equiv \grave{t}^2(mod\,n)$$

$$\acute{k} = \grave{k}//2$$

$\square$

Definition A.2.3 Invariant Relation. Given an algorithm's inputs and a set of variables that are used in the algorithm, an *invariant relation* is a set of one or more equations that are true prior to entering a loop and remain true in every iteration of the loop. $\diamond$

Example A.2.4 Invariant Relation for Fast Exponentiation. We claim that the invariant relation in the fast algorithm is $bt^k = a^m(mod\,n)$. We will prove that his is indeed true below. $\square$

Theorem A.2.5 The Invariant Relation Theorem. *Given a loop within an algorithm, if R is a relation with the properties*

(a) R is true before entering the loop

(b) the truth of R is maintained in any iteration of the loop

(c) the condition for exiting the loop will always be reached in a finite number of iterations.

then R will be true upon exiting the loop.

Proof. The condition that the loop ends in a finite number of iterations lets us apply mathematical induction with the induction variable being the number of iterations. We leave the details to the reader. ■

We can verify the correctness of the fast exponentiation algorithm using the Invariant Relation Theorem. First we note that prior to entering the loop, $bt^k = 1a^m = a^m(mod\,n)$. Assuming the relation is true at the start of any iteration, that is $\grave{b}\grave{t}^k = a^m (mod\, n)$, then

$$\acute{b}\acute{t}^k \equiv (\grave{b}\grave{t}^{k\, mod\, 2})(\grave{t}^2)^{k//2}(mod\, n)$$
$$\equiv \grave{b}\grave{t}^{2(k//2)+k\, mod\, 2}(mod\, n)$$
$$\equiv \grave{b}\grave{t}^k (mod\, n)$$
$$\equiv a^m (mod\, n)$$

Finally, the value of k will decrease to zero in a finite number of steps because the number of binary digits of k decreases by one with each iteration. At the end of the loop,

$$b = bt^0 = bt^k \equiv a^m (mod\, n)$$

which verifies the correctness of the algorithm.

A.2.3 Exercises for Part 2 of the Algorithms Appendix

1. How are the pre and post values in the slow exponentiation algorithm related? What is the invariant relation between the variables in the slow algorithm?

2. Verify the correctness of the following algorithm to compute the greatest common divisor of two integers that are not both zero.

```
def gcd(a,b):
        r0=a
        r1=b
        while r1 !=0:
                t= r0 % r1
                r0=r1
                r1=t
        return r0

gcd(1001,154)   #test
```

Hint. The invariant of this algorithm is $gcd(r0, r1) = gcd(a, b)$.

3. Verify the correctness of the Binary Conversion Algorithm in Chapter 1.

Appendix B

Python and SageMath

SageMath (originally Sage) is a computer algebra system that is built on top of Python, which is a popular general-purpose programming language. In this appendix we highlight a few features of Python through a series of SageMath cells. Pure Python code can generally be evaluated in these cells and most of what you see here is just Python. There are exceptions. For example, Sage-Math has enhanced capabilities to work with sets. In Python, the expression `set([0,1,2,3])` is a set of four integers, and certain basic set operations can be performed on these types of expressions. This is a valid expression in Sage-Math too, but a different SageMath expression, `Set([0,1,2,3])`, with a capital S, has enhanced properties. For example, we can create the power set of the SageMath expression, which we do in the discussion of iterators.

B.1 Python Iterators

All programming languages allow for looping. A common form of loop is one in which a series of instructions are executed for each value of some index variable, commonly for values between two integers. Python allows a bit more generality by having structures called "iterators" over which looping can be done. An iterator can be as simple as a list, such as `[0,1,2,3]`, but also can be a power set of a finite set, as we see below, or the keys in a dictionary, which is described in the next section.

B.1.1 Counting Subsets

Suppose we want to count the number of subsets of $\{0, 1, 2, ..., 9\}$ that contain no adjacent elements. First, we will define our universe and its power set. The plan will be to define a function that determines whether a subset is "valid" in the sense that it contains no adjacent elements. Then we will iterate over the subsets, counting the valid ones. We know that the number of all subsets will be 2 raised to the number of elements in U, which would be $2^{10} = 1024$, but let's check.

```
U=Set(range(10))
power_set=U.subsets()
len(power_set)
```

1024

The validity check in this case is very simple. For each element, k, of a set, B, we ask whether its successor, $k + 1$, is also in the set. If we never get an answer of "True" then we consider the set valid. This function could be edited to define validity in other ways to answer different counting questions. It's always a good idea to test your functions, so we try two tests, one with a valid set and one with an invalid one.

```
def valid(B):
    v=true
    for k in B:
        if k+1 in B:
            v=false
            break
    return v
[valid(Set([1,3,5,9])),valid(Set([1,2,4,9]))]
```

```
[True, False]
```

Finally we do the counting over our power set, incrementing the count variable with each valid set.

```
count=0
for B in power_set:
    if valid(B):
        count+=1
count
```

144

B.2 Dictionaries

B.2.1 Colors of Fruits

In Python and SageMath, a dictionary is a convenient data structure for establishing a relationship between sets of data. From the point of view of this text, we can think of a dictionary as a concrete realization of a relation between two sets or on a single set. A dictionary resembles a function in that there is a set of data values called the keys, and for each key, there is a value. The value associated with a key can be almost anything, but it is most commonly a list.

To illustrate the use of dictionaries, we will define a relationship between colors and fruits. The keys will be a set of colors and values associated with each color will be a list of fruits that can take on that color. We will demonstrate how to initialize the dictionary and how to add to it. The following series of assignments have no output, so we add a print statement to verify that this cell is completely evaluated.

```
fruit_color={}
fruit_color['Red']=['apple','pomegranate','blood_orange']
fruit_color['Yellow']=['banana','apple','lemon']
fruit_color['Green']=['apple','pear','grape','lime']
fruit_color['Purple']=['plum','grape']
fruit_color['Orange']=['orange','pineapple']
print 'done'
```

We distinguish a color from a fruit by capitalizing colors but not fruit. The keys of this dictionary are the colors:

```
fruit_color.keys()
```

```
['Purple', 'Orange', 'Green', 'Yellow', 'Red']
```

As an afterthough, we might add the information that a raspberry is red as follows. You have to be careful in that if 'Red' isn't already in the dictionary, it doesn't have a value. This is why we need an if statement.

```
if 'Red' in fruit_color:
        fruit_color['Red']=fruit_color['Red']+['raspberry']
else:
        fruit_color['Red']=['raspberry']
fruit_color['Red']
```

```
['apple', 'pomegranate', 'blood_orange', 'raspberry',
    'raspberry']
```

A dictionary is iterable, with an iterator taking on values that are the keys. Here we iterate over the our dictionary to output lists consisting of a color followed by a list of fruits that come in that color.

```
for fruit in fruit_color:
    print [fruit,fruit_color[fruit]]
```

```
['Purple', ['plum', 'grape']]
['Orange', ['orange', 'pineapple']]
['Green', ['apple', 'pear', 'grape', 'lime']]
['Yellow', ['banana', 'apple', 'lemon']]
['Red', ['apple', 'pomegranate', 'blood_orange','raspberry']]
```

We can view a graph of this relation between colors and fruits, but the default view is a bit unconventional.

```
DiGraph(fruit_color).plot()
```

With a some additional coding we can line up the colors and fruits in their own column. First we set the positions of colors on the left with all x-coordinates equal to -5 using another dictionary called vertex_pos.

```
vertex_pos={}
k=0
for c in fruit_color.keys():
    vertex_pos[c]=(-5,k)
    k+=1
vertex_pos
```

```
{'Purple': (-5, 0), 'Orange': (-5, 1), 'Green': (-5, 2),
    'Red': (-5, 4), 'Yellow': (-5, 3)}
```

Next, we place the fruit vertices in another column with x-coordinates all equal to 5. In order to do this, we first collect all the fruit values into one set we call fruits.

```
fruits=Set([ ])
for v in fruit_color.values():
    fruits=fruits.union(Set(v))
k=0
for f in fruits:
    vertex_pos[f]=(5,k)
    k+=1
vertex_pos
```

```
{'blood_orange': (5, 0), 'grape': (5, 1), 'apple': (5, 2),
    'Purple': (-5, 0), 'plum': (5, 10), 'pomegranate': (5, 3),
    'pear': (5, 4), 'Yellow': (-5, 3), 'orange': (5, 7),
    'Green': (-5, 2), 'pineapple': (5, 6), 'Orange': (-5, 1),
    'lemon': (5, 8), 'raspberry': (5, 9), 'banana': (5, 5),
    'Red': (-5, 4), 'lime': (5, 11)}
```

Now the graph looks like a conventional graph for a relation between two different sets. Notice that it's not a function

```
DiGraph(fruit_color).plot(pos=vertex_pos,vertex_size=1)
```

Appendix C

Hints and Solutions to Selected Exercises

For the most part, solutions are provided here for odd-numbered exercises.

1 · Set Theory

1.1 · Set Notation and Relations

1.1.3 · Exercises for Section 1.1

1.1.3.1. Answer. These answers are not unique.

(a) $8, 15, 22, 29$

(b) apple, pear, peach, plum

(c) $1/2, 1/3, 1/4, 1/5$

(d) $-8, -6, -4, -2$

(e) $6, 10, 15, 21$

1.1.3.3. Answer.

(a) $\{2k + 1 \mid k \in \mathbb{Z}, 2 \leqslant k \leqslant 39\}$

(b) $\{x \in \mathbb{Q} \mid -1 < x < 1\}$

(c) $\{2n \mid n \in \mathbb{Z}\}$

(d) $\{9n \mid n \in \mathbb{Z}, -2 \leq n\}$

1.1.3.5. Answer.

(a) True

(b) False

(c) True

(d) True

(e) False

(f) True

(g) False

(h) True

1.2 · Basic Set Operations

1.2.4 · Exercises for Section 1.2

1.2.4.1. Answer.

(a) $\{2, 3\}$

(b) $\{0, 2, 3\}$

(c) $\{0, 2, 3\}$

(d) $\{0, 1, 2, 3, 5, 9\}$

(e) $\{0\}$

(f) $\emptyset$

(g) $\{1, 4, 5, 6, 7, 8, 9\}$

(h) $\{0, 2, 3, 4, 6, 7, 8\}$

(i) $\emptyset$

(j) $\{0\}$

1.2.4.3. Answer. These are all true for any sets A, B, and C.

1.2.4.5. Answer.

(a) $\{1,4\} \subseteq A \subseteq \{1,2,3,4\}$

(b) $\{2\} \subseteq A \subseteq \{1,2,4,5\}$

(c) $A = \{2,4,5\}$

1.2.4.7. Answer.

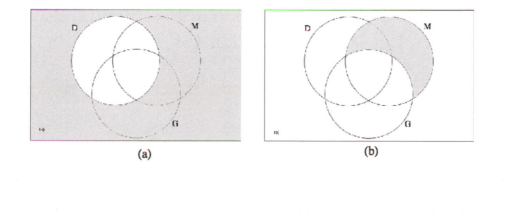

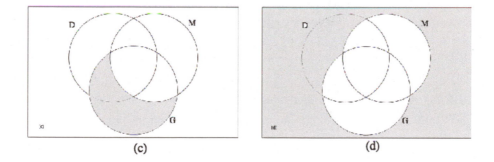

Figure C.0.1

1.3 · Cartesian Products and Power Sets
1.3.4 · EXERCISES FOR SECTION 1.3

1.3.4.1. Answer.

(a) $\{(0,2),(0,3),(2,2),(2,3),(3,2),(3,3)\}$

(b) $\{(2,0),(2,2),(2,3),(3,0),(3,2),(3,3)\}$

(c) $\{(0,2,1),(0,2,4),(0,3,1),(0,3,4),(2,2,1),(2,2,4),$
$(2,3,1),(2,3,4),(3,2,1),(3,2,4),(3,3,1),(3,3,4)\}$

(d) $\emptyset$

(e) $\{(0,1),(0,4),(2,1),(2,4),(3,1),(3,4)\}$

(f) $\{(2,2),(2,3),(3,2),(3,3)\}$

(g) $\{(2,2,2),(2,2,3),(2,3,2),(2,3,3),(3,2,2),(3,2,3),(3,3,2),(3,3,3)\}$

(h) $\{(2,\emptyset),(2,\{2\}),(2,\{3\}),(2,\{2,3\}),(3,\emptyset),(3,\{2\}),(3,\{3\}),(3,\{2,3\})\}$

1.3.4.3. Answer. $\{a,b\},\{a,c\},\{a,d\},\{b,c\},\{b,d\}$ and $\{c,d\}$

1.3.4.5. Answer. There are n singleton subsets, one for each element.

1.3.4.7. Answer.

(a) $\{+00,+01,+10,+11,-00,-01,-10,-11\}$

(b) 16 and 512

1.3.4.9. Answer. They are equal when $A = B$.

1.4 · Binary Representation of Positive Integers
1.4.3 · Exercises for Section 1.4

1.4.3.1. Answer.

(a) 11111

(b) 100000

(c) 1010

(d) 1100100

1.4.3.3. Answer.

(a) 18

(b) 19

(c) 42

(d) 1264

1.4.3.5. Answer. There is a bit for each power of 2 up to the largest one needed to represent an integer, and you start counting with the zeroth power. For example, 2017 is between $2^{10} = 1024$ and $2^{11} = 2048$, and so the largest power needed is 2^{10}. Therefore there are 11 bits in binary 2017.

(a) 11

(b) 12

(c) 13

(d) 51

1.4.3.7. Answer. A number must be a multiple of four if its binary representation ends in two zeros. If it ends in k zeros, it must be a multiple of 2^k.

1.5 · Summation Notation and Generalizations
1.5.3 · Exercises for Section 1.5

1.5.3.1. Answer.

(a) 24

(b) 6

(c) $3,7,15,31$

(d) $1,4,9,16$

1.5.3.3. Answer.

(a) $\frac{1}{1(1+1)} + \frac{1}{2(2+1)} + \frac{1}{3(3+1)} + \cdots + \frac{1}{n(n+1)} = \frac{n}{n+1}$

(b) $\frac{1}{1(2)} + \frac{1}{2(3)} + \frac{1}{3(4)} = \frac{1}{2} + \frac{1}{6} + \frac{1}{12} = \frac{3}{4} = \frac{3}{3+1}$

(c) $1 + 2^3 + 3^3 + \cdots + n^3 = \left(\frac{1}{4}\right)n^2(n+1)^2 \quad 1 + 8 + 27 = 36 = \left(\frac{1}{4}\right)(3)^2(3+1)^2$

1.5.3.5. Answer. $(x+y)^3 = \binom{3}{0}x^3 + \binom{3}{1}x^2y + \binom{3}{2}xy^2 + \binom{3}{3}y^n$

1.5.3.7. Answer.

(a) $\{x \in \mathbb{Q} \mid 0 < x \leq 5\}$

(b) $\{x \in \mathbb{Q} \mid -5 < x < 5\} = B_5$

(c) $\emptyset$

(d) $\{x \in \mathbb{Q} \mid -1 < x < 1\} = B_1$

1.5.3.9. Answer.
 (a) 36   (b) 105

2 · Combinatorics

2.1 · Basic Counting Techniques - The Rule of Products

2.1.3 · Exercises

2.1.3.1. Answer. If there are m horses in race 1 and n horses in race 2 then there are $m \cdot n$ possible daily doubles.

2.1.3.3. Answer. $72 = 4 \cdot 6 \cdot 3$

2.1.3.5. Answer. $720 = 6 \cdot 5 \cdot 4 \cdot 3 \cdot 2 \cdot 1$

2.1.3.7. Answer. If we always include the blazer in the outfit we would have 6 outfits. If we consider the blazer optional then there would be 12 outfits. When we add a sweater we have the same type of choice. Considering the sweater optional produces 24 outfits.

2.1.3.9. Answer.

 (a) $2^8 = 256$

 (b) $2^4 = 16$. Here we are concerned only with the first four bits, since the last four must be the same.

 (c) $2^7 = 128$, you have no choice in the last bit.

2.1.3.11. Answer.
 (a) 16   (b) 30

2.1.3.13. Answer.
 (a) 3   (b) 6

2.1.3.15. Answer. 18

2.1.3.17. Answer.

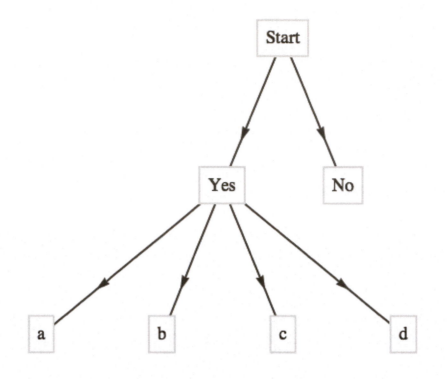

Figure C.0.2: Solution to 17(a)

(a) See Figure C.0.2

(b) 5^6

2.1.3.19. Answer. $2^{n-1} - 1$ and $2^n - 2$

2.2 · Permutations
2.2.2 · Exercises

2.2.2.1. Answer. $P(1000, 3)$

2.2.2.3. Answer. With repetition: $26^8 \approx 2.0883 \times 10^{11}$
Without repetition: $P(26, 8) \approx 6.2991 \ 10^{10}$

2.2.2.5. Answer. $15!$

2.2.2.7. Answer.

(a) $P(15, 5) = 360360$

(b) $2 \cdot 14 \cdot 13 \cdot 12 \cdot 11 = 48048$

2.2.2.9. Answer. $2 \cdot P(3, 3) = 12$

2.2.2.11. Answer.

(a) $P(4, 2) = 12$

(b) $P(n; 2) = n(n - 1)$

(c) Case 1: $m > n$. Since the coordinates must be different, this case is impossible.

Case 2: $m \leqslant n.P(n; m)$.

2.3 · Partitions of Sets and the Law of Addition
2.3.3 · Exercises for Section 2.3

2.3.3.1. Answer. $\{\{a\}, \{b\}, \{c\}\}, \{\{a, b\}, \{c\}\}, \{\{a, c\}, \{b\}\}, \{\{a\}, \{b, c\}\}, \{\{a, b, c\}\}$

2.3.3.3. Answer. No. By this definition it is possible that an element of A might belong to two of the subsets.

2.3.3.5. Answer. The first subset is all the even integers and the second is all the odd integers. These two sets do not intersect and they cover the integers completely.

2.3.3.7. Answer. Since 17 participated in both activities, 30 of the tennis players only played tennis and 25 of the swimmers only swam. Therefore, $17 + 30 + 25 = 72$ of those who were surveyed participated in an activity and so 18 did not.

2.3.3.9. Solution. We assume that $|A_1 \cup A_2| = |A_1| + |A_2| - |A_1 \cap A_2|$.

$$
\begin{aligned}
|A_1 \cup A_2 \cup A_3| &= |(A_1 \cup A_2) \cup A_3| \quad Why? \\
&= |A_1 \cup A_2| + |A_3| - |(A_1 \cup A_2) \cap A_3| \quad Why? \\
&= |(A_1 \cup A_2| + |A_3| - |(A_1 \cap A_3) \cup (A_2 \cap A_3)| \quad Why? \\
&= |A_1| + |A_2| - |A_1 \cap A_2| + |A_3| \\
&\quad - (|A_1 \cap A_3| + |A_2 \cap A_3| - |(A_1 \cap A_3) \cap (A_2 \cap A_3)| \quad Why? \\
&= |A_1| + |A_2| + |A_3| - |A_1 \cap A_2| - |A_1 \cap A_3| \\
&\quad - |A_2 \cap A_3| + |A_1 \cap A_2 \cap A_3| \quad Why?
\end{aligned}
$$

The law for four sets is

$$
\begin{aligned}
|A_1 \cup A_2 \cup A_3 \cup A_4| &= |A_1| + |A_2| + |A_3| + |A_4| \\
&\quad - |A_1 \cap A_2| - |A_1 \cap A_3| - |A_1 \cap A_4| \\
&\quad - |A_2 \cap A_3| - |A_2 \cap A_4| - |A_3 \cap A_4| \\
&\quad + |A_1 \cap A_2 \cap A_3| + |A_1 \cap A_2 \cap A_4| \\
&\quad + |A_1 \cap A_3 \cap A_4| + |A_2 \cap A_3 \cap A_4| \\
&\quad - |A_1 \cap A_2 \cap A_3 \cap A_4|
\end{aligned}
$$

Derivation:

$$
\begin{aligned}
|A_1 \cup A_2 \cup A_3 \cup A_4| &= |(A_1 \cup A_2 \cup A_3) \cup A_4| \\
&= |(A_1 \cup A_2 \cup A_3| + |A_4| - |(A_1 \cup A_2 \cup A_3) \cap A_4| \\
&= |(A_1 \cup A_2 \cup A_3| + |A_4| \\
&\quad - |(A_1 \cap A_4) \cup (A_2 \cap A_4) \cup (A_3 \cap A_4)| \\
&= |A_1| + |A_2| + |A_3| - |A_1 \cap A_2| - |A_1 \cap A_3| \\
&\quad - |A_2 \cap A_3| + |A_1 \cap A_2 \cap A_3| + |A_4| - |A_1 \cap A_4| \\
&\quad + |A_2 \cap A_4| + |A_3 \cap A_4| - |(A_1 \cap A_4) \cap (A_2 \cap A_4)| \\
&\quad - |(A_1 \cap A_4) \cap (A_3 \cap A_4)| - |(A_2 \cap A_4) \cap (A_3 \cap A_4)| \\
&\quad + |(A_1 \cap A_4) \cap (A_2 \cap A_4) \cap (A_3 \cap A_4)| \\
&= |A_1| + |A_2| + |A_3| + |A_4| - |A_1 \cap A_2| - |A_1 \cap A_3| \\
&\quad - |A_2 \cap A_3| - |A_1 \cap A_4| - |A_2 \cap A_4| \quad - |A_3 \cap A_4| \\
&\quad + |A_1 \cap A_2 \cap A_3| + |A_1 \cap A_2 \cap A_4| \\
&\quad + |A_1 \cap A_3 \cap A_4| + |A_2 \cap A_3 \cap A_4| \\
&\quad - |A_1 \cap A_2 \cap A_3 \cap A_4|
\end{aligned}
$$

2.3.3.11. Answer. Partition the set of fractions into blocks, where each block contains fractions that are numerically equivalent. Describe how you would determine whether two fractions belong to the same block. Redefine the rational numbers to be this partition. Each rational number is a set of fractions.

2.4 · Combinations and the Binomial Theorem

2.4.4 · Exercises

2.4.4.1. Answer. $\binom{10}{3} \cdot \binom{25}{4} = 1,518,000$

2.4.4.2. Hint. Think of the set of positions that contain a 1 to turn this is into a question about sets.

Solution. (a) $\binom{8}{3}$ (b) $2^8 - \left(\binom{8}{0} + \binom{8}{1} \right)$

2.4.4.3. Answer. $\binom{10}{7} + \binom{10}{8} + \binom{10}{9} + \binom{10}{10} = 120 + 45 + 10 + 1 = 176$

2.4.4.5. Hint. Think of each path as a sequence of instructions to go right (R) and up (U).

Answer. Each path can be described as a sequence or R's and U's with exactly six of each. The six positions in which R's could be placed can be selected from the twelve positions in the sequence $\binom{12}{6}$ ways. We can generalize this logic and see that there are $\binom{m+n}{m}$ paths from $(0,0)$ to (m,n).

2.4.4.7. Answer.

(a) $C(52,5) = 2,598,960$

(b) $\binom{52}{5} \cdot \binom{47}{5} \cdot \binom{42}{5} \cdot \binom{37}{5}$

2.4.4.9. Answer. $\binom{4}{2} \cdot \binom{48}{3} = 6 \cdot 17296 = 103776$

2.4.4.11. Answer. $\binom{12}{3} \cdot \binom{9}{4} \cdot \binom{5}{5}$

2.4.4.13. Answer.

(a) $\binom{10}{2} = 45$

(b) $\binom{10}{3} = 120$

2.4.4.15. Answer. Assume $|A| = n$. If we let $x = y = 1$ in the Binomial Theorem, we obtain $2^n = \binom{n}{0} + \binom{n}{1} + \cdots + \binom{n}{n}$, with the right side of the equality counting all subsets of A containing $0, 1, 2, \ldots, n$ elements. Hence $|P(A)| = 2^{|A|}$

2.4.4.17. Hint. $9998 = 10000 - 2$

Answer. $10000^3 - 3 \cdot 2 \cdot 10000^2 + 3 \cdot 2^2 \cdot 10000 - 2^3 = 999,400,119,992.$

3 · Logic

3.1 · Propositions and Logical Operators

3.1.3 · Exercises for Section 3.1

3.1.3.1. Answer.

(a) $d \wedge c$

(c) $\neg(d \wedge s)$

(b) $s \vee \neg c$

(d) $\neg s \wedge \neg c$

3.1.3.3. Answer.

(a) $2 > 5$ and 8 is an even integer. False.

(b) If $2 \leqslant 5$ then 8 is an even integer. True.

(c) If $2 \leqslant 5$ and 8 is an even integer then 11 is a prime number. True.

(d) If $2 \leqslant 5$ then either 8 is an even integer or 11 is not a prime number. True.

(e) If $2 \leqslant 5$ then either 8 is an odd integer or 11 is not a prime number. False.

(f) If 8 is not an even integer then $2 > 5$. True.

3.1.3.5. Answer. Only the converse of d is true.

3.2 · Truth Tables and Propositions Generated by a Set

3.2.3 · Exercises for Section 3.2

3.2.3.1. Answer.

(a)
p	$p \vee p$
0	0
1	1

(b)
p	$\neg p$	$p \wedge p$
0	1	0
1	0	0

(c)
p	$\neg p$	$p \wedge (\neg p)$
0	1	1
1	0	1

(d)
p	$p \wedge p$
0	0
1	1

3.2.3.3. Answer.

(a) $\neg(p \wedge r) \vee s$

(b) $(p \vee q) \wedge (r \vee q)$

3.2.3.5. Answer. $2^4 = 16$ rows.

3.3 · Equivalence and Implication

3.3.5 · Exercises for Section 3.3

3.3.5.1. Answer. $a \Leftrightarrow e, d \Leftrightarrow f, g \Leftrightarrow h$

3.3.5.3. Solution. No. In symbolic form the question is: Is $(p \rightarrow q) \Leftrightarrow (q \rightarrow p)$?

p	q	$p \rightarrow q$	$q \rightarrow p$	$(p \rightarrow q) \leftrightarrow (q \rightarrow p)$
0	0	1	1	1
0	1	1	0	0
1	0	0	1	0
1	1	1	1	1

This table indicates that an implication is not always equivalent to its converse.

3.3.5.5. Solution. Let x be any proposition generated by p and q. The truth table for x has 4 rows and there are 2 choices for a truth value for x for each row, so there are $2 \cdot 2 \cdot 2 \cdot 2 = 2^4$ possible propositions.

3.3.5.7. Answer. $0 \to p$ and $p \to 1$ are tautologies.

3.4 · The Laws of Logic

3.4.2 · Exercises for Section 3.4

3.4.2.1. Answer. Let $s =$ I will study,$t =$ I will learn. The argument is: $((s \to t) \wedge (\neg t)) \to (\neg s)$, call the argument a.

s	t	$s \to t$	$(s \to t) \wedge (\neg t)$	a
0	0	1	1	1
0	1	1	0	1 .
1	0	0	0	1
1	1	1	0	1

Since a is a tautology, the argument is valid.

3.4.2.3. Answer. In any true statement S, replace; $\wedge$ with $\vee$, $\vee$ with $\wedge$, 0 with 1, 1 with 0, $\Leftarrow$ with $\Rightarrow$, and $\Rightarrow$ with $\Leftarrow$. Leave all other connectives unchanged.

3.5 · Mathematical Systems and Proofs

3.5.4 · Exercises for Section 3.5

3.5.4.1. Answer.

(a)

p	q	$(p \vee q) \wedge \neg q$	$((p \vee q) \wedge \neg q) \to p$
0	0	0	1
0	1	0	1
1	0	1	1
1	1	0	1

(b)

p	q	$(p \to q) \wedge \neg q$	$\neg p$	$(p \to q) \wedge (\neg q)$
0	0	1	1	1
0	1	0	1	1
1	0	0	0	1
1	1	0	0	1

3.5.4.3. Answer.

(a) (1) Direct proof:

  (2) $d \to (a \vee c)$

  (3) d

  (4) $a \vee c$

  (5) $a \to b$

  (6) $\neg a \vee b$

  (7) $c \to b$

  (8) $\neg c \vee b$

  (9) $(\neg a \vee b) \wedge (\neg c \vee b)$

  (10) $(\neg a \wedge \neg c) \vee b$

(11) $\neg(a \vee c) \vee b$

(12) b □

Indirect proof:

(1) $\neg b$ Negated conclusion

(2) $a \rightarrow b$ Premise

(3) $\neg a$ Indirect Reasoning (1), (2)

(4) $c \rightarrow b$ Premise

(5) $\neg c$ Indirect Reasoning (1), (4)

(6) $(\neg a \wedge \neg c)$ Conjunctive (3), (5)

(7) $\neg(a \vee c)$ DeMorgan's law (6)

(8) $d \rightarrow (a \vee c)$ Premise

(9) $\neg d$ Indirect Reasoning (7), (8)

(10) d Premise

(11) $\nvdash$ (9), (10)   □

(b) Direct proof:

(1) $(p \rightarrow q) \wedge (r \rightarrow s)$

(2) $p \rightarrow q$

(3) $(p \rightarrow t) \wedge (s \rightarrow u)$

(4) $q \rightarrow t$

(5) $p \rightarrow t$

(6) $r \rightarrow s$

(7) $s \rightarrow u$

(8) $r \rightarrow u$

(9) $p \rightarrow r$

(10) $p \rightarrow u$

(11) $p \rightarrow (t \wedge u)$ Use $(x \rightarrow y) \wedge (x \rightarrow z) \Leftrightarrow x \rightarrow (y \wedge z)$

(12) $\neg(t \wedge u) \rightarrow \neg p$

(13) $\neg(t \wedge u)$

(14) $\neg p$ □

Indirect proof:

(1) p

(2) $p \rightarrow q$

(3) q

(4) $q \rightarrow t$

(5) t

(6) $\neg(t \wedge u)$

(7) $\neg t \vee \neg u$

(8) $\neg u$

(9) $s \rightarrow u$

(10) $\neg s$

(11) $r \to s$

(12) $\neg r$

(13) $p \to r$

(14) r

(15) 0 ☐

(c) Direct proof:

(1) $\neg s \lor p$ Premise

(2) s Added premise (conditional conclusion)

(3) $\neg(\neg s)$ Involution (2)

(4) p Disjunctive simplification (1), (3)

(5) $p \to (q \to r)$ Premise

(6) $q \to r$ Detachment (4), (5)

(7) q Premise

(8) r Detachment (6), (7) ☐

Indirect proof:

(1) $\neg(s \to r)$ Negated conclusion

(2) $\neg(\neg s \lor r)$ Conditional equivalence (I)

(3) $s \land \neg r$ DeMorgan (2)

(4) s Conjunctive simplification (3)

(5) $\neg s \lor p$ Premise

(6) $s \to p$ Conditional equivalence (5)

(7) p Detachment (4), (6)

(8) $p \to (q \to r)$ Premise

(9) $q \to r$ Detachment (7), (8)

(10) q Premise

(11) r Detachment (9), (10)

(12) $\neg r$ Conjunctive simplification (3)

(13) 0 \quad Conjunction (11), (12) ☐

(d) Direct proof:

(1) $p \to q$

(2) $q \to r$

(3) $p \to r$

(4) $p \lor r$

(5) $\neg p \lor r$

(6) $(p \lor r) \land (\neg p \lor r)$

(7) $(p \land \neg p) \lor r$

(8) $0 \lor r$

(9) r☐

Indirect proof:

(1) $\neg r$ Negated conclusion

(2) $p \vee r$ Premise

(3) p (1), (2)

(4) $p \to q$ Premise

(5) q Detachment (3), (4)

(6) $q \to r$ Premise

(7) r Detachment (5), (6)

(8) 0 (1), (7) □

3.5.4.5. Answer.

(a) Let W stand for "Wages will increase," I stand for "there will be inflation," and C stand for "cost of living will increase." Therefore the
argument is: $W \to I$, $\neg I \to \neg C$, $W \Rightarrow C$. The argument is invalid. The
easiest way to see this is through a truth table. Let x be the conjunction of

all premises.

W	I	C	$\neg I$	$\neg C$	$W \to I$	$\neg I \to \neg C$	x	$x \to C$
0	0	0	1	1	1	0	0	1
0	0	1	1	0	1	1	0	1
0	1	0	0	1	1	1	0	1
0	1	1	0	0	1	1	0	1
1	0	0	1	1	0	0	0	1
1	0	1	1	0	0	1	0	1
1	1	0	0	1	1	1	1	1
1	1	1	0	0	1	1	1	0

(b) Let r stand for "the races are fixed," c stand for "casinos are crooked," t
stand for "the tourist trade will decline," and p stand for "the police will
be happy." Therefore, the argument is:

$$(r \vee c) \to t, t \to p, \neg p \to \neg r.$$

The argument is valid. Proof:

(1) $t \to p$ Premise

(2) $\neg p$ Premise

(3) $\neg t$ Indirect Reasoning (1), (2)

(4) $(r \vee c) \to t$ Premise

(5) $\neg(r \vee c)$ Indirect Reasoning (3), (4)

(6) $(\neg r) \wedge (\neg c)$ DeMorgan (5)

(7) $\neg r$ Conjunction simplification (6) □

3.5.4.7. Answer.

$p_1 \to p_k$ and $p_k \to p_{k+1}$ implies $p_1 \to p_{k+1}$. It takes two
steps to get to $p_1 \to p_{k+1}$ from $p_1 \to p_k$ This means it takes $2(100 - 1)$ steps to
get to $p_1 \to p_{100}$ (subtract 1 because $p_1 \to p_2$ is stated as a premise). A final
step is needed to apply detachment to imply p_{100}

3.6 · Propositions over a Universe

3.6.3 · Exercises for Section 3.6

3.6.3.1. Answer.

(a) $\{\{1\}, \{3\}, \{1, 3\}, \emptyset\}$

(b) $\{\{3\}, \{3, 4\}, \{3, 2\}, \{2, 3, 4\}\}$

(c) $\{\{1\}, \{1,2\}, \{1,3\}, \{1,4\}, \{1,2,3\}, \{1,2,4\}, \{1,3,4\}, \{1,2,3,4\}\}$

(d) $\{\{2\}, \{3\}, \{4\}, \{2,3\}, \{2,4\}, \{3,4\}\}$

(e) $\{A \subseteq U : |A| = 2\}$

3.6.3.2. Solution.

(a) (i) $T_p = \{2,3,5,7,11,13,17,19,23,29,31\}$

(ii) $T_q = \{1,3,9,27,81,\dots\}$

(iii) $T_r = \{1,3,9,27\}$

(b) $r \Rightarrow q$

3.6.3.3. Answer.   There are $2^3 = 8$ subsets of U, allowing for the possibility of 2^8 nonequivalent propositions over U.

3.6.3.5. Answer.   Two possible answers: s is odd and $(s-1)(s-3)(s-5)(s-7)=0$

3.6.3.7. Solution. b and c

3.7 · Mathematical Induction
3.7.4 · Exercises for Section 3.7

3.7.4.1. Answer.   We wish to prove that $P(n) : 1+3+5+\cdots+(2n-1) = n^2$ is true for $n \geqslant 1$. Recall that the nth odd positive integer is 2n - 1.

Basis: for $n = 1$, $P(n)$ is $1 = 1^2$, which is true

Induction: Assume that for some $n \geqslant 1, P(n)$ is true. Then:

$$1+3+\cdots+(2(n+1)-1) = (1+3+\cdots+(2n-1)) + (2(n+1)-1)$$
$$= n^2 + (2n+1) \quad \text{by } P(n) \text{ and basic algebra}$$
$$= (n+1)^2 \quad \blacksquare$$

3.7.4.3. Answer.   Proof:

- Basis: $1 = 1(2)(3)/6 = 1$

- Induction:

$$\sum_{1}^{n+1} k^2 = \sum_{1}^{n} k^2 + (n+1)^2$$
$$= \frac{n(n+1)(2n+1)}{6} + (n+1)^2$$
$$= \frac{(n+1)(2n^2+7n+6)}{6}$$
$$= \frac{(n+1)(n+2)(2n+3)}{6} \quad \blacksquare$$

3.7.4.5. Answer. Basis: For $n = 1$, we observe that $\frac{1}{(1 \cdot 2)} = \frac{1}{(1+1)}$

Induction: Assume that for some $n \geqslant 1$, the formula is true. Then:

$$\frac{1}{(1 \cdot 2)} + \cdots + \frac{1}{n(n+1)} + \frac{1}{(n+1)(n+2)} = \frac{n}{n+1} + \frac{1}{(n+1)(n+2)}$$

$$= \frac{(n+2)n}{(n+1)(n+2)} + \frac{1}{(n+1)(n+2)}$$

$$= \frac{(n+1)^2}{(n+1)(n+2)}$$

$$= \frac{n+1}{n+2} \quad \blacksquare$$

3.7.4.6. Solution. Basis: ($n = 2$) Proven with a truth table already.

Induction: Assume the generalized DeMorgan's Law with n propositions is true for some $n \geq 2$.

$$\neg(p_1 \wedge p_2 \wedge \cdots \wedge p_n \wedge p_{n+1}) \Leftrightarrow \neg((p_1 \wedge p_2 \wedge \cdots \wedge p_n) \wedge p_{n+1})$$

$$\Leftrightarrow \neg(p_1 \wedge p_2 \wedge \cdots \wedge p_n) \vee (\neg p_{n+1})$$

$$\Leftrightarrow ((\neg p_1) \vee (\neg p_2) \vee \cdots \vee (\neg p_n)) \vee (\neg p_{n+1})$$

$$\Leftrightarrow (\neg p_1) \vee (\neg p_2) \vee \cdots \vee (\neg p_n) \vee (\neg p_{n+1})$$

3.7.4.7. Answer. Let A_n be the set of strings of zeros and ones of length n (we assume that $|A_n| = 2^n$ is known). Let E_n be the set of the "even" strings, and E_n^c = the odd strings. The problem is to prove that for $n \geqslant 1$, $|E_n| = 2^{n-1}$. Clearly, $|E_1| = 1$, and, if for some $n \geqslant 1, |E_n| = 2^{n-1}$, it follows that $|E_{n+1}| = 2^n$ by the following reasoning.

We partition E_{n+1} according to the first bit: $E_{n+1} = \{1s \mid s \in E_n^c\} \cup \{0s \mid s \in E_n\}$

Since $\{1s \mid s \in E_n^c\}$ and $\{0s \mid s \in E_n\}$ are disjoint, we can apply the addition law. Therefore,

$$|E_{n+1}| = |E_n^c| + |E_n|$$

$$= 2^{n-1} + (2^n - 2^{n-1}) = 2^n. \quad \blacksquare$$

3.7.4.9. Answer. Assume that for n persons $(n \geqslant 1)$, $\frac{(n-1)n}{2}$ handshakes take place. If one more person enters the room, he or she will shake hands with n people,

$$\frac{(n-1)n}{2} + n = \frac{n^2 - n + 2n}{2}$$

$$= \frac{n^2 + n}{2} = \frac{n(n+1)}{2}$$

$$= \frac{((n+1) - 1)(n+1)}{2}$$

Also, for $n = 1$, there are no handshakes, which matches the conjectured formula:

$$\frac{(1-1)(1)}{2} = 0 \quad \blacksquare.$$

3.7.4.11. Solution. Let $p(n)$ be "$a_1 + a_2 + \cdots + a_n$ has the same value no matter how it is evaluated."

Basis: $a_1 + a_2 + a_3$ may be evaluated only two ways. Since $+$ is associative,

$(a_1 + a_2) + a_3 = a_1 + (a_2 + a_3)$. Hence, $p(3)$ is true.

Induction: Assume that for some $n \geq 3$, $p(3), p(4), \ldots, p(n)$ are all true. Now consider the sum $a_1 + a_2 + \cdots + a_n + a_{n+1}$. Any of the n additions in this expression can be applied last. If the jth addition is applied last, we have $c_j = (a_1 + a_2 + \cdots + a_j) + (a_{j+1} + \cdots + a_{n+1})$. No matter how the expression to the left and right of the jth addition are evaluated, the result will always be the same by the induction hypothesis, specifically $p(j)$ and $p(n+1-j)$. We now can prove that $c_1 = c_2 = \cdots = c_n$. If $i < j$,

$$\begin{aligned}
c_i &= (a_1 + a_2 + \cdots + a_i) + (a_{i+1} + \cdots + a_{n+1}) \\
&= (a_1 + a_2 + \cdots + a_i) + ((a_{i+1} + \cdots + a_j) + (a_{j+1} + \cdots + a_{n+1})) \\
&= ((a_1 + a_2 + \cdots + a_i) + ((a_{i+1} + \cdots + a_j)) + (a_{j+1} + \cdots + a_{n+1}) \\
&= ((a_1 + a_2 + \cdots + a_j)) + (a_{j+1} + \cdots + a_{n+1}) \\
&= c_j \qquad \square
\end{aligned}$$

3.7.4.12. Hint. The number of times the rules are applied should be the integer that you do the induction on.

3.7.4.13. Hint. Let $p(m)$ be the proposition that $x^{m+n} = x^m x^n$ for all $n \geq 1$.

Solution. For $m \geq 1$, let $p(m)$ be $x^{n+m} = x^n x^m$ for all $n \geq 1$. The basis for this proof follows directly from the basis for the definition of exponentiation.

Induction: Assume that for some $m \geq 1$, $p(m)$ is true. Then

$$\begin{aligned}
x^{n+(m+1)} &= x^{(n+m)+1} && \text{by associativity of integer addition} \\
&= x^{n+m} x^1 && \text{by recursive definition} \\
&= x^n x^m x^1 && \text{induction hypothesis} \\
&= x^n x^{m+1} && \text{recursive definition} \quad \square
\end{aligned}$$

3.8 · Quantifiers

3.8.5 · Exercises for Section 3.8

3.8.5.1. Answer.

(a) $(\forall x)(F(x) \to C(x))$

(b) There are objects in the sea which are not fish.

(c) Every fish lives in the sea.

3.8.5.3. Answer.

(a) There is a book with a cover that is not blue.

(b) Every mathematics book that is published in the United States has a blue cover.

(c) There exists a mathematics book with a cover that is not blue.

(d) There exists a book that appears in the bibliography of every mathematics book.

(e) $(\forall x)(B(x) \to M(x))$

(f) $(\exists x)(M(x) \wedge \neg U(x))$

(g) $(\exists x)((\forall y)(\neg R(x, y))$

3.8.5.5. Answer. The equation $4u^2 - 9 = 0$ has a solution in the integers. (False)

3.8.5.6. Hint. Your answer will depend on your choice of a universe

3.8.5.7. Answer.

(a) Every subset of U has a cardinality different from its complement. (True)

(b) There is a pair of disjoint subsets of U both having cardinality 5. (False)

(c) $A - B = B^c - A^c$ is a tautology. (True)

3.8.5.9. Answer. $(\forall a)_\mathbb{Q}(\forall b)_\mathbb{Q}(a + b$ is a rational number.$)$

3.8.5.10. Hint. You will need three quantifiers.

3.8.5.11. Answer. Let $I = \{1, 2, 3, \ldots, n\}$

(a) $(\exists x)_I\,(x \in A_i)$

(b) $(\forall x)_I\,(x \in A_i)$

3.9 · A Review of Methods of Proof
3.9.3 · Exercises for Section 3.9

3.9.3.1. Answer. The given statement can be written in if ... , then ... format as: If x and y are two odd positive integers, then $x + y$ is an even integer.

    Proof: Assume x and y are two positive odd integers. It can be shown that $x + y = 2 \cdot$ (some positive integer).
    x odd and positive $\Rightarrow x = 2m + 1$ for some $m \geq 0$,
    y odd and positive $\Rightarrow y = 2n + 1$ for some $n \geq 0$.
    Then,

$$x + y = (2m + 1) + (2n + 1) = 2((m + n) + 1) = 2 \cdot \text{(some positive integer)}$$

Therefore, $x + y$ is a positive even integer. $\square$

3.9.3.3. Answer. Proof: (Indirect) Assume to the contrary, that $\sqrt{2}$ is a rational number. Then there exists $p, q \in \mathbb{Z}, (q \neq 0)$ where $\frac{p}{q} = \sqrt{2}$ and where $\frac{p}{q}$ is in lowest terms, that is, p and q have no common factor other than 1.

$$\frac{p}{q} = \sqrt{2} \Rightarrow \frac{p^2}{q^2} = 2$$
$$\Rightarrow p^2 = 2q^2$$
$$\Rightarrow p^2 \text{ is an even integer}$$
$$\Rightarrow p \text{ is an even integer (see Exercise 2)}$$
$$\Rightarrow 4 \text{ is a factor of } p^2$$
$$\Rightarrow q^2 \text{ is even}$$
$$\Rightarrow q \text{ is even}$$

    Hence both p and q have a common factor, namely 2, which is a contradiction. $\square$

3.9.3.5. Answer. Proof: (Indirect) Assume $x, y \in \mathbb{R}$ and $x + y \leq 1$. Assume to the contrary that $\left(x \leq \frac{1}{2} \text{ or } y \leq \frac{1}{2}\right)$ is false, which is equivalent to $x >$

$\frac{1}{2}$ and $y > \frac{1}{2}$. Hence $x + y > \frac{1}{2} + \frac{1}{2} = 1$. This contradicts the assumption that $x + y \leqslant 1$. $\square$

4 · More on Sets

4.1 · Methods of Proof for Sets

4.1.5 · Exercises for Section 4.1

4.1.5.1. Answer.

(a) Assume that $x \in A$ (condition of the conditional conclusion $A \subseteq C$). Since $A \subseteq B$, $x \in B$ by the definition of $\subseteq$. $B \subseteq C$ and $x \in B$ implies that $x \in C$. Therefore, if $x \in A$, then $x \in C$. $\square$

(b) (Proof that $A - B \subseteq A \cap B^c$) Let x be in $A - B$. Therefore, x is in A, but it is not in B; that is, $x \in A$ and $x \in B^c \Rightarrow x \in A \cap B^c$. $\square$

(c) $(\Rightarrow)$Assume that $A \subseteq B$ and $A \subseteq C$. Let $x \in A$. By the two premises,$x \in B$ and $x \in C$. Therefore, by the definition of intersection, $x \in B \cap C$. $\square$

(d) $(\Rightarrow)$(Indirect) Assume that B^c is not a subset of A^c . Therefore, there exists $x \in B^c$ that does not belong to A^c. $x \notin A^c \Rightarrow x \in A$. Therefore, $x \in A$ and $x \notin B$, a contradiction to the assumption that $A \subseteq B$. $\square$

4.1.5.3. Answer.

(a) If $A = \mathbb{Z}$ and $B = \emptyset$, $A - B = \mathbb{Z}$, while $B - A = \emptyset$.

(b) If $A = \{0\}$ and $B = \{1\}$, $(0,1) \in A \times B$, but $(0,1)$ is not in $B \times A$.

(c) Let $A = \emptyset$, $B = \{0\}$, and $C = \{1\}$.

(d) If $A = \{1\}$, $B = \{1\}$, and $C = \emptyset$, then the left hand side of the identity is $\{1\}$ while the right hand side is the empty set. Another example is $A = \{1, 2\}$, $B = \{1\}$, and $C = \{2\}$.

4.1.5.5. Solution. Proof: Let $p(n)$ be

$$A \cap (B_1 \cup B_2 \cup \cdots \cup B_n) = (A \cap B_1) \cup (A \cap B_2) \cup \cdots \cup (A \cap B_n).$$

Basis: We must show that $p(2) : A \cap (B_1 \cup B_2) = (A \cap B_1) \cup (A \cap B_2)$ is true. This was done by several methods in section 4.1.

Induction: Assume for some $n \geq 2$ that $p(n)$ is true. Then

$$\begin{aligned}
A \cap (B_1 \cup B_2 \cup \cdots \cup B_{n+1}) &= A \cap ((B_1 \cup B_2 \cup \cdots \cup B_n) \cup B_{n+1}) \\
&= (A \cap (B_1 \cup B_2 \cup \cdots \cup B_n)) \cup (A \cap B_{n+1}) \quad \text{by } p(2) \\
&= ((A \cap B_1) \cup \cdots \cup (A \cap B_n)) \cup (A \cap B_{n+1}) \quad \text{by the induction hypothesis} \\
&= (A \cap B_1) \cup \cdots \cup (A \cap B_n) \cup (A \cap B_{n+1}) \quad \square
\end{aligned}$$

4.1.5.6. Answer. The statement is false. The sets $A = \{1, 2\}$, $B = \{2, 3\}\backslash$ and $C = \{3, 4\}\backslash$ provide a counterexample. Looking ahead to Chapter 6, we would say that the relation of being non-disjoint is not transitive 6.3.3

4.2 · Laws of Set Theory

4.2.4 · Exercises for Section 4.2

4.2.4.1. Answer.

(a)

(b)

A	B	A^c	B^c	$A \cup B$	$(A \cup B)^c$	$A^c \cap B^c$
0	0	1	1	0	1	1
0	1	1	0	1	0	0
1	0	0	1	1	0	0
1	1	0	0	1	0	0

The last two columns are the same so the two sets must be equal.

(c)

$$x \in A \cup A \Rightarrow (x \in A) \vee (x \in A) \quad \text{by the definition of } \cap$$
$$\Rightarrow x \in A \quad \text{by the idempotent law of logic}$$

Therefore, $A \cup A \subseteq A$.

$$x \in A \Rightarrow (x \in A) \vee (x \in A) \quad \text{by conjunctive addition}$$
$$\Rightarrow x \in A \cup A$$

Therefore, $A \subseteq A \cup A$ and so we have $A \cup A = A$.

4.2.4.3. Answer. For all parts of this exercise, a reason should be supplied for each step. We have supplied reasons only for part a and left them out of the other parts to give you further practice.

(a)

$$A \cup (B - A) = A \cup (B \cap A^c) \text{ by Exercise 1 of Section 4.1}$$
$$= (A \cup B) \cap (A \cup A^c) \text{ by the distributive law}$$
$$= (A \cup B) \cap U \text{ by the null law}$$
$$= (A \cup B) \text{ by the identity law } \square$$

(b)

$$A - B = A \cap B^c$$
$$= B^c \cap A$$
$$= B^c \cap (A^c)^{c\cdot}$$
$$= B^c - A^c$$

(c) Select any element, $x \in A \cap C$. One such element exists since $A \cap C$ is not empty.

$$x \in A \cap C \Rightarrow x \in A \wedge x \in C$$
$$\Rightarrow x \in B \wedge x \in C$$
$$\Rightarrow x \in B \cap C$$
$$\Rightarrow B \cap C \neq \emptyset \quad \square$$

(d)

$$A \cap (B - C) = A \cap (B \cap C^c)$$
$$= (A \cap B \cap A^c) \cup (A \cap B \cap C^c)$$
$$= (A \cap B) \cap (A^c \cup C^c)$$
$$= (A \cap B) \cap (A \cup C)^c$$
$$= (A - B) \cap (A - C) \quad \square$$

(e)

$$A - (B \cup C) = A \cap (B \cup C)^c$$
$$= A \cap (B^c \cap C^c)$$
$$= (A \cap B^c) \cap (A \cap C^c)$$
$$= (A - B) \cap (A - C) \quad \square$$

4.2.4.5. Hierarchy of Set Operations. Answer.
(a) $A \cup ((B^c) \cap C)$ (b) $(A \cap B) \cup (C \cap B)$ (c) $(A \cup B) \cup (C^c)$

4.3 · Minsets

4.3.3 · Exercises for Section 4.3

4.3.3.1. Answer.

(a) $\{1\}, \{2, 3, 4, 5\}, \{6\}, \{7, 8\}, \{9, 10\}$

(b) 2^5 , as compared with 2^{10}. $\{1, 2\}$ is one of the 992 sets that can't be generated.

4.3.3.3. Answer. $B_1 = \{00, 01, 10, 11\}$ and $B_2 = \{0, 00, 01\}$ generate minsets $\{00, 01\}, \{0\}, \{10, 11\}$, and $\{\lambda, 1\}$. Note: λ is the null string, which has length zero.

4.3.3.5. Answer.

(a) $B_1 \cap B_2 = \emptyset$, $B_1 \cap B_2^c = \{0, 2, 4\}$, $B_1^c \cap B_2 = \{1, 5\}$, $B_1^c \cap B_2^c = \{3\}$

(b) 2^3, since there are 3 nonempty minsets.

4.3.3.7. Answer. Let $a \in A$. For each i, $a \in B_i$, or $a \in B_i{}^c$, since $B_i \cup B_i{}^c = A$ by the complement law. Let $D_i = B_i$ if $a \in B_i$, and $D = B_i{}^c$ otherwise. Since a is in each D_i, it must be in the minset $D_1 \cap D_2 \cdots \cap D_n$. Now consider two different minsets $M_1 = D_1 \cap D_2 \cdots \cap D_n$, and $M_2 = G_1 \cap G_2 \cdots \cap G_n$, where each D_i and G_i is either B_i or $B_i{}^c$. Since these minsets are not equal, $D_i \neq G_i$, for some i. Therefore, $M_1 \cap M_2 = D_1 \cap D_2 \cdots \cap D_n \cap G_1 \cap G_2 \cdots \cap G_n = \emptyset$, since two of the sets in the intersection are disjoint. Since every element of A is in a minset and the minsets are disjoint, the nonempty minsets must form a partition of A. $\square$

4.4 · The Duality Principle

4.4.2 · Exercises for Section 4.4

4.4.2.1. Answer.

(a) $A \cap (B \cup A) = A$

(b) $A \cap ((B^c \cap A) \cup B)^c = \emptyset$

(c) $(A \cap B^c)^c \cup B = A^c \cup B$

4.4.2.3. Answer.

(a) $(p \wedge \neg(\neg q \wedge p) \vee q) \Leftrightarrow 0$

(b) $(\neg(p \vee (\neg q))) \wedge q \Leftrightarrow ((\neg p) \wedge q)$

4.4.2.5. Answer. The maxsets are:

- $B_1 \cup B_2 = \{1, 2, 3, 5\}$

- $B_1 \cup B_2{}^c = \{1, 3, 4, 5, 6\}$

- $B_1{}^c \cup B_2 = \{1, 2, 3, 4, 6\}$

- $B_1{}^c \cup B_2{}^c = \{2, 4, 5, 6\}$

They do not form a partition of A since it is not true that the intersection of any two of them is empty. A set is said to be in **maxset normal form** when it is expressed as the intersection of distinct nonempty maxsets or it is the universal set U.

5 · Introduction to Matrix Algebra

5.1 · Basic Definitions and Operations

5.1.4 · Exercises for Section 5.1

5.1.4.1. Answer. For parts c, d and i of this exercise, only a verification is needed. Here, we supply the result that will appear on both sides of the equality.

(a) $AB = \begin{pmatrix} -3 & 6 \\ 9 & -13 \end{pmatrix}$ $BA = \begin{pmatrix} 2 & 3 \\ -7 & -18 \end{pmatrix}$

(b) $\begin{pmatrix} 1 & 0 \\ 5 & -2 \end{pmatrix}$

(c) $\begin{pmatrix} 3 & 0 \\ 15 & -6 \end{pmatrix}$

(d) $\begin{pmatrix} 18 & -15 & 15 \\ -39 & 35 & -35 \end{pmatrix}$

(e) $\begin{pmatrix} -12 & 7 & -7 \\ 21 & -6 & 6 \end{pmatrix}$

(f) $B + 0 = B$

(g) $\begin{pmatrix} 0 & 0 \\ 0 & 0 \end{pmatrix}$

(h) $\begin{pmatrix} 0 & 0 \\ 0 & 0 \end{pmatrix}$

(i) $\begin{pmatrix} 5 & -5 \\ 10 & 15 \end{pmatrix}$

5.1.4.3. Answer. $\begin{pmatrix} 1/2 & 0 \\ 0 & 1/3 \end{pmatrix}$

5.1.4.5. Answer. $A^3 = \begin{pmatrix} 1 & 0 & 0 \\ 0 & 8 & 0 \\ 0 & 0 & 27 \end{pmatrix}$ $A^{15} = \begin{pmatrix} 1 & 0 & 0 \\ 0 & 32768 & 0 \\ 0 & 0 & 14348907 \end{pmatrix}$

5.1.4.7. Answer.

(a) $Ax = \begin{pmatrix} 2x_1 + 1x_2 \\ 1x_1 - 1x_2 \end{pmatrix}$ equals $\begin{pmatrix} 3 \\ 1 \end{pmatrix}$ if and only if both of the equalities $2x_1 + x_2 = 3$ and $x_1 - x_2 = 1$ are true.

(b) (i) $A = \begin{pmatrix} 2 & -1 \\ 1 & 1 \end{pmatrix}$ $x = \begin{pmatrix} x_1 \\ x_2 \end{pmatrix}$ $B = \begin{pmatrix} 4 \\ 0 \end{pmatrix}$

(c) $A = \begin{pmatrix} 1 & 1 & 2 \\ 1 & 2 & -1 \\ 1 & 3 & 1 \end{pmatrix}$ $x = \begin{pmatrix} x_1 \\ x_2 \\ x_3 \end{pmatrix}$ $B = \begin{pmatrix} 1 \\ -1 \\ 5 \end{pmatrix}$

(d) $A = \begin{pmatrix} 1 & 1 & 0 \\ 0 & 1 & 0 \\ 1 & 0 & 3 \end{pmatrix}$ $x = \begin{pmatrix} x_1 \\ x_2 \\ x_3 \end{pmatrix}$ $B = \begin{pmatrix} 3 \\ 5 \\ 6 \end{pmatrix}$

5.2 · Special Types of Matrices

5.2.3 · Exercises for Section 5.2

5.2.3.1. Answer.

(a) $\begin{pmatrix} -1/5 & 3/5 \\ 2/5 & -1/5 \end{pmatrix}$

(d) $A^{-1} = A$

(b) No inverse exists.

(c) $\begin{pmatrix} 1 & 3 \\ 0 & 1 \end{pmatrix}$

(e) $\begin{pmatrix} 1/3 & 0 & 0 \\ 0 & 2 & 0 \\ 0 & 0 & -1/5 \end{pmatrix}$

5.2.3.3. Answer. Let A and B be n by n invertible matrices.

$$
\begin{aligned}
\left(B^{-1}A^{-1}\right)(AB) &= \left(B^{-1}\right)\left(A^{-1}(AB)\right) \\
&= \left(B^{-1}\right)\left(\left(A^{-1}A\right)B\right) \\
&= \left(\left(B^{-1}\right)IB\right) \\
&= B^{-1}(B) \\
&= I
\end{aligned}
$$

Similarly, $(AB)\left(B^{-1}A^{-1}\right) = I$.

By Theorem 5.2.6, $B^{-1}A^{-1}$ is the only inverse of AB. If we tried to invert AB with $A^{-1}B^{-1}$, we would be unsuccessful since we cannot rearrange the order of the matrices.

5.2.3.5. Linearity of Determinants. Answer. $1 = \det I = \det\left(AA^{-1}\right) = \det A \ \det A^{-1}$. Now solve for $\det A^{-1}$.

5.2.3.7. Answer. Basis: $(n = 1) : \det A^1 = \det A = (\det A)^1$

Induction: Assume $\det A^n = (\det A)^n$ for some $n \geq 1$.

$$
\begin{aligned}
\det A^{n+1} &= \det\left(A^n A\right) && \text{by the definition of exponents} \\
&= \det\left(A^n\right)\det(A) && \text{by exercise 5} \\
&= (det A)^n (\det A) && \text{by the induction hypothesis} \\
&= (\det A)^{n+1}
\end{aligned}
$$

5.2.3.9. Answer.

(a) Assume $A = BDB^{-1}$

    Basis: $(m = 1)$: $A^1 = A = BD^1 B^{-1}$ is given.

    Induction: Assume that for some positive integer m, $A^m = BD^m B^{-1}$

$$
\begin{aligned}
A^{m+1} = A^m A &\\
&= (BD^m B^{-1})(BDB^{-1}) \quad \text{by the induction hypothesis}\\
&= (BD^m (B^{-1}B)(DB^{-1}) \quad \text{by associativity}\\
&= BD^m DB^{-1} \quad \text{by the definition of inverse}\\
&= BD^{m+1}B^{-1} \quad \square
\end{aligned}
$$

(b) $A^{10} = BD^{10}B^{-1} = \begin{pmatrix} -9206 & 15345 \\ -6138 & 10231 \end{pmatrix}$

5.3 · Laws of Matrix Algebra

5.3.3 · Exercises for Section 5.3

5.3.3.1. Answer.

(a) Let A and B be m by n matrices. Then $A + B = B + A$,

(b) Let A, B, and C be m by n matrices. Then $A + (B + C) = (A + B) + C$.

(c) Let A and B be m by n matrices, and let $c \in \mathbb{R}$. Then $c(A+B) = cA + cB$,

(d) Let A be an m by n matrix, and let $c_1, c_2 \in \mathbb{R}$. Then $(c_1 + c_2)A = c_1 A + c_2 A$.

(e) Let A be an m by n matrix, and let $c_1, c_2 \in \mathbb{R}$. Then $c_1 (c_2 A) = (c_1 c_2) A$

(f) Let $\mathbf{0}$ be the zero matrix, of size m by n, and let A be a matrix of size n by r. Then $\mathbf{0}A = \mathbf{0} = $ the m by r zero matrix.

(g) Let A be an m by n matrix, and $0 = $ the number zero. Then $0A = 0 = $ the m by n zero matrix.

(h) Let A be an m by n matrix, and let $\mathbf{0}$ be the m by n zero matrix. Then $A + \mathbf{0} = A$.

(i) Let A be an m by n matrix. Then $A + (-1)A = \mathbf{0}$, where $\mathbf{0}$ is the m by n zero matrix.

(j) Let A, B, and C be m by n, n by r, and n by r matrices respectively. Then $A(B + C) = AB + AC$.

(k) Let A, B, and C be m by n, r by m, and r by m matrices respectively. Then $(B + C)A = BA + CA$.

(l) Let A, B, and C be m by n, n by r, and r by p matrices respectively. Then $A(BC) = (AB)C$.

(m) Let A be an m by n matrix, I_m the m by m identity matrix, and I_n the n by n identity matrix. Then $I_m A = A I_n = A$

(n) Let A be an n by n matrix. Then if A^{-1} exists, $\left(A^{-1}\right)^{-1} = A$.

(o) Let A and B be n by n matrices. Then if A^{-1} and B^{-1} exist, $(AB)^{-1} = B^{-1}A^{-1}$.

5.3.3.3. Answer.

(a) $AB + AC = \begin{pmatrix} 21 & 5 & 22 \\ -9 & 0 & -6 \end{pmatrix}$

(b) $A^{-1} = \begin{pmatrix} 1 & 2 \\ 0 & -1 \end{pmatrix} = A$

(c) $A(B + C) = AB + BC$, which is given in part (a).

(d) $\left(A^2\right)^{-1} = (AA)^{-1} = (A^{-1}A) = I^{-1} = I$ by part c

5.4 · Matrix Oddities
5.4.2 · Exercises for Section 5.4

5.4.2.1. Answer. In elementary algebra (the algebra of real numbers), each of the given oddities does not exist.

- AB may be different from BA. Not so in elementary algebra, since $ab = ba$ by the commutative law of multiplication.

- There exist matrices A and B such that $AB = \mathbf{0}$, yet $A \neq \mathbf{0}$ and $B \neq \mathbf{0}$. In elementary algebra, the only way $ab = 0$ is if either a or b is zero. There are no exceptions.

- There exist matrices A, $A \neq \mathbf{0}$, yet $A^2 = \mathbf{0}$. In elementary algebra, $a^2 = 0 \Leftrightarrow a = 0$.

- There exist matrices $A^2 = A$. where $A \neq \mathbf{0}$ and $A \neq I$. In elementary algebra, $a^2 = a \Leftrightarrow a = 0$ or 1.

- There exist matrices A where $A^2 = I$ but $A \neq I$ and $A \neq -I$. In elementary algebra, $a^2 = 1 \Leftrightarrow a = 1$ or -1.

5.4.2.3. Answer.

(a) $\det A \neq 0 \Rightarrow A^{-1}$ exists, and if you multiply the equation $A^2 = A$ on both sides by A^{-1}, you obtain $A = I$.

(b) Counterexample: $A = \begin{pmatrix} 1 & 0 \\ 0 & -1 \end{pmatrix}$

5.4.2.5. Answer.

(a) $A^{-1} = \begin{pmatrix} 1/3 & 1/3 \\ 1/3 & -2/3 \end{pmatrix}$ $x_1 = 4/3$, and $x_2 = 1/3$

(b) $A^{-1} = \begin{pmatrix} 1 & -1 \\ 1 & -2 \end{pmatrix}$ $x_1 = 4$, and $x_2 = 4$

(c) $A^{-1} = \begin{pmatrix} 1/3 & 1/3 \\ 1/3 & -2/3 \end{pmatrix}$ $x_1 = 2/3$, and $x_2 = -1/3$

(d) $A^{-1} = \begin{pmatrix} 1/3 & 1/3 \\ 1/3 & -2/3 \end{pmatrix}$ $x_1 = 0$, and $x_2 = 1$

(e) The matrix of coefficients for this system has a zero determinant; therefore, it has no inverse. The system cannot be solved by this method. In fact, the system has no solution.

6 · Relations

6.1 · Basic Definitions

6.1.4 · Exercises for Section 6.1

6.1.4.1. Answer.

(a) $(2, 4), (2, 8)$

(b) $(2, 3), (2, 4), (5, 8)$

(c) $(1, 1), (2, 4)$

6.1.4.3. Answer.

(a) $r = \{(1, 2), (2, 3), (3, 4), (4, 5)\}$

(b) $r^2 = \{(1, 3), (2, 4), (3, 5)\} = \{(x, y) : y = x + 2, x, y \in A\}$

(c) $r^3 = \{(1, 4), (2, 5)\} = \{(x, y) : y = x + 3, x, y \in A\}$

6.1.4.5. Answer.

(a) When $n = 3$, there are 27 pairs in the relation.

(b) Imagine building a pair of disjoint subsets of S. For each element of S there are three places that it can go: into the first set of the ordered pair, into the second set, or into neither set. Therefore the number of pairs in the relation is 3^n, by the product rule.

6.2 · Graphs of Relations on a Set

6.2.2 · Exercises for Section 6.2

6.2.2.1. Answer.

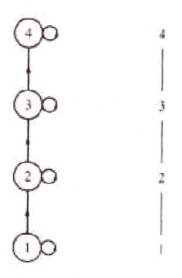

Figure C.0.3

6.2.2.3. Answer. See Figure C.0.4

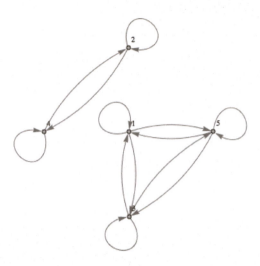

Figure C.0.4: Digraph of the relation t

6.3 · Properties of Relations

6.3.4 · Exercises for Section 6.3

6.3.4.1. Answer.

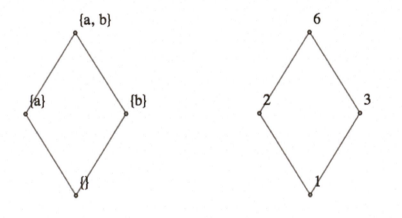

Figure C.0.5

(a) See Figure C.0.5.

(b) The graphs are the same if we disregard the names of the vertices.

6.3.4.2. Hint. Here is a Hasse diagram for the part (a).

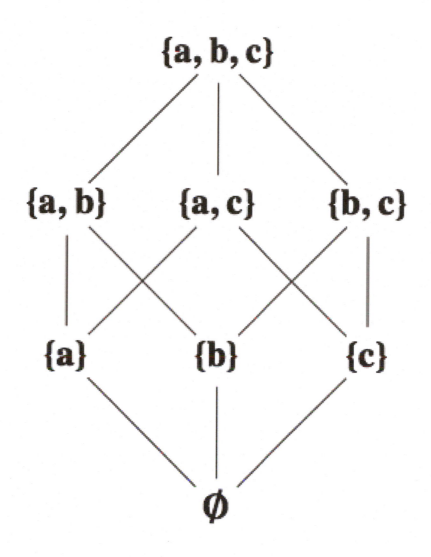

Figure C.0.6

6.3.4.3. Answer.

Part	reflexive?	symmetric?	antisymmetric?	transitive?
i	yes	no	no	yes
ii	yes	no	yes	yes
iii	no	yes	no	yes
iv	no	yes	yes	yes
v	yes	yes	no	yes
vi	yes	no	yes	yes
vii	no	no	no	no

Table C.0.7: Properties of relations in number 3

(i) See Table C.0.7

(ii) Graphs ii and vi show partial ordering relations. Graph v is of an equivalence relation.

6.3.4.5. Answer.

(a) No, since $\mid 1 - 1 \mid = 0 \neq 2$, for example

(b) Yes, because $\mid i - j \mid = \mid j - i \mid$.

(c) No, since $\mid 2 - 4 \mid = 2$ and $\mid 4 - 6 \mid = 2$, but $\mid 2 - 6 \mid = 4 \neq 2$, for example.

(d) See Figure C.0.8

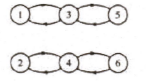

Figure C.0.8

6.3.4.7. Equivalence Classes. Answer.

(a)

(b) $c(0) = \{0\}, c(1) = \{1, 2, 3\} = c(2) = c(3)$

(c) $c(0) \cup c(1) = A$ and $c(0) \cap c(1) = \emptyset$

(d) Let A be any set and let r be an equivalence relation on A. Let a be any element of A. $a \in c(a)$ since r is reflexive, so each element of A is in some equivalence class. Therefore, the union of all equivalence classes equals A. Next we show that any two equivalence classes are either identical or disjoint and we are done. Let $c(a)$ and $c(b)$ be two equivalence classes, and assume that $c(a) \cap c(b) \neq \emptyset$. We want to show that $c(a) = c(b)$. To show that $c(a) \subseteq c(b)$, let $x \in c(a)$. $x \in c(a) \Rightarrow arx$. Also, there exists an element, y, of A that is in the intersection of $c(a)$ and $c(b)$ by our assumption. Therefore,

$$ary \wedge bry \Rightarrow ary \wedge yrb \quad r \text{ is symmetric}$$
$$\Rightarrow arb \quad \text{transitivity of } r$$

Next,

$$arx \wedge arb \Rightarrow xra \wedge arb$$
$$\Rightarrow xrb$$
$$\Rightarrow brx$$
$$\Rightarrow x \in c(b)$$

Similarly, $c(b) \subseteq c(a)$. $\square$

6.3.4.9. Answer.

(a) Equivalence Relation, $c(0) = \{0\}, c(1) = \{1\}, c(2) = \{2,3\} = c(3), c(4) = \{4,5\} = c(5)$, and $c(6) = \{6,7\} = c(7)$

(b) Not an Equivalence Relation.

(c) Equivalence Relation, $c(0) = \{0,2,4,6\} = c(2) = c(4) = c(6)$ and $c(1) = \{1,3,5,7\} = c(3) = c(5) = c(7)$

6.3.4.11. Answer.

(a) The proof follows from the biconditional equivalence in Table 3.4.4.

(b) Apply the chain rule.

(c) See Figure C.0.9.

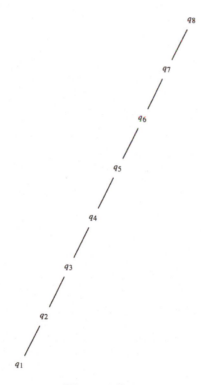

Figure C.0.9

6.4 · Matrices of Relations

6.4.3 · Exercises for Section 6.4

6.4.3.1. Answer.

(a)
$$\begin{array}{c} \\ 1 \\ 2 \\ 3 \\ 4 \end{array} \begin{array}{ccc} 4 & 5 & 6 \\ \left(\begin{array}{ccc} 0 & 0 & 0 \\ 1 & 0 & 0 \\ 0 & 1 & 0 \\ 0 & 0 & 1 \end{array}\right) \end{array} \text{ and } \begin{array}{c} \\ 4 \\ 5 \\ 6 \end{array} \begin{array}{ccc} 6 & 7 & 8 \\ \left(\begin{array}{ccc} 0 & 0 & 0 \\ 1 & 0 & 0 \\ 0 & 1 & 0 \end{array}\right) \end{array}$$

(b) $r_1 r_2 = \{(3,6),(4,7)\}$

(c)
$$\begin{array}{c} \\ 1 \\ 2 \\ 3 \\ 4 \end{array} \begin{array}{ccc} 6 & 7 & 8 \\ \left(\begin{array}{ccc} 0 & 0 & 0 \\ 0 & 0 & 0 \\ 1 & 0 & 0 \\ 0 & 1 & 0 \end{array}\right) \end{array}$$

6.4.3.3. Answer.

$$R : xry \text{ if and only if } |x - y| = 1$$
$$S : xsy \text{ if and only if } x \text{ is less than } y.$$

Table C.0.10

6.4.3.5. Hint.   Consider the possible matrices.

Answer.   The diagonal entries of the matrix for such a relation must be 1. When the three entries above the diagonal are determined, the entries below are also determined. Therefore, there are 2^3 fitting the description.

6.4.3.7. Answer.

(a)
$$\begin{array}{c} \\ 1 \\ 2 \\ 3 \\ 4 \end{array} \begin{array}{cccc} 1 & 2 & 3 & 4 \\ \left(\begin{array}{cccc} 0 & 1 & 0 & 0 \\ 1 & 0 & 1 & 0 \\ 0 & 1 & 0 & 1 \\ 0 & 0 & 1 & 0 \end{array}\right) \end{array} \text{ and } \begin{array}{c} \\ 1 \\ 2 \\ 3 \\ 4 \end{array} \begin{array}{cccc} 1 & 2 & 3 & 4 \\ \left(\begin{array}{cccc} 1 & 0 & 1 & 0 \\ 0 & 1 & 0 & 1 \\ 1 & 0 & 1 & 0 \\ 0 & 1 & 0 & 1 \end{array}\right) \end{array}$$

(b) $PQ = \begin{array}{c} \\ 1 \\ 2 \\ 3 \\ 4 \end{array} \begin{array}{cccc} 1 & 2 & 3 & 4 \\ \left(\begin{array}{cccc} 0 & 1 & 0 & 0 \\ 1 & 0 & 1 & 0 \\ 0 & 1 & 0 & 1 \\ 0 & 0 & 1 & 0 \end{array}\right) \end{array} P^2 = \begin{array}{c} \\ 1 \\ 2 \\ 3 \\ 4 \end{array} \begin{array}{cccc} 1 & 2 & 3 & 4 \\ \left(\begin{array}{cccc} 0 & 1 & 0 & 0 \\ 1 & 0 & 1 & 0 \\ 0 & 1 & 0 & 1 \\ 0 & 0 & 1 & 0 \end{array}\right) \end{array} = Q^2$

6.4.3.9. Answer.

(a) Reflexive: $R_{ij} = R_{ij}$ for all i, j, therefore $R_{ij} \le R_{ij}$

Antisymmetric: Assume $R_{ij} \le S_{ij}$ and $S_{ij} \le R_{ij}$ for all $1 \le i, j \le n$. Therefore, $R_{ij} = S_{ij}$ for all $1 \le i, j \le n$ and so $R = S$

Transitive: Assume R, S, and T are matrices where $R_{ij} \le S_{ij}$ and $S_{ij} \le T_{ij}$, for all $1 \le i, j \le n$. Then $R_{ij} \le T_{ij}$ for all $1 \le i, j \le n$, and so $R \le T$.

(b)

$$\begin{aligned} \left(R^2\right)_{ij} &= R_{i1} R_{1j} + R_{i2} R_{2j} + \cdots + R_{in} R_{nj} \\ &\le S_{i1} S_{1j} + S_{i2} S_{2j} + \cdots + S_{in} S_{nj} \quad . \\ &= \left(S^2\right)_{ij} \Rightarrow R^2 \le S^2 \end{aligned}$$

To verify that the converse is not true we need only one example. For $n = 2$, let $R_{12} = 1$ and all other entries equal 0, and let S be the zero matrix. Since R^2 and S^2 are both the zero matrix, $R^2 \leq S^2$, but since $R_{12} > S_{12}$, $R \leq S$ is false.

(c) The matrices are defined on the same set $A = \{a_1, a_2, \ldots, a_n\}$. Let $c(a_i), i = 1, 2, \ldots, n$ be the equivalence classes defined by R and let $d(a_i)$ be those defined by S. Claim: $c(a_i) \subseteq d(a_i)$.

$$a_j \in c(a_i) \Rightarrow a_i r a_j$$
$$\Rightarrow R_{ij} = 1 \Rightarrow S_{ij} = 1$$
$$\Rightarrow a_i s a_j$$
$$\Rightarrow a_j \in d(a_i)$$

6.5 · Closure Operations on Relations

6.5.3 · Exercises for Section 6.5

6.5.3.3. Answer.

(a) See graphs below.

(b) For example, $0s^+4$ and using S one can go from 0 to 4 using a path of length 3.

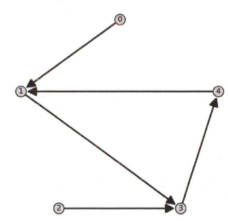

Figure C.0.11: Digraph of $\mathcal{S}$

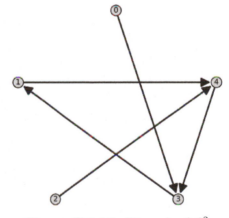

Figure C.0.12: Digraph of $\mathcal{S}^2$

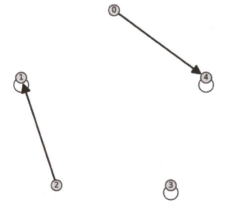

Figure C.0.13: Digraph of $\mathcal{S}^3$

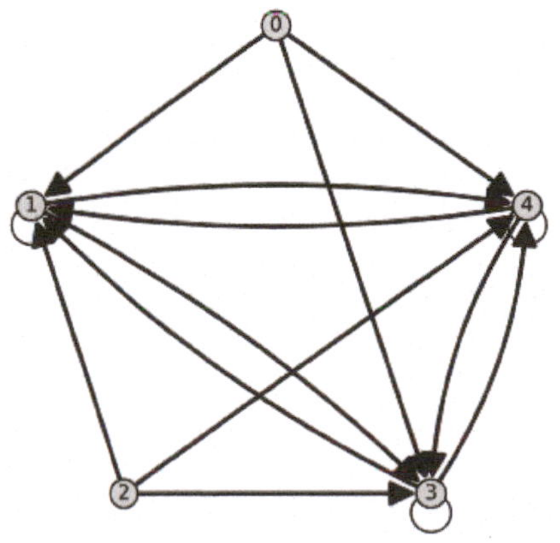

Figure C.0.14: Digraph of $\mathcal{S}^+$

6.5.3.5. Answer. Definition: Reflexive Closure. Let r be a relation on A. The reflexive closure of r is the smallest reflexive relation that contains r.

Theorem: The reflexive closure of r is the union of r with $\{(x,x) : x \in A\}$

6.5.3.7. Answer.

(a) By the definition of transitive closure, r^+ is the smallest relation which contains r; therefore, it is transitive. The transitive closure of r^+, $(r^+)^+$, is the smallest transitive relation that contains r^+. Since r^+ is transitive, $(r^+)^+ = r^+$.

(b) The transitive closure of a symmetric relation is symmetric, but it may not be reflexive. If one element is not related to any elements, then the transitive closure will not relate that element to others.

7 · Functions

7.1 · Definition and Notation

7.1.5 · Exercises for Section 7.1

7.1.5.1. Answer.

(a) Yes   (c) No   (e) Yes

(b) Yes   (d) No

7.1.5.3. Answer.

(a) Range of $f = f(A) = \{a,b,c,d\} = B$

(b) Range of $g = g(A) = \{a,b,d\}$

(c) Range of $L = L(A) = \{1\}$

7.1.5.7. Answer. For each of the $|A|$ elements of A, there are $|B|$ possible images, so there are $|B| \cdot |B| \cdot \ldots \cdot |B| = |B|^{|A|}$ functions from A into B.

7.2 · Properties of Functions

7.2.3 · Exercises for Section 7.2

7.2.3.1. Answer. The only one-to-one function and the only onto function is f.

7.2.3.3. Answer.

(a) f_1 is onto but not one-to-one: $f_1(0) = f_1(1)$.

(b) f_2 is one-to-one and onto.

(c) f_3 is one-to-one but not onto.

(d) f_4 is onto but not one-to-one.

(e) f_5 is one-to-one but not onto.

(f) f_6 is one-to-one but not onto.

7.2.3.5. Answer. Let $X = \{\text{socks selected}\}$ and $Y = \{\text{pairs of socks}\}$ and define $f : X \to Y$ where $f(x) =$ the pair of socks that x belongs to . By the Pigeonhole principle, there exist two socks that were selected from the same pair.

7.2.3.7. Answer.

(a) $f(n) = n$, for example

(b) $f(n) = 1$, for example

   (c) None exist.

   (d) None exist.

7.2.3.9. Answer.

   (a) Use $s : \mathbb{N} \to \mathbb{P}$ defined by $s(x) = x + 1$.

   (b) Use the function $f : \mathbb{N} \to \mathbb{Z}$ defined by $f(x0 = x/2$ if x is even and $f(x) = -(x+1)/2$ if x is odd.

   (c) The proof is due to Georg Cantor (1845-1918), and involves listing the rationals through a definite procedure so that none are omitted and duplications are avoided. In the first row list all nonnegative rationals with denominator 1, in the second all nonnegative rationals with denominator 2, etc. In this listing, of course, there are duplications, for example, $0/1 = 0/2 = 0$, $1/1 = 3/3 = 1$, $6/4 = 9/6 = 3/2$, etc. To obtain a list without duplications follow the arrows in Figure C.0.15, listing only the circled numbers.

   We obtain: $0, 1, 1/2, 2, 3, 1/3, 1/4, 2/3, 3/2, 4/1, \ldots$ Each nonnegative rational appears in this list exactly once. We now must insert in this list the negative rationals, and follow the same scheme to obtain:

$$0, 1, -1, 1/2, -1/2, 2, -2, 3, -3, 1/3, -1/3, \ldots$$

   which can be paired off with the elements of $\mathbb{N}$.

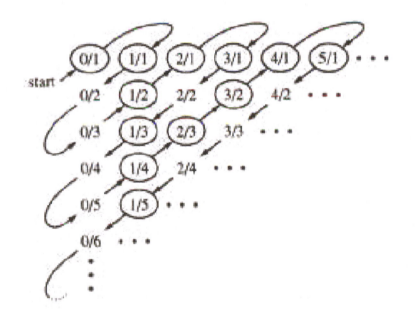

Figure C.0.15: Enumeration of the rational numbers.

7.2.3.11. Answer. Let f be any function from A into B. By the Pigeonhole principle with $n = 1$, there exists an element of B that is the image of at least two elements of A. Therefore, f is not an injection.

7.2.3.13. Answer. The proof is indirect and follows a technique called the Cantor diagonal process. Assume to the contrary that the set is countable, then the elements can be listed: $n_1, n_2, n_3, \ldots$ where each n_i is an infinite sequence of 0s and 1s. Consider the array:

$$n_1 = n_{11}n_{12}n_{13}\cdots$$
$$n_2 = n_{21}n_{22}n_{23}\cdots$$
$$n_3 = n_{31}n_{32}n_{33}\cdots$$
$$\vdots$$

We assume that this array contains all infinite sequences of 0s and 1s. Consider the sequence s defined by $s_i = \begin{cases} 0 & \text{if } n_{ii} = 1 \\ 1 & \text{if } n_{ii} = 0 \end{cases}$

Notice that s differs from each n_i in the ith position and so cannot be in the list. This is a contradiction, which completes our proof.

7.3 · Function Composition

7.3.4 · Exercises for Section 7.3

7.3.4.1. Answer.

(a) $g \circ f : A \to C$ is defined by $(g \circ f)(k) = \begin{cases} + & \text{if } k = 1 \text{ or } k = 5 \\ - & \text{otherwise} \end{cases}$

(b) No, since the domain of f is not equal to the codomain of g.

(c) No, since f is not surjective.

(d) No, since g is not injective.

7.3.4.3. Answer.

(a) The permutations of A are $i, r_1, r_2, f_1, f_2,$ and f_3, defined in Table 11.1.1.

(b)

g	g^{-1}	g^2
i	i	i
r_1	r_2	r_2
r_2	r_1	r_1
f_1	f_1	i
f_2	f_2	i
f_3	f_3	i

(c) If f and g are permutations of A, then they are both injections and their composition, $f \circ g$, is a injection, by Theorem 7.3.6. By Theorem 7.3.7, $f \circ g$ is also a surjection; therefore, $f \circ g$ is a bijection on A, a permutation.

(d) Proof by induction: Basis: ($n = 1$). The number of permutations of A is one, the identity function, and $1! = 1$.

Induction: Assume that the number of permutations on a set with n elements, $n \geq 1$, is $n!$. Furthermore, assume that $|A| = n + 1$ and that A contains an element called σ. Let $A' = A - \{\sigma\}$. We can reduce the definition of a permutation, f, on A to two steps. First, we select any one of the $n!$ permutations on A'. (Note the use of the induction hypothesis.) Call it g. This permutation almost completely defines a permutation on A that we will call f. For all a in A', we start by defining $f(a)$ to be $g(a)$. We may be making some adjustments, but define it that way for now. Next, we select the image of σ, which can be done $n + 1$ different ways, allowing for any value in A. To keep our function bijective, we must adjust f as follows: If we select $f(\sigma) = y \neq \sigma$, then we must find the element, z, of A such that $g(z) = y$, and redefine the image of z to $f(z) = \sigma$. If we had selected $f(\sigma) = \sigma$, then there is no adjustment needed. By the rule of products, the number of ways that we can define f is $n!(n+1) = (n+1)!$ $\square$

7.3.4.7. Answer.

(a) $f \circ g(n) = n + 3$

(b) $f^3(n) = n + 15$

(c) $f \circ h(n) = n^2 + 5$

7.3.4.9. Hint. You have seen a similar proof in matrix algebra.

7.3.4.11. Answer. If $f : A \to B$ and f has an inverse, then that inverse is unique.

Proof: Suppose that g and h are both inverses of f, both having domain B and codomain A.

$$
\begin{aligned}
g &= g \circ i_B \\
&= g \circ (f \circ h) \\
&= (g \circ f) \circ h \\
&= i_A \circ h \\
&= h \quad \Rightarrow g = h \quad \square
\end{aligned}
$$

7.3.4.12. Hint. See Exercise 3 of Section 5.4.

7.3.4.13. Answer. Let x, x' be elements of A such that $g \circ f(x) = g \circ f(x')$; that is, $g(f(x)) = g(f(x'))$. Since g is injective, $f(x) = f(x')$ and since f is injective, $x = x'$. $\square$

Let x be an element of C. We must show that there exists an element of A whose image under $g \circ f$ is x. Since g is surjective, there exists an element of B, y, such that $g(y) = x$. Also, since f is a surjection, there exists an element of A, z, such that $f(z) = y$, $g \circ f(z) = g(f(z)) = g(y) = x$. $\square$

7.3.4.15. Answer. Basis: $(n = 2)$: $(f_1 \circ f_2)^{-1} = f_2^{-1} \circ f_1^{-2}$ by exercise 10.
Induction: Assume $n \geq 2$ and

$$
(f_1 \circ f_2 \circ \cdots \circ f_n)^{-1} = f_n^{-1} \circ \cdots \circ f_2^{-1} \circ f_1^{-1}
$$

and consider $(f_1 \circ f_2 \circ \cdots \circ f_{n+1})^{-1}$.

$$
\begin{aligned}
(f_1 \circ f_2 \circ \cdots \circ f_{n+1})^{-1} &= ((f_1 \circ f_2 \circ \cdots \circ f_n) \circ f_{n+1})^{-1} \\
&= f_{n+1}^{-1} \circ (f_1 \circ f_2 \circ \cdots \circ f_n)^{-1} \\
&\qquad \text{by the basis} \\
&= f_{n+1}^{-1} \circ \left(f_n^{-1} \circ \cdots \circ f_2^{-1} \circ f_1^{-1} \right) \\
&\qquad \text{by the induction hypothesis} \\
&= f_{n+1}^{-1} \circ \cdots \circ f_2^{-1} \circ f_1^{-1} \quad . \square
\end{aligned}
$$

8 · Recursion and Recurrence Relations
8.1 · The Many Faces of Recursion
8.1.8 · Exercises for Section 8.1

8.1.8.1. Answer.

$$\binom{7}{2} = \binom{6}{2} + \binom{6}{1}$$
$$= \binom{5}{2} + \binom{5}{1} + \binom{5}{1} + \binom{5}{0}$$
$$= \binom{5}{2} + 2\binom{5}{1} + 1$$
$$= \binom{4}{2} + \binom{4}{1} + 2(\binom{4}{1} + \binom{4}{0}) + 1$$
$$= \binom{4}{2} + 3\binom{4}{1} + 3$$
$$= \binom{3}{2} + \binom{3}{1} + 3(\binom{3}{1} + \binom{3}{0}) + 3$$
$$= \binom{3}{2} + 4\binom{3}{1} + 6$$
$$= \binom{2}{2} + \binom{2}{1} + 4(\binom{2}{1} + \binom{2}{0}) + 6$$
$$= 5\binom{2}{1} + 11$$
$$= 5(\binom{1}{1} + \binom{1}{0}) + 11$$
$$= 21$$

8.1.8.3. Answer.

(a) $p(x)$ in telescoping form: $((((x + 3)x - 15)x + 0)x + 1)x - 10$

(b) $p(3) = ((((3 + 3)3 - 15)3 - 0)3 + 1)3 - 10 = 74$

8.1.8.5. Answer. The basis is not reached in a finite number of steps if you try to compute $f(x)$ for a nonzero value of x.

8.2 · Sequences

8.2.3 · Exercises for Section 8.2

8.2.3.1. Answer. Basis: $B(0) = 3 \cdot 0 + 2 = 2$, as defined.
Induction: Assume: $B(k) = 3k + 2$ for some $k \geq 0$.

$$\begin{aligned} B(k + 1) &= B(k) + 3 \\ &= (3k + 2) + 3 \quad \text{by the induction hypothesis} \\ &= (3k + 3) + 2 \\ &= 3(k + 1) + 2 \quad \text{as desired} \end{aligned}$$

8.2.3.3. Answer. Imagine drawing line k in one of the infinite regions that it passes through. That infinite region is divided into two infinite regions by line k. As line k is drawn through every one of the $k - 1$ previous lines, you enter another region that line k divides. Therefore k regions are divided and the number of regions is increased by k. From this observation we get $P(5) = 16$.

8.2.3.5. Answer. For n greater than zero, $M(n) = M(n - 1) + 1$, and $M(0) = 0$.

8.3 · Recurrence Relations

8.3.5 · Exercises for Section 8.3

8.3.5.1. Answer. $S(k) = 2 + 9^k$

8.3.5.3. Answer. $S(k) = 6(1/4)^k$

8.3.5.5. Answer. $S(k) = k^2 - 10k + 25$

8.3.5.7. Answer. $S(k) = (3 + k)5^k$

8.3.5.9. Answer. $S(k) = (12 + 3k) + \left(k^2 + 7k - 22\right)2^{k-1}$

8.3.5.11. Answer. $P(k) = 4(-3)^k + 2^k - 5^{k+1}$

8.3.5.13. Answer.

(a) The characteristic equation is $a^2 - a - 1 = 0$, which has solutions $\alpha = \left(1 + \sqrt{5}\right)/2$ and $\beta = \left(1 - \sqrt{5}\right)/2$, It is useful to point out that $\alpha + \beta = 1$ and $\alpha - \beta = \sqrt{5}$. The general solution is $F(k) = b_1 \alpha^k + b_2 \beta^k$. Using the initial conditions, we obtain the system: $b_1 + b_2 = 1$ and $b_1 \alpha + b_2 \beta = 1$. The solution to this system is $b_1 = \alpha/(\alpha - \beta) = \left(5 + \sqrt{5}\right)/2\sqrt{5}$ and $b_2 = \beta/(\alpha - \beta) = \left(5 - \sqrt{5}\right)/2\sqrt{5}$

Therefore the final solution is

$$F(n) = \frac{\alpha^{n+1} - \beta^{n+1}}{\alpha - \beta}$$
$$= \frac{\left((1 + \sqrt{5})/2\right)^{n+1} - \left((1 - \sqrt{5})/2\right)^{n+1}}{\sqrt{5}}$$

(b) $C_r = F(r + 1)$

8.3.5.15. Answer.

(a) For each two-block partition of $\{1, 2, \ldots, n - 1\}$, there are two partitions we can create when we add n, but there is one additional two-block partition to count for which one block is $\{n\}$. Therefore, $D(n) = 2D(n - 1) + 1$ for $n \geq 2$ and $D(1) = 0$.

(b) $D(n) = 2^{n-1} - 1$

8.4 · Some Common Recurrence Relations
8.4.5 · Exercises for Section 8.4

8.4.5.1. Answer.

(a) $S(n) = 1/n!$

(b) $U(k) = 1/k$, an improvement.

(c) $T(k) = (-3)^k k!$, no improvement.

8.4.5.3. Answer.

(a) $T(n) = 3\left(\lfloor \log_2 n \rfloor + 1\right)$

(b) $T(n) = 2$

(c) $V(n) = \lfloor \log_8 n \rfloor + 1$

8.4.5.4. Hint. Prove by induction on r.

8.4.5.5. Answer. The indicated substitution yields $S(n) = S(n + 1)$. Since $S(0) = T(1)/T(0) = 6$, $S(n) = 6$ for all n. Therefore $T(n + 1) = 6T(n) \Rightarrow$

$T(n) = 6^n$.

8.4.5.7. Answer.

(a) A good approximation to the solution of this recurrence relation is based on the following observation: n is a power of a power of two; that is, n is 2^m, where $m = 2^k$, then $Q(n) = 1 + Q\left(2^{m/2}\right)$. By applying this recurrence relation k times we obtain $Q(n) = k$. Going back to the original form of n, $\log_2 n = 2^k$ or $\log_2(\log_2 n) = k$. We would expect that in general, $Q(n)$ is $\lfloor \log_2(\log_2 n) \rfloor$. We do not see any elementary method for arriving at an exact solution.

(b) Suppose that n is a positive integer with $2^{k-1} \leq n < 2^k$. Then n can be written in binary form, $(a_{k-1}a_{k-2}\cdots a_2a_1a_0)_{\text{two}}$ with $a_{k-1} = 1$ and $R(n)$ is equal to the sum $\sum\limits_{i=0}^{k-1}(a_{k-1}a_{k-2}\cdots a_i)_{\text{two}}$. If $2^{k-1} \leq n < 2^k$, then we can estimate this sum to be between $2n - 1$ and $2n + 1$. Therefore, $R(n) \approx 2n$.

8.5 · Generating Functions

8.5.7 · Exercises for Section 8.5

8.5.7.1. Answer.

(a) $1, 0, 0, 0, 0, \ldots$

(b) $5(1/2)^k$

(c) $1, 1, 0, 0, 0, \ldots$

(d) $3(-2)^k + 3 \cdot 3^k$

8.5.7.3. Answer.

(a) $1/(1 - 9z)$

(b) $(2 - 10z)/\left(1 - 6z + 5z^2\right)$

(c) $1/\left(1 - z - z^2\right)$

8.5.7.5. Answer.

(a) $3/(1 - 2z) + 2/(1 + 2z), 3 \cdot 2^k + 2(-2)^k$

(b) $10/(1 - z) + 12/(2 - z), 10 + 6(1/2)^k$

(c) $-1/(1 - 5z) + 7/(1 - 6z), 7 \cdot 6^k - 5^k$

8.5.7.7. Answer.

(a) $11k$

(b) $(5/3)k(k + 1)(2k + 1) + 5k(k + 1)$

(c) $\sum\limits_{j=0}^{k}(j)(10(k - j)) = 10k\sum\limits_{j=0}^{k}j - 10\sum\limits_{j=0}^{k}j^2 = 5k^2(k + 1) - (5k(k + 1)(2k + 1)/6) = (5/3)k(k + 1)(2k + 1)$

(d) $k(k + 1)(2k + 7)/12$

8.5.7.9. Answer. Coefficients of z^0 through z^5 in $(1+5z)(2+4z)(3+3z)(4+$

$2z)(5 + z)$

k	Number of ways of getting a score of k
0	120
1	1044
2	2724
3	2724
4	1044
5	120

9 · Graph Theory
9.1 · Graphs - General Introduction
9.1.5 · Exercises for Section 9.1

9.1.5.1. Answer. In Figure 9.1.10, computer b can communicate with all other computers. In Figure 9.1.11, there are direct roads to and from city b to all other cities.

9.1.5.3. Answer.

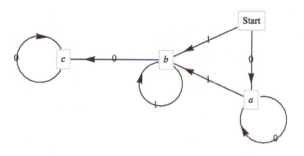

Figure C.0.16: Solution to exercise 3 of Section 9.1

9.1.5.5. Answer. The maximum number of edges would be $\binom{8}{2} = \frac{(7)(8)}{2} = 28$.

9.1.5.7. Answer.

(a) $\binom{n}{2} = \frac{(n-1)n}{2}$

(b) $n - 1$, each vertex except the champion vertex has an indegree of 1 and the champion vertex has an indegree of zero.

9.1.5.9. Answer.

(a) Not graphic - if the degree of a graph with seven vertices is 6, it is connected to all other vertices and so there cannot be a vertex with degree zero.

(b) Graphic. One graph with this degree sequence is a cycle of length 6.

(c) Not Graphic. The number of vertices with odd degree is odd, which is impossible.

(d) Graphic. A "wheel graph" with one vertex connected to all other and the others connected to one another in a cycle has this degree sequence.

(e) Graphic. Pairs of vertices connected only to one another.

(f) Not Graphic. With two vertices having maximal degree, 5, every vertex would need to have a degree of 2 or more, so the 1 in this sequence makes it non-graphic.

9.2 · Data Structures for Graphs

9.2.3 · Exercises for Section 9.2

9.2.3.1. Answer.

(a) A rough estimate of the number of vertices in the "world airline graph" would be the number of cities with population greater than or equal to 100,000. This is estimated to be around 4,100. There are many smaller cities that have airports, but some of the metropolitan areas with clusters of large cities are served by only a few airports. 4,000-5,000 is probably a good guess. As for edges, that's a bit more difficult to estimate. It's certainly not a complete graph. Looking at some medium sized airports such as Manchester, NH, the average number of cities that you can go to directly is in the 50-100 range. So a very rough estimate would be $\frac{75 \cdot 4500}{2} = 168,750$. This is far less than $4,500^2$, so an edge list or dictionary of some kind would be more efficient.

(b) The number of ASCII characters is 128. Each character would be connected to $\binom{8}{2} = 28$ others and so there are $\frac{128 \cdot 28}{2} = 3,584$ edges. Comparing this to the $128^2 = 16,384$, an array is probably the best choice.

(c) The Oxford English Dictionary as approximately a half-million words, although many are obsolete. The number of edges is probably of the same order of magnitude as the number of words, so an edge list or dictionary is probably the best choice.

9.2.3.3. Answer. Each graph is isomorphic to itself. In addition, G_2 and G_4 are isomorphic; and G_3, G_5, and G_6 are isomorphic to one another.

9.3 · Connectivity

9.3.5 · Exercises for Section 9.3

9.3.5.1. Answer.

k	1	2	3	4	5	6	
$V[k]$.found	T	T	T	F	F	T	($* =$ undefined)
$V[k]$.from	2	5	6	$*$	$*$	5	
DepthSet	2	1	2	$*$	$*$	1	

9.3.5.3. Answer. If the number of vertices is n, there can be $\frac{(n-1)(n-2)}{2}$ vertices with one vertex not connected to any of the others. One more edge and connectivity is assured.

9.3.5.5. Answer. Basis: $(k = 1)$ Is the relation r^1, defined by vr^1w if there is a path of length 1 from v to w? Yes, since vrw if and only if an edge, which is a path of length 1, connects v to w.

Induction: Assume that vr^kw if and only if there is a path of length k from v to w. We must show that $vr^{k+1}w$ if and only if there is a path of length $k+1$ from v to w.

$$vr^{k+1}w \Rightarrow vr^ky \text{ and } yrw \text{ for some vertex } y$$

By the induction hypothesis, there is a path of length k from v to y. And by the basis, there is a path of length one from y to w. If we combine these two paths, we obtain a path of length $k + 1$ from v to w. Of course, if we start with a path of length $k + 1$ from v to w, we have a path of length k from v to some vertex y and a path of length 1 from y to w. Therefore, vr^ky and $yrw \Rightarrow vr^{k+1}w$.

9.4 · Traversals: Eulerian and Hamiltonian Graphs

9.4.3 · Exercises for Section 9.4

9.4.3.1. Answer. Using a recent road map, it appears that an Eulerian circuit exists in New York City, not including the small islands that belong to the city. Lowell, Massachusetts, is located at the confluence of the Merrimack and Concord rivers and has several canals flowing through it. No Eulerian path exists for Lowell.

9.4.3.3. Answer. Gray Code for the 4-cube:

$$G_4 = \begin{pmatrix} 0000 \\ 0001 \\ 0011 \\ 0010 \\ 0110 \\ 0111 \\ 0101 \\ 0100 \\ 1100 \\ 1101 \\ 1111 \\ 1110 \\ 1010 \\ 1011 \\ 1001 \\ 1000 \end{pmatrix}$$

9.4.3.5. Answer. Any bridge between two land masses will be sufficient. To get an Eulerian circuit, you must add a second bridge that connects the two land masses that were not connected by the first bridge.

9.4.3.7. Answer. Let $G = (V, E)$ be a directed graph. G has an Eulerian circuit if and only if G is connected and $indeg(v) = outdeg(v)$ for all $v \in V$. There exists an Eulerian path from v_1 to v_2 if and only if G is connected, $indeg(v_1) = outdeg(v_1) - 1$, $indeg(v_2) = outdeg(v_2) + 1$, and for all other vertices in V the indegree and outdegree are equal.

9.4.3.8. Hint. Prove by induction on the number of edges.

9.4.3.9. Answer. A round-robin tournament graph is rarely Eulerian. It will be Eulerian if it has an odd number of vertices and each vertex (team) wins exactly as many times as it loses. Every round-robin tournament graph has a Hamiltonian path. This can be proven by induction on the number of vertices.

9.5 · Graph Optimization

9.5.5 · Exercises for Section 9.5

9.5.5.1. Answer. The circuit would be Boston, Providence, Hartford, Concord, Montpelier, Augusta, Boston. It does matter where you start. If you start in Concord, for example, your mileage will be higher.

9.5.5.3. Answer.

(a) Optimal cost $= 2\sqrt{2}$. Phase 1 cost $= 2.4\sqrt{2}$. Phase 2 cost $= 2.6\sqrt{2}$.

(b) Optimal cost $= 2.60$. Phase 1 cost $= 3.00$. Phase 2 cost $2\sqrt{2}$.

(c) $A = (0.0, 0.5), B = (0.5, 0.0), C = (0.5, 1.0), D = (1.0, 0.5)$

There are 4 points; so we will divide the unit square into two strips.

- Optimal Path: (B, A, C, D) Distance $= 2\sqrt{2}$
- Phase I Path: (B, A, C, D) Distance $= 2\sqrt{2}$
- Phase II Path: (A, C, B, D) \quad \quadDistance $= 2 + \sqrt{2}$

(d) $A = (0, 0), B = (0.2, 0.6), C = (0.4, 0.1), D = (0.6, 0.8), E = (0.7, 0.5)$

There are 5 points; so we will divide the unit square into three strips.

- Optimal Path: (A, B, D, E, C) Distance $= 2.31$
- Phase I Path: (A, C, B, C, E) Distance $= 2.57$
- Phase II Path: (A, B, D, E, C) Distance $= 2.31$

9.5.5.5. Answer.

(a) $f(c, d) = 2$, $f(b, d) = 2$, $f(d, k) = 5$, $f(a, g) = 1$, and $f(g, k) = 1$.

(b) There are three possible flow-augmenting paths. s, b, d, k with flow increase of 1. s, a, d, k with flow increase of 1, and s, a, g, k with flow increase of 2.

(c) The new flow is never maximal, since another flow-augmenting path will always exist. For example, if s, b, d, k is used above, the new flow can be augmented by 2 units with s, a, g, k.

9.5.5.7. Answer.

(a) Value of maximal flow $= 31$.

(b) Value of maximal flow $= 14$.

(c) Value of maximal flow $= 14$. See Table C.0.17 for one way to got this flow.

Step	Flow-augmenting path	Flow added
1	Source, A, Sink	2
2	Source, C, B, Sink	3
3	Source, E, D, Sink	4
4	Source, A, B, Sink	1
5	Source, C, D, Sink	2
6	Source, A, B, C, D, Sink	2

Table C.0.17

9.5.5.9. Hint. Count the number of comparisons of distances that must be done.

Answer. To locate the closest neighbor among the list of k other points on the unit square requires a time proportional to k. Therefore the time required for the closest-neighbor algorithm with n points is proportional to $(n - 1) + (n - 2) + \cdots + 2 + 1$, which is proportional to n^2. Since the strip algorithm takes a time proportional to $n(\log n)$, it is much faster for large values of n.

9.6 · Planarity and Colorings

9.6.3 · Exercises for Section 9.6

9.6.3.1. Answer. Theorem 9.6.12 can be applied to infer that if $n \geqslant 5$, then K_n is nonplanar. A K_4 is the largest complete planar graph.

9.6.3.3. Answer.

(a) 4	(c) 3	(e) 2
(b) 3	(d) 3	(f) 4

9.6.3.5. Answer. The chromatic number is n since every vertex is connected to every other vertex.

9.6.3.7. Answer. Suppose that G' is not connected. Then G' is made up of 2 components that are planar graphs with less than k edges, G_1 and G_2. For $i = 1, 2$ let $v_i, r_i,$ and e_i be the number of vertices, regions and edges in G_i. By the induction hypothesis, $v_i + r_i - e_i = 2$ for $i = 1, 2$.

One of the regions, the infinite one, is common to both graphs. Therefore, when we add edge e back to the graph, we have $r = r_1 + r_2 - 1$, $v = v_1 + v_2$, and $e = e_1 + e_2 + 1$.

$$\begin{aligned} v + r - e &= (v_1 + v_2) + (r_1 + r_2 - 1) - (e_1 + e_2 + 1) \\ &= (v_1 + r_1 - e_1) + (v_2 + r_2 - e_2) - 2 \\ &= 2 + 2 - 2 \\ &= 2 \end{aligned}$$

9.6.3.9. Answer. Since $|E| + E^c = \frac{n(n-1)}{2}$, either E or E^c has at least $\frac{n(n-1)}{4}$ elements. Assume that it is E that is larger. Since $\frac{n(n-1)}{4}$ is greater than $3n - 6$ for $n \geqslant 11$, G would be nonplanar. Of course, if E^c is larger, then G' would be nonplanar by the same reasoning. Can you find a graph with ten vertices such that it is planar and its complement is also planar?

9.6.3.11. Answer. Suppose that (V, E) is bipartite (with colors red and blue), $|E|$ is odd, and $(v_1, v_2, \ldots, v_{2n+1}, v_1)$ is a Hamiltonian circuit. If v_1 is red, then v_{2n+1} would also be red. But then $\{v_{2n+1}, v_1\}$ would not be in E, a contradiction.

9.6.3.13. Answer. Draw a graph with one vertex for each edge, If two edges in the original graph meet at the same vertex, then draw an edge connecting the corresponding vertices in the new graph.

10 · Trees
10.1 · What Is a Tree?
10.1.3 · Exercises for Section 10.1

10.1.3.1. Answer. The number of trees are: (a) 1, (b) 3, and (c) 16. The trees that connect V_c are:

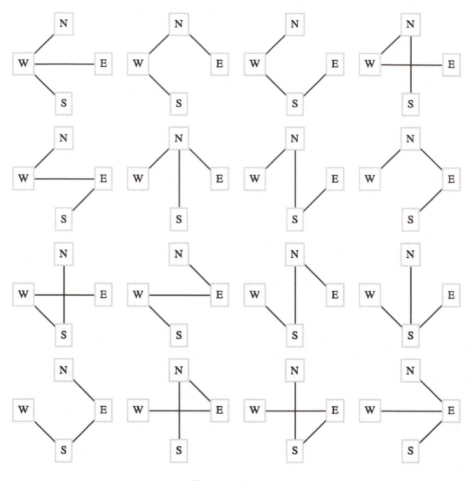

Figure C.0.18

10.1.3.3. Hint. Use induction on $|E|$.

10.1.3.5. Answer.

(a) Assume that (V, E) is a tree with $|V| \geq 2$, and all but possibly one vertex in V has degree two or more.

$$2|E| = \sum_{v \in V} \deg(v) \geq 2|V| - 1 \Rightarrow |E| \geq |V| - \frac{1}{2}$$

$$\Rightarrow |E| \geq |V|$$
$$\Rightarrow (V, E) \text{ is not a tree.}$$

(b) The proof of this part is similar to part a in that we can infer $2|E| \geq 2|V| - 1$, using the fact that a non-chain tree has at least one vertex of degree three or more.

10.2 · Spanning Trees

10.2.4 · Exercises for Section 10.2

10.2.4.1. Answer. It might not be most economical with respect to Objective 1. You should be able to find an example to illustrate this claim. The new system can always be made most economical with respect to Objective 2 if the old system were designed with that objective in mind.

10.2.4.3. Answer. In the figure below, $\{1,2\}$ is not a minimal bridge between $L = \{1,4\}$ and $R = \{2,3\}$, but it is part of the minimal spanning tree for this graph.

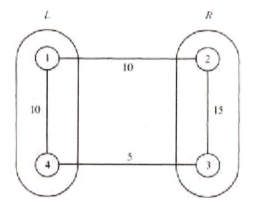

Figure C.0.19

10.2.4.5. Answer.

(a) Edges in one solution are: $\{8,7\}, \{8,9\}, \{8,13\}, \{7,6\}, \{9,4\}, \{13,12\}, \{13,14\}, \{6,11\}, \{6,1\}, \{1,2\}, \{4,$

(b) Vertices 8 and 9 are centers of the graph. Starting from vertex 8, a minimum diameter spanning tree is $\{\{8,3\}, \{8,7\}, \{8,13\}, \{8,14\}, \{8,9\}, \{3,2\}, \{3,4\}, \{7,6\}, \{13,12\}, \{13,1$. The diameter of the tree is 7.

10.3 · Rooted Trees

10.3.4 · Exercises for Section 10.3

10.3.4.1. Answer. Locate any simple path of length d and locate the vertex in position $\lceil d/2 \rceil$ on the path. The tree rooted at that vertex will have a depth of $\lceil d/2 \rceil$, which is minimal.

10.3.4.3. Answer.

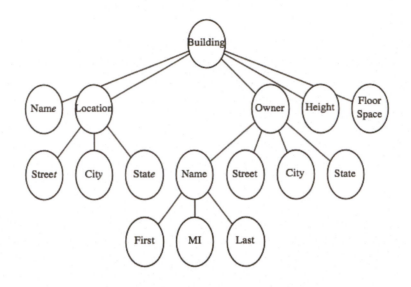

Figure C.0.20

10.4 · Binary Trees

10.4.6 · Exercises for Section 10.4

10.4.6.1. Answer.

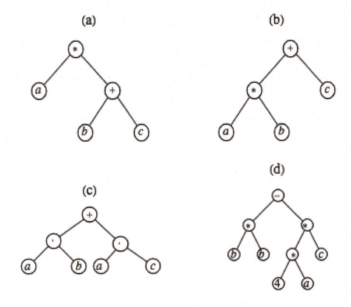

Figure C.0.21

(e)

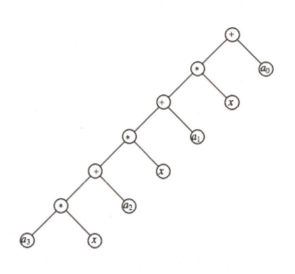

Figure C.0.22

10.4.6.3. Answer.

	Preorder	Inorder	Postorder
(a)	$\cdot a + bc$	$a \cdot b + c$	$abc + \cdot$
(b)	$+ \cdot abc$	$a \cdot b + c$	$ab \cdot c+$
(c)	$+ \cdot ab \cdot ac$	$a \cdot b + a \cdot c$	$ab \cdot ac \cdot +$

10.4.6.5. Answer.

(a)

(b)

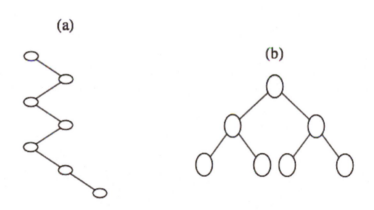

Figure C.0.23

10.4.6.7. Answer.   Solution 1:

Basis: A binary tree consisting of a single vertex, which is a leaf, satisfies the equation leaves = internal vertices + 1

Induction:Assume that for some $k \geq 1$, all full binary trees with k or fewer vertices have one more leaf than internal vertices. Now consider any full binary tree with $k + 1$ vertices. Let T_A and T_B be the left and right subtrees of the tree which, by the definition of a full binary tree, must both be full. If i_A and i_B are the numbers of internal vertices in T_A and T_B, and j_A and j_B are the numbers of leaves, then $j_A = i_A + 1$ and $j_B = i_B + 1$. Therefore, in the whole tree,

$$\begin{aligned}
\text{the number of leaves} &= j_A + j_B \\
&= (i_A + 1) + (i_B + 1) \\
&= (i_A + i_B + 1) + 1 \\
&= (\text{number of internal vertices}) + 1
\end{aligned}$$

Solution 2:

Imagine building a full binary tree starting with a single vertex. By continuing to add leaves in pairs so that the tree stays full, we can build any full binary tree. Our starting tree satisfies the condition that the number of leaves is one more than the number of internal vertices . By adding a pair of leaves to a full binary tree, an old leaf becomes an internal vertex, increasing the number of internal vertices by one. Although we lose a leaf, the two added leaves create a net increase of one leaf. Therefore, the desired equality is maintained.

Appendix D

Notation

The following table defines the notation used in this book. Page numbers or references refer to the first appearance of each symbol.

Symbol	Description	Page
$x \in A$	x is an element of A	1
$x \notin A$	x is not an element of A	1
$\|A\|$	The number of elements in a finite set A.	2
$A \subseteq B$	A is a subset of B.	3
$\emptyset$	the empty set	3
$A \cap B$	The intersection of A and B.	4
$A \cup B$	The union of A and B.	5
$B - A$	The complement of A relative to B.	6
A^c	The complement of A relative to the universe.	6
$A \oplus B$	The symmetric difference of A and B.	7
$A \times B$	The cartesian product of A with B.	10
$\mathcal{P}(A)$	The power set of A, the set of all subsets of A.	11
$n!$	n factorial, the product of the first n positive integers	25
$\binom{n}{k}$	n choose k, the number of k element subsets of an n element set.	34
$p \wedge q$	the conjunction, p and q	41
$p \vee q$	the disjunction, p or q	41
$\neg p$	the negation of p, "not p"	41
$p \rightarrow q$	The conditional proposition If p then q.	42
$p \leftrightarrow q$	The biconditional proposition p if and only if q	43
1	symbol for a tautology	47
0	symbol for a contradiction	47
$r \iff s$	r is logically equivalent to s	48
$r \Rightarrow s$	r implies s	48
$p \mid q$	the Sheffer Stroke of p and q	49
T_p	the truth set of p	58
$(\exists n)_U(p(n))$	The statement that $p(n)$ is true for at least one value of n	68
$(\forall n)_U(p(n))$	The statement that $p(n)$ is always true.	68
$\mathbf{0}_{m \times n}$	the m by n zero matrix	94
I_n	The $n \times n$ identity matrix	97

(Continued on next page)

Symbol	Description	Page
A^{-1}	A inverse, the multiplicative inverse of A	97
$\det A$ or $|A|$	The determinant of A	98
$a \mid b$	a divides b, or a divides evenly into b	106
xsy	x is related to y through the relation s	106
rs	the composition of relation r with relation s	107
$a \equiv_m b$	a is congruent to b modulo m	117
$a \equiv b(\mathrm{mod}\ m)$	a is congruent to b modulo m	117
$c(a)$	the equivalence class of a under r	121
r^+	The transitive closure of r	127
$f : A \to B$	A function, f, from A into B	132
B^A	The set of all functions from A into B	133
$f(a)$	The image of a under f	133
$f(X)$	Range of function $f : X \to Y$	133
$|A| = n$	A has cardinality n	137
$(g \circ f)(x) = g(f(x))$	The composition of g with f	141
$f \circ f = f^2$	the "square" of a function.	142
i or i_A	The identitiy function (on a set A)	142
f^{-1}	The inverse of function f read "f inverse"	142
$log_b a$	Logarithm, base b of a	170
		176
$S \uparrow$	S pop	179
$S \downarrow$	S push	179
$S * T$	Convolution of sequences S and T	179
$S \uparrow p$	Multiple pop operation on S	180
$S \downarrow p$	Multiple push operation on S	180
K_n	A complete undirected graph with n vertices	193
$deg(v), indeg(v), outdeg(v)$	degree, indegree and outdegree of vertex v	198
Q_n	the n-cube	218
$V(f)$	The value of flow f	231
		237
P_n	a path graph of length n	237
$\chi(G)$	the chromatic number of G	241
C_n	A cycle with n edges.	246
$\grave{x}, \acute{x}$	pre and post values of a variable x	279

References

Many of the references listed here were used in preparing the original 1980's version of this book. In most cases, the mathematics that they contain is still worth reading for further background. Many can be found online, in university libraries or used bookstores. A few references more current references have been added.

[1] Allenby, R.B.J.T, *Rings, Fields and Groups*, Edward Arnold, 1983.

[2] Appel, K., and W. Haken, *Every Planar Map Is 4-colorable*, Bull, Am. Math. Soc. no. 82 (1976): 711–12.
     This has historical significance in that it announced the first correct proof of the Four Color Theorems

[3] Austin, A. Keith, *An Elementary Approach to NP-Completeness* American Math. Monthly 90 (1983): 398-99.

[4] Beardwood, J., J. H. Halton, and J. M. Hammersley, *The Shortest Path Through Many Points* Proc. Cambridge Phil. Soc. no. 55 (1959): 299–327.

[5] Ben-Ari, M, *Principles of Concurrent Programming*, Englewood Cliffs, NJ: Prentice-Hall, 1982.

[6] Berge, C, *The Theory of Graphs and Its Applications*, New York: Wiley, 1962.

[7] Bogart, Kenneth P, *Combinatorics Through Guided Discovery*, 2005.
     This book may be freely downloaded and redistributed under the terms of the GNU Free Documentation License (FDL), as published by the Free Software Foundation.

[8] Busacker, Robert G., and Thomas L. Saaty, *Finite Graphs and Networks*, New York: McGraw-Hill, 1965.

[9] Connell, Ian, *Modern Algebra, A Constructive Introduction*, New York: North-Holland, 1982.

[10] Denning, Peter J., Jack B. Dennis, and Joseph L. Qualitz, *Machines, Languages, and Computation*, Englewood Cliffs, NJ: Prentice-Hall, 1978.

[11] Denning, Peter J, *Multigrids and Hypercubes*. American Scientist 75 (1987): 234-238.

[12] Dornhoff, L. L., and F. E. Hohn, *Applied Modern Algebra*, New York: Macmillan, 1978.

[13] Ford, L. R., Jr., and D. R. Fulkerson, *Flows in Networks*, Princeton, NJ: Princeton Univesity Press, 1962.

[14] Fraleigh, John B, *A First Course in Abstract Algebra*, 3rd ed. Reading,

MA: Addison-Wesley, 1982.

[15] Gallian, Joseph A, *Contemporary Abstract Algebra*, D.C. Heath, 1986.

[16] Gallian, Joseph A, *Group Theory and the Design of a Letter-Facing Machine*, American Math. Monthly 84 (1977): 285-287.

[17] Hamming, R. W, *Coding and Information Theory*, Englewood Cliffs, NJ: Prentice-Hall, 1980.

[18] Hill, F. J., and G. R. Peterson, *Switching Theory and Logical Design*, 2nd ed. New York: Wiley, 1974.

[19] Hofstadter, D. R, *Godel, Escher, Bach: An Eternal Golden Braid*, New York: Basic Books, 1979.

[20] Hohn, F. E, *Applied Boolean Algebra*, 2nd ed. New York: Macmillan, 1966.

[21] Hopcroft, J. E., and J. D. Ullman, *Formal Languages and Their Relation to Automata*, Reading, MA: Addison-Wesley, 1969.

[22] Hu, T. C, *Combinatorial Algorithms*, Reading, MA: Addison-Wesley, 1982.

[23] Knuth, D. E, *The Art of Computer Programming. Vol. 1, Fundamental Algorithms*, 2nd ed. Reading, MA: Addison-Wesley, 1973.

[24] Knuth, D. E, *The Art of Computer Programming. Vol. 2, Seminumerical Algorithms*, 2nd ed., Reading, MA: Addison-Wesley, 1981.

[25] Knuth, D. E, *The Art of Computer Programming. Vol. 3, Sorting and Searching*, Reading, MA: Addison-Wesley, 1973.

[26] Knuth, D. E, *The Art of Computer Programming. Vol. 4A, Combinatorial Algorithms, Part 1*, Upper Saddle River, New Jersey: Addison-Wesley, 2011.
https://www-cs-faculty.stanford.edu/~knuth/taocp.html

[27] Kulisch, U. W., and Miranker, W. L, *Computer Arithmetic in Theory and Practice*, New York: Academic Press, 1981.

[28] Lipson, J. D, *Elements of Algebra and Algebraic Computing*, Reading, MA: Addison-Wesley, 1981.

[29] Liu, C. L, *Elements of Discrete Mathematics*, New York: McGraw-Hill, 1977.

[30] O'Donnell, *Analysis of Boolean Functions*.
A book about Fourier analysis of boolean functions that is being developed online in a blog.

[31] Ore, O, *Graphs and Their Uses*, New York: Random House, 1963.

[32] Parry, R. T., and H. Pferrer, *The Infamous Traveling-Salesman Problem: A Practical Approach* Byte 6 (July 1981): 252-90.

[33] Pless, V, *Introduction to the Theory of Error-Correcting Codes*, New York: Wiley-Interscience, 1982.

[34] Purdom, P. W., and C. A. Brown, *The Analysis of Algorithms*, Holt, Rinehart, and Winston, 1985.

[35] Quine, W. V, *The Ways of Paradox and Other Essays*, New York: Random House, 1966.

[36] Ralston, A, *The First Course in Computer Science Needs a Mathematics Corequisite*, Communications of the ACM 27-10 (1984): 1002-1005.

[37] Solow, Daniel, *How to Read and Do Proofs*, New York: Wiley, 1982.

[38] Sopowit, K. J., E. M. Reingold, and D. A. Plaisted *The Traveling Salesman Problem and Minimum Matching in the Unit Square.*SIAM J. Computing, 1983,**12**, 144–56.

[39] Standish, T. A, *Data Structure Techniques*, Reading, MA: Addison-Wesley, 1980.

[40] Stoll, Robert R, Sets, *Logic and Axiomatic Theories*, San Francisco: W. H. Freeman, 1961.

[41] Strang, G, *Linear Algebra and Its Applications*, 2nd ed. New York: Academic Press, 1980.

[42] Tucker, Alan C, *Applied Combinatorics*, 2nd ed. New York: John Wiley and Sons, 1984.

[43] Wand, Mitchell, *Induction, Recursion, and Programming*, New York: North-Holland, 1980.

[44] Warshall, S, *A Theorem on Boolean Matrices* Journal of the Association of Computing Machinery, 1962, 11-12.

[45] Weisstein, Eric W. *Strassen Formulas*, MathWorld--A Wolfram Web Resource, http://mathworld.wolfram.com/StrassenFormulas.html.

[46] Wilf, Herbert S, *Some Examples of Combinatorial Averaging*, American Math. Monthly 92 (1985).

[47] Wilf, Herbert S. *generatingfunctionology*, A K Peters/CRC Press, 2005 The 1990 edition of this book is available at https://www.math.upenn.edu/~wilf/DownldGF.html

[48] Winograd, S, *On the Time Required to Perform Addition*, J. Assoc. Comp. Mach. 12 (1965): 277-85.

[49] Wilson, R., *Four Colors Suffice - How the Map Problem Was Solved*Princeton, NJ: Princeton U. Press, 2013.

Index